PLANT PRODUCTION ON THE THRESHOLD OF A NEW CENTURY

Developments in Plant and Soil Sciences

VOLUME 61

Plant Production on the Threshold of a New Century

Proceedings of the International Conference at the Occasion of
the 75th Anniversary of the Wageningen Agricultural University,
Wageningen, The Netherlands, held June 28 – July 1, 1993

Edited by

P. C. STRUIK

Department of Agronomy,
Wageningen Agricultural University, Wageningen, The Netherlands

W. J. VREDENBERG

Department of Plant Physiology,
Wageningen Agricultural University, Wageningen, The Netherlands

J. A. RENKEMA

Department of Farm Management,
Wageningen Agricultural University, Wageningen, The Netherlands

and

J. E. PARLEVLIET

Department of Plant Breeding,
Wageningen Agricultural University, Wageningen, The Netherlands

SPRINGER SCIENCE+BUSINESS MEDIA, B.V.

Library of Congress Cataloging-in-Publication Data

Plant production on the threshold of a new century   proceedings of
  the international conference at the occasion of the 75th anniversary
  of the Wageningen Agricultural University, held June 28 - July 1,
  1993, Wageningen, The Netherlands / edited by P.C. Struik ... [et
  al.].
      p.   cm. -- (Developments in plant and soil sciences , v. 61)
    Includes index.
    ISBN 978-0-7923-2903-9   ISBN 978-94-011-1158-4 (eBook)
    DOI 10.1007/978-94-011-1158-4
    1. Crops--Congresses. 2. Cropping systems--Congresses. 3. Farm
  management--Congresses. 4. Crops--Physiology--Congresses.
  I. Struik, P. C. (Paul Christiaan), 1954-   .
  II. Landbouwuniversiteit Wageningen. III. Series.
  SB16.P58   1994
  630--dc20   94-16568
ISBN 978-0-7923-2903-9

Printed on acid-free paper

TABLE OF CONTENTS

SESSIONS ON PLANT AND CROP PHYSIOLOGICAL ASPECTS (WITH SPECIAL REFERENCE TO STRESS)

VIII

PREFACE

This volume contains the proceedings of an international conference on plant production on the threshold of a new century, held June 28 - July 1, 1993, in Wageningen, The Netherlands, at the occasion of the 75th Anniversary of the Wageningen Agricultural University. Oral presentations are published in the order of appearance during the conference, reflecting the original structure of the theme. Summaries of the poster presentations are compiled at the end of this volume.

The book describes problems and frontier developments in the different sectors of plant production at the farming system level, the production system level and the crop and plant levels, thus aiming at an integration of developments in basic plant sciences, in crop sciences and in socio-economic sciences contributing to a sustainable plant production.

The conference theme highlights the challenge that the agricultural sciences have to face to make agriculture more sustainable, making full use of the potentials but avoiding negative side-effects of agricultural production on landscape, nature, environment, biodiversity, etc. and contributing to the quality of life in general, but in rural areas in particular. In the approach of the conference, there was room for generalists and specialists in different fields. By the approach the scientists present were challenged to:

* formulate goals and constraints in policy, economy, production, environment and land use;
* indicate how these goals and constraints could be translated into possible or desirable farming styles and cropping systems;
* describe how the fundamental plant sciences could contribute to the realisation of these farming styles and cropping systems.

We gratefully acknowledge the financial support of our sponsors, the excellent performance of the organizing committee, the hospitality of the Wageningen International Conference Centre, the administrative support of the Office of the Sector Plant Production of the Wageningen Agricultural University, the support of the home institutions of the organizers and the rewarding interaction with the participants.

The preparation of the proceedings would have been impossible without the skillful service of the Text Processing Service of the Wageningen Agricultural University and of Kluwer Academic Publishers.

The editors
P.C. Struik
W.J. Vredenberg
J.A. Renkema
J.E. Parlevliet

ORGANISATION OF CONGRESS

Sponsors: ROFA-MAVI
                        HOEK LOOS
                        CSM SUIKER
                        CIBA-GEIGY
                        Wageningen Agricultural University

Programme committee: J.A. Renkema
                        P.C. Struik
                        W.J. Vredenberg

Organising committee: H.H. Bartelink
                        A. Hink
                        G.H.M. Kronenberg
                        W.J.M. Lommen
                        J.L. Meulenbroek
                        J.H. van Niejenhuis
                        J.E. Parlevliet
                        P.A. van de Pol

CROP PRODUCTION ON THE THRESHOLD OF A NEW CENTURY

J.C. ZADOKS
Wageningen Agricultural University
Department of Phytopathology
P.O. Box 8025
6700 EE Wageningen
The Netherlands

ABSTRACT. New aspects in crop production of possible interest in the 21st century are highlighted. Emphasis is on technology development, gene technology, ecotechnology and agrotechnology, at the molecular, crop and agroecosystem level, respectively, with some bias towards crop protection and plant breeding.

1. What's new in crop production?

What's new in crop production? Well, more or less everything. We realise that growers have to cope with various abiotic constraints leading to well-defined production levels (De Wit, 1982). We realise that, even at the no-constraint level, we rarely obtain the theoretical yield (Zadoks and Schein, 1979) for reasons still difficult to fathom.

We also realise that present day production systems in western countries have too many side effects to remain politically acceptable. The new look is 'sustainable' or 'durable'. Durability in the modern political sense (World Commission, 1987) shies away from continuous top production with high inputs and complete neglect of anything outside the production perimeter, and considers external effects. Durability is a road sign, indicating the direction to be taken but without mention of the distance to go (Zadoks, 1993).

Following the roadsign towards durability we will pass the crossroad of global climatic change. Struggling forward we will have to climb the hill of overproduction in some parts of the world, or to descend into the valley of underproduction in other parts. Overproduction and underproduction have physical, economical and social dimensions, leading to grave political decisions. How selfish should we remain as individuals and nations, and as regions such as the European Community? How much solidarity (De Wit *et al.*, 1987) should we feel with less affluent individuals, nations and regions, be it only out of sheer self-interest (Zadoks, 1993)?

1

P. C. Struik et al. (eds.), Plant Production on the Threshold of a New Century, 1–8.

If the political dimension of future agriculture is so undecided, so wobbly, what can we do as research scientists? Durability is riding the crest of the political fashion wave. The message to agriculture and agricultural science is loud and clear, and that message will reverberate far into the 21st century. We can translate the message into operational terms by means of the minimax principle (Netherlands Scientific Council, 1992), to produce maximum output using minimum inputs.

Outputs are Ecu's, kilograms of product, but also product quality, production quality, nature values and other externalities, either translated into Ecu's or measured with their own intrinsic yardsticks. Production means are soil, water and air, which should be pure or purified. This troika will be complemented by nature and biodiversity, to be maintained or improved. The economic inputs are capital, labour and knowledge. The present congress deals with the latter input, science and technology of crop production.

2. The role of technology

Science and technology are as two sides of a coin. Science without technology is like an empty shell, whereas technology without science cannot make much progress. Economists see technology as the motor of economic development. With so much change in the air, we have to provide new technology to implement the desired change.

Within the domain of technology, biotechnology recently attained a position of prominence. In line with the three systems levels of this congress, I will describe three sub-sets of biotechnology, to be called gene technology, ecotechnology and agrotechnology. Gene technology deals with the workings of genes in plants. Ecotechnology deals with crops at the systems level of field or greenhouse. Agrotechnology refers to the farming systems level.

In plant breeding we manipulate genes exploiting whatever genetic variation we can access. The gene technology relevant here uses genetic modification to extend the natural range of variation and to recombine genes which cannot be recombined by classical methods. New combinations can be constructed by molecular techniques and literally inserted as a 'construct' into a plant genome (De Wit, 1993).

In the agricultural context of this conference I define eco-technology as the technology to manipulate agroecosystems at crop level in such a way that a set target is attained. A common target is to keep populations of harmful agents below a nuisance threshold by means of supervised, biological or integrated control.

Agrotechnology integrates all available techniques into techno-logy packages that serve objectives at farm level. With the term agro-technology I want to emphasize that these technology packages have to be designed explicitly by agronomic designers (Goewie, 1993).

3. Plant breeding and gene technology

Plant breeding has contributed much to the success of agriculture and will continue to do so, now supported by molecular biology. Tradition-ally, resistance breeding utilised high-level monogenic resistance. This I call first order resistance. Second order resistance is the so-called partial resistance, often of polygenic nature (Jacobs and Parlevliet, 1993).

Here, I postulate a third order, enhanced resistance, in which enhancement refers to resistance, biological control, or even both. To mention a few possibilities:

a. Reduced hairiness in cucumber is a selectable character which facilitates predators and parasitoids to find their prey (Van Lenteren, 1990).

b. Some plants 'cry for help' when attacked by a phytophage, so that a predator or parasitoid will easily find the infested plant; genetic variability seems to exist for the loudness of the outcry materialised as semiochemicals (Dicke et al., 1990).

c. Biological control of foliar pathogens by fungivorous fungi may be easier when partial resistance slows the pathogen down, as seems to be the case in cucumber mildew (Verhaar and Hijwegen, 1993).

d. Partial resistance may be chemically enhanced and could thus facilitate biocontrol of fungal pathogens.

e. Plant infection by fungi may enhance resistance to other fungi; again, genetic variability seems to exist in the strenth of induced resistance (Martinelli et al., 1993).

New questions will appear, such as breeding for tolerance to various physical and chemical stresses (e.g. air pollution), tolerance to low inputs, and tolerance to increased CO_2 levels. It is my firm opinion that durable agriculture begins with breeding suitable plant types.

4. Ecotechnology

Ecotechnology was christened when Elsevier produced its journal 'Ecological Engineering, Journal of Ecotechnology' in 1992. It refers to the designing, monitoring or constructing of ecosystems, 'for the mutual benefit of humans and nature'. The engineering aspect appeals to me. My feeling is that durability can only be implemented with target-specific technology. The implication is that for every situation and every objective a specific agro-ecosystem must be designed by professional designers, sitting at their drawing boards or punching their computers, after having made a thorough field study.

Such is the case for biological control of pests and diseases in protected cultivation (Van Lenteren, 1990) where varieties, greenhouse climate, parasitoids and predators, entomovorous and fungivorous fungi and viruses, and various management methods have to be fine-tuned together. The approach is increasingly successful, in part because Dutch growers perceive biocontrol as 'high tech' and are proud of it (Spaan and Van der Ploeg, 1992).

In the open crops this fine tuning is far more difficult, at least in areas with a clear alternation of the growing season with either a cold or a dry season. Nevertheless, various possibilities avail, such as cultural methods, suppressive crops, maybe disease-suppressive soil amendments, and various other techniques. Progress has been made on experimental farms (Vereijken, 1989, 1992; Zadoks, 1989b) and, more important, on hundreds of commercial farms (Wijnands, 1992).

I make a special plea for habitat management, that is management of the surroundings of fields. There is nothing new to the idea. The effective control of black stem rust on wheat through barberry (*Berberis vulgaris*) eradication is a classic case (Eriksson and Henning, 1896; Stakman and Harrar, 1957). Whereas in much of North-west Europe aphids have become a major pest of cereals, this was not the case in Southern Germany, supposedly because of the many roughages which hosted predators, among which coccinellids and syrphids (Ohnesorgen, pers. comm.). Similarly, predatory spiders, easily killed by tillage operations and insecticide spraying, survive in roughages and rapidly recolonize a treated field (Everts, 1990; Thomas et al., 1990).

Computer modeling helps to understand the complexities of di- and tritrophic systems (Rabbinge, 1976). Computer-based decision support systems such as (semi-)expert systems and optimisation-type models (Rossing, 1993) can be instrumental in the management of intricate ecosystems. The Dutch example, which spread over parts of Europe, is EPIPRE for supervised control of pests and diseases in wheat (Zadoks, 1989a).

5. Economics

No new ecosystem design is acceptable to growers if it is not at least as reliable and at least as profitable as any current system (Van Lenteren, 1993). It is the scientist's duty to demonstrate that the proposed ecosystem design fulfils that requirement. In addition, he or she has to incorporate the external effects of the proposed innovation in all considerations.

In Western Europe farmers were so greedy for kilograms that they forgot about the economics of inputs. They neglected externalities. In crop protection, EPIPRE with its economic module convinced farmers of the idea that a maximisation of kilogrammes per hectare does not equal maximisation of financial returns per hectare. One conclusion was that many farmers used supra-optimal inputs (Zadoks and Schein, 1979).

The Farming Systems Experiment indicated that returns of reduced input farming, systematically tested on the so-called Integrated Farm, were no less than of the current high input farm (Zadoks, 1989b). Meanwhile, many more data became available, from similar farming systems experiments, from commercial farms (Wijnands et al., 1992), and from various model calculations (De Koeijer and Wossink, 1992; Wossink, 1993). They all point in the same direction, reduction of inputs is economically feasible, does not necessarily lead to lower returns, and has positive external effects.

New approaches, reaching beyond the classical dose-reponse relationships are needed if we are to design and engineer new agro-ecosystems. If all ecological and economical objectives are explicitized, a 'trade off' between objectives can be envisaged without necessarily expressing all objectives in Ecu's.

6. Agrotechnology

Agrotechnology refers to the highest systems level of our concern. It is more than crop husbandry but less than agronomy. The term techno-logy is not customary in Europe, but in the Third World the term 'Green Revolution Technology' is current. Agrotechnology should provide growers with durable agro-ecosystems. The Green Revolution, which swept over the Western World shortly after World War II (De Wit et al., 1987) and over South-east Asia some twenty years later, should become greener (Swaminathan, 1990).

To obtain region and nation wide durability, according to whatever criterion, local solutions must be found. 'Think nationally but plan and act locally' (Swaminathan, 1991). Whereas West European arable agriculture tends to become more and more uniform, I expect it to diversify after the year 2000 with local solutions for local durability constraints.

The Netherlands as a small nation have lost considerable agri-cultural diversity on the one hand, but on the other hand have gained much diversity by growing over 600 crops. New concepts are applied, such as 'closed cropping', with negligible emission of chemicals. Recirculation of organic matter is being improved, composting of organic refuse becomes general, and disease-suppressive compost is being designed (Hoitink et al., 1993).

Classical mixed farming, mainly to produce manure for arable crops, is nearly lost. But it may reappear, e.g. by cooperation of neighbouring arable and dairy cattle farmers. Whereas multiple cropping is customary, often with a cash crop followed by a green manure crop, intercropping as in the tropics is seldom except maybe in biological horticulture. In the temperate climate intercropping may have advantages similar to those in the tropics for plant nutrition and crop protection (Theunissen et al., 1992), but new machinery should be designed for large scale field application and product processing.

In international circles, thinking about agriculture is changing too. It is as if the high tide of individual disciplines is over. New terms appear to emphasize integration of disciplines coined according to the format of Integrated Pest Management, such as Integrated Nutrient Management and Integrated Crop Management (FAO, 1991). Such terms point to fresh opportunities for agrotechnology as an engineering science, to design new constructs. The concept of Integrated Crop Management tallies with the concept of Plant Health, avoiding weeds, pests and diseases by Good Agricultural Practice.

Two concepts were proposed to attain the objectives of durable crop production. The first, Good Agricultural Practice (GAP), is a benevolent but somewhat vague concept. The second is the concept of Best Technical Means (BTM; Netherlands Scientific Council, 1992). There is nothing vague in it because BTM can be calculated once the objectives are set, such as Production oriented, Environment oriented and Land use oriented BTM, at European, regional or farm level.

7. Winding up

Our challenge is to prepare crop production for the 21st century. We will have to proceed in the direction of durability, even if and when this concept will have been forgotten by our politicians. We will have to reduce our material inputs into crop production if we are going to reduce undesirable externalities.

New technology, even within a GAP or BTM context, will not be effective without changing the minds of people, first the scientists designing the new technology, second the farmers applying it, and third the consumers accepting it. The ideas behind the technology have to be 'internalized', to quote Vereijken (pers. comm.). Extension campaigns to this purpose can be effective (Van de Fliert, 1993).

Agriculture will thus become an ever more interesting profession. The material inputs are to be reduced, but that is possible only if we increase the non-material inputs, knowledge, know-how and determination. We will have to step up our efforts in disciplinary, inter-disciplinary and integrative research. A coherent Agricultural Knowledge and Innovation System (AKIS; Röling, 1989) is conditional for success.

For agricultural scientists, life will become ever more fasci-nating. For design and construction of new agroecosystems more rather than less professionals will be needed to usher crop production into a new phase.

8. References

Dicke, M., Sabelis, M.W., Takabayashi, J., Bruin, J. and Posthumus, M.A. (1990) 'Plant strategies of manipulating predator-prey interactions through allelochemicals: Prospects for application in pest control'. Journal Chemical Ecology 16, 3091-3118.

Eriksson, J. and Henning F. (1896) 'Die Getreideroste. Ihre Geschichte und Natur sowie Massregeln gegen dieselben'. Norstedt & Soener, Stockholm, 463 pp.

Everts, J.W. (1990) 'Sensitive indicators of side-effects of pesticides on the epigeal fauna of arable land'. Ph.D. Thesis, Wageningen, 114 pp.

FAO (1991) FAO/Netherlands Conference on Agriculture and the Environment, 's-Hertogenbosch, The Netherlands, 15-19 April 1991. Sustainable crop production and protection: background document no. 2. Rome, FAO. 27 pp.

Fliert, E. van de (1993) 'Integrated Pest Management: Farmer field schools generate sustainable practices'. A case study in Central Java evaluating IPM training. Wageningen Agricultural University Papers 93-3. 304 pp.

Goewie, E.A. (1993) 'Ecologische landbouw: Een duurzaam perspectief'? Inaugural Lecture, Wageningen. 29 pp.

Hoitink, H.A.J., Boehm, M.J. and Hadar, Y. (1993) 'Mechanisms of suppression of soilborne plant pathogens in compost-amended substrates', in H.A.J. Hoitink and H.M. Keener (eds.), Science and engineering of composting: design, environmental, microbiological and utilization aspects. Renaissance Press, Worthington (Ohio), pp. 601-621.

Jacobs, Th. and Parlevliet J.E. (eds., 1993) 'Durability of disease resistance'. Kluwer Academic Publishers, Dordrecht NL.

Koeijer, T.J. de, and Wossink, G.A.A. (1992) 'Milieu-economische modellering voor de akkerbouw'. Verslag Vakgroep Agrarische Bedrijfseconomie, Wageningen. 59 pp.

Lenteren, J.C. van (1990) 'Biological control in a tritrophic system approach', in D.C. Peeters et al. (eds.), Proceedings aphid-plant interactions: populations to molecules, Oklahoma Agric. Exp. Station, Stillwater, #177. pp. 3-28.

Lenteren, J.C. van (1993) 'Integrated Pest Management: inescapable trend', in Zadoks, J.C. (ed.) Modern Crop Protection: developments and perspectives. Pudoc, Wageningen (in press).

Martinelli, J.A., Brown, J.K.M. and Wolfe, M.S. (1993) 'Effects of barley genotype on induced resistance to powdery mildew'. Plant Pathology 42, 195-202.

Netherlands Scientific Council for Government Policy (1992) 'Ground for Choices'. Four perspectives for the rural areas in the European Community. SDU Publishers, The Hague. 144 pp.

Rabbinge, R. (1976) 'Biological control of fruit-tree red spider mite'. Pudoc, Wageningen. 234 pp.

Röling, N.G. (1989) 'The agricultural research-technology transfer interface: a knowledge systems perspective'. ISNAR, The Hague, The Netherlands, Linkage Theme Paper 6, 27 pp.

Rossing, W.A.H. (1993) 'On damage, uncertainty and risk in supervised control'. Aphids and brown rust in winter wheat as an example. Ph.D. Thesis, Wageningen (in press).

Spaan, J.H. and Ploeg, J.D. van der (1992) 'Toppers en tuinders. Bedrijfsstijlen in de glastuinbouw: een verkenning'. Vakgroep Agrarische Ontwikkeling LUW, Wageningen. 103 pp.

Stakman, E.C. and Harrar, J.G. (1957) 'Principles of plant pathology'. Ronald Press, New York. 581 pp.

Swaminathan, M.S. (1990) 'The Green Revolution and small-farm agriculture'. CIMMYT 1990 Annual Report, pp. 12-15.

Swaminathan, M.S. (1991) 'From Stockholm to Rio de Janeiro'. The road to sustainable agriculture. Monograph No. 4. M.S. Swaminathan Research Foundation, Madras. 68 pp.

Theunissen, J., Booij, C.J.H., Schelling, G. and Noorlander, J. (1992) 'Intercropping white cabbage with clover'. Bulletin OILB/SROP = IOBC/WPRS Bulletin 15/4: 104-114.

8

Thomas, C.F.G., Hol, E.H.A. and Everts, J.W. (1990) 'Modelling the diffusion component of dispersal during recovery of a population of linyphiid spiders from exposure to an insecticide'. Functional Ecology 4, 357-368.

Vereijken, P. (1989) 'The DFS farming systems experiment', in J.C. Zadoks (ed.), Development of farming systems. Evaluation of the five-year period 1980-1984. Pudoc, Wageningen, pp 1-8.

Vereijken, P. (1992) 'A methodic way to more sustainable farming systems'. Netherlands Journal of Agricultural Science 40, 209-223.

Verhaar, M.A. and Hijwegen, T. (1993) 'Biologische bestrijding van komkommermeeldauw (Sphaerotheca fuliginea) met behulp van mycoparasieten'. Gewasbescherming 23, 149.

Wijnands, F.G. (1992). 'Evaluation and introduction of integrated arable farming in practice'. Netherlands Journal of Agricultural Science 40, 239-249.

Wijnands, F.G., Janssens, S.R.M., Asperen, P. van and Bon, K.B. van (1992) 'Innovatiebedrijven geïntegreerde akkerbouw. Opzet en eerste resultaten'. PAGV Verslag nr. 144, Lelystad, 88 pp.

Wit, C.T. de (1982) 'La productivité des paturages sahéliens', in F.T.W. Penning de Vries, M.A. Djiteye (eds.). La productivité des paturages sahéliens. Une étude des sols, des végétations et de l'exploitation de cette ressource naturelle. Agricultural Research Report #918. Pudoc, Wageningen. pp 20-35.

Wit, C.T. de, Huisman, H. and Rabbinge, R. (1987) 'Agriculture and its environment: Are there other ways?' Agricultural Systems 23, 211-236.

Wit, P.J.G.M. de (1993) 'Moleculaire gewasbescherming. Schimmelgen bewaakt tomaat'. Natuur en Techniek 61, 46-57.

World Commission on Environment and Development (1987) 'Our common future'. Oxford University Press, Oxford (UK). 400 pp.

Wossink, A. (1993) 'Analysis of future agricultural change: A farm economics approach applied to Dutch arable farming'. Ph.D. Thesis, Wageningen. 221 pp.

Zadoks, J.C. (1989a) 'EPIPRE, a computer-based decision support system for pest and disease control in wheat: Its development and implementation in Europe'. Plant Disease Epidemiology 2, 3-29.

Zadoks, J.C. (ed., 1989b) 'Development of farming systems. Evaluation of the five-year period 1980-1984'. Pudoc, Wageningen. 90 pp.

Zadoks, J.C. (1993) 'Speurtocht naar duurzaamheid. Verleden, heden en toekomst van de gewasbescherming'. Lecture at the occasion of the 75th anniversary of the Wageningen Agricultural University, 9 maart 1993. Brochure, Wageningen. 25 pp.

Zadoks, J.C. and Schein, R.D. (1979) 'Epidemiology and plant disease management'. Oxford University Press, New York. 427 pp.

REFLECTIONS ON THE ECONOMIC AND ECOLOGICAL SUSTAINABILITY OF MODERN PLANT PRODUCTION

G. WEINSCHENCK
Inst. f. landwirtschaftliche Betriebslehre
Universität Hohenheim
70593 Stuttgart
Germany

ABSTRACT. The paper tries to define sustainability, distinguishing between strong and week sustainability and asks for the responsibility of science for sustainability of technical and economic development.

1. Introduction

Does Carleton deserve a monument? Gottfried Benn, the well-known German poet and novelist, asked this question about 25 years ago. (Benn, 1968). Carleton is the breeder of the famous Kubanka and Charkow durum wheat varieties which revolutionised wheat cultivation in the United States and Canada at the beginning of our century. He was not - as one would assume - the boss of one of the big plant breeding societies which dominate the seed business in the USA to day. He was also not a scientist but a simple state agent, who was motivated by the misery of his farmers, whom he served as a consultant. The new varieties spread quickly and brought a period of wealth to American farmers and to the nation. Already in 1907, less than 10 years after their introduction, the annual value of the harvest of the new varieties was estimated at 30 Mill. dollars, and in 1914 half of the total wheat harvest of the USA originated from Carleton's varieties.

Carleton himself died in poverty. He remained the little state agent with an annual salary of 2000 dollars and he borrowed money from one of the large firms - which owed all its wealth to Carleton's wheat - in order to cover the medical costs of his daughter's sickness. The daughter died in 1918 and was burnt, since Carleton could not pay for an ordinary funeral. Carleton was not able to repay the money he had borrowed and was dismissed from the civil service. He had to leave the United States and died in poverty, forgotten by the farmers and traders who were busy accumulating wealth by cultivating his varieties.

Little more than 10 years later, poverty caught the farmers again. Wheat production had expanded too fast, markets were flooded,

9

P. C. Struik et al. (eds.), Plant Production on the Threshold of a New Century, 9–22.

prices decreased and the landscape was destroyed since the farmers, anxious to accumulate as much profit as fast as possible, neglected basic crop rotation rules. The development ended in the well-known American erosion and market disaster, which John Steinbeck has described so dramatically in his "Grapes of Wrath". After a period of 30 years of wealth, farmers became poorer than they had been before the introduction of the new varieties.

The question, whether Carleton deserves a monument - a late Nobel prize, like his more fortunate colleague Borlaug, in our days - is principally important when asked about the value of technical progress in general. Gottfried Benn presents two answers. The one is "yes". It is sufficient to serve the needs of the living generation. Carleton could not know that people would misuse his invention destroying the markets as well as the landscape. Anyway, since he did not have the power to avoid the misuse of his inventions, he is not responsible for it. The second answer is "yes but". It realises the consequences knowing that everything has contradiction within itself, and that even the apparent short run benefits of a new variety of wheat in a period of hunger and poverty can change into destruction. It demands that if Carleton should get a monument at all, the inscription must remind us of the ambiguity of all things, even the things which seem beneficial at present, beyond any doubts. "Eternity" Benn proposes is "where flowers are like poison and serpents, like dragon-flies delicate and mortifying in one". (Benn, 1968, p. 772).

2. Weak and strong sustainability

The first answer sounds reasonable and realistic, though maybe a bit simple and superficial. The second answer is perhaps a bit too philosophical.

In our prosaic days, one would simply ask for the sustainability of the introduction of Carleton's varieties. This is as ambiguous as the second answer, taking into account that we live in a dual world divided into poor and rich societies, in which people suffer from hunger while farmers produce more than the markets can absorb. The concept of sustainability is not clearly defined. One can distinguish weak (anthropocentric) and strong (biocentric) sustainability (Batie, 1989, Weinschenck, 1991).

Weak sustainability is based on the concept of intergenerational equity. Strong sustainability includes, additionally, harmony with nature and the dignity of natural life. The concept of intergenerational equity is generally accepted. It demands that the current generation does not compromise the ability of future generations to meet their material needs and to enjoy a healthy environment.

Strong sustainability demands, in addition, fairness to the non-human elements of natural life which share the universe with human beings. Strong sustainability is debated. It is too narrow, even serious advocates of an environmental policy say, considering that plants and animals are necessary resources to satisfy human needs. The biocentric answer is: Fairness to the non-human elements of natural life does not mean that human being must not use them to satisfy their needs. Humans are even allowed to kill them. What counts is the way animals are kept and the urgency of the purpose for which they are killed.

2.1. CHARACTERISTICS OF AGRICULTURAL PRODUCTION

Strong sustainability is of particular importance in agriculture and it is in agriculture where it is discussed most passionately due to the particular characteristics of production.

Modern eco-economists like Immler (1990) suggest to reconsider the physiocratic theory, which suggests that nature is the only true productive element in the economy. In agriculture this is evident. Milk is produced by cows. The farmers only work with cows by shaping their environment and supplying feed in order to make the cows produce. Naturally cows are on the farm in order to produce, however this is not their only purpose. Besides of milk production they have an end in themselves. The farmer cannot produce wheat, he puts the seed in the earth and then has to wait for the nature to make wheat grow. He influences only the conditions under which the seed will be transformed into wheat.

The farmer uses the landscape, but agriculture is landscape as well. It is using and sharing the resources of the landscape with other natural elements which are part of the ecosystem, having value in itself.

If mankind wants to increase the productivity of agricultural efforts, it intervenes mainly in three ways:
- changing the characteristics of animals and plants by breeding new varieties,
- transforming the landscape by displacing natural elements which hinder or do not serve the productive objectives,
- increasing the use of capital by adjusting their equipment to the requirements of high labour productivity, frequently disregarding crop rotation rules and the conditions of animal husbandry true to kind.

These kinds of intervention change the characteristics of the animals and plants used in agricultural production and the aesthetic value of the landscape as well as the conditions of life essential for maintaining the diversity of varieties in the landscape and the quality of the life of animals on the farm.

2.2. THE VALUE OF SUSTAINABILITY

The concept of sustainability puts limits to these interventions. With respect to the principle, there is no difference between weak and strong sustainability.

Even a strictly anthropocentric view is not indifferent to the maintenance of the beauty of the landscape and the diversity of

species living in it. The anthropocentric point of view is concerned with the kinds of landscape in which future generations will live, whether they will be surrounded by useful plants and animals only and how their food will be produced. For the biocentric point of view, these concerns go without saying. Hence, both concepts include the same subjects, namely
- the material and aesthetic needs of present and future gene- rations,
- the preservation of the conditions of life for the animals used in agriculture, as well as for the natural elements sharing the use of the landscape with agriculture,
- the protection of the dignity of life with respect to the creatures which bio-technical engineering might be able to construct in the future.

Though agreeing with respect to the items which ought to be included in the concept of sustainability, the anthropocentric and the bio- centric versions differ with respect to the values attached to the items. As a consequence they differ with respect to the limits in shaping the organization of a farm or the structure of an economy.

In the anthropocentric version values are estimated according to the benefits which the observation of sustainability provides to the present generations. Sustainability is observed only to the extent at which these benefits exceed the opportunity costs measured by the value of lost material production.

The strong (biocentric) version of sustainability is based on the intrinsic value of nature. Nature has to be respected for its own sake. It has to be respected, not exploited. The use of intrinsic values implies the concept of "enoughness" and implies greatly reduced rates of economic growth, at least for developed nations (Sachs, 1989). It is a minimization concept that implies minimizing the use of the nature environment observing that nature has an end in itself.

2.3. SOCIAL SUSTAINABILITY

Technical changes not only affect ecological structures, they also affect social structures as we all know. The misery of European agricultural policy results, at least partly, from the fact, that it does not want to accept the social structures which follow from adjustment to the changed technical and market conditions. The green revolution was criticized for its social implications inspite of its undeniable merits with respect to the provision of food. Other examples could be easily added. They show that policy attaches intrinsic values not only to ecological phenomena but also to social structures, like full employment or the maintenance of family farms in agriculture. It seems necessary, therefore, to complement the dis- course on environmental sustainability by introducing the concept of social sustainability.

The social disasters which occur if the rules of ecological sustainability are not properly obeyed, are obvious as for instance in Carleton's case. However, social changes may also occur within the limits of ecological sustainability, affecting the benefits of technical changes which have induced the social changes. There are good reasons to assume that the biotechnological progress, which is in the pipeline of modern plant production, will have not only ecological but also social consequences which could endanger the social sustainability, especially in developing countries where 50 - 80 % of the active population is still working in agriculture. What for instance will happen, if biotechniques succeed in producing food, not in the country side but in factories at a fraction of the current cost and labour input?

The concept of sustainability demands a discussion of ecological and social issues simultaneously.

2.4. SUSTAINABILITY, FARMING SYSTEMS THEORY AND ECONOMIC THEORY

Sustainability has always been considered an essential element in farming systems theory, at least in Europe. Consideration was, however, restricted to the production system, predominantly aimed at the sustainability of soil fertility. Brinkmanns's concept of the farm as an organismus can be considered as an early version of a sustainable farming system. It does not consider a single field but the farming system as a whole, of which the sustainability is based on an equilibrium between fertility production, by "fertility producing plants" and manure and fertility consuming plants like cereals, sugar beets etc. (Brinkmann, 1922, 1950). Sustainable agriculture is more than a sustainable farming system, it includes the maintenance and enhancement of the quality of the environment of which it is a part.

The structure of a farm-household landscape system is shown in Chart 1.

Its general structure can be described as follows: Given are
1) The production functions including
    a) the traditional inputs land, labour, capital, including fertilizer, herbicides,
    b) the "ecological" inputs like the quality of the soil and the landscape, the weather.

$$(1) \quad Y_t = f(W_t, S_t, Z_t, P_t, L_t, T)$$
Y = Yields
W = Weather variable
S = Structure of the soil (pH-value, physical structure)
Z = Chemical structure of the soil
$Z = B + x_1, x_2, \ldots, x_n$
    B = Existing nutients + inputs from the air
    $x_1, x_2, \ldots, x_n$ traditional inputs like fertilizer herbicides
P = Expectation of plant diseases, weeds etc
L = Ecological quality of the landscape
T = Technical progress.

14

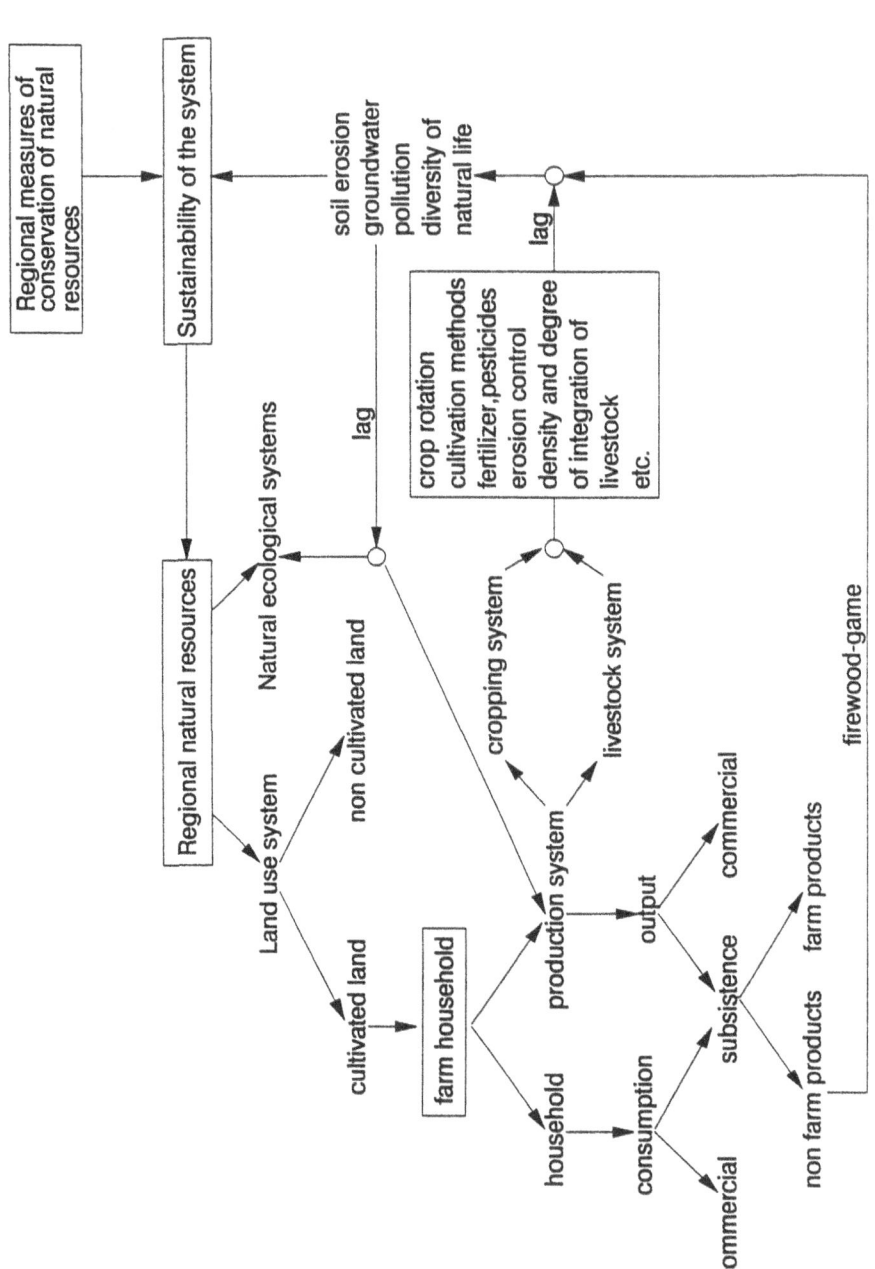

Chart 1. Structure of a farm-household landscape system.

S is not an exogenous variable, but determined by the history of the system as follows:

(2) $S_t = f(W_{t-n}, U_{t-n}, A_{t-n}, X_{t-n}, P_{t-n}, L_{t-n})$

$t-n = t$ and previous years with the vectors

W = Weather
U = Utilization of land
A = Cultivation methods
X = "Traditional" inputs (fertilizer, herbicides etc)
P = Plant diseases and weeds
L = Structure and quality of the landscape.

2) Fixed inputs.
3) Prices for products and means of production.
4) The objective function containing economic success and ecological quality.
   (Lanzer and Paris, 1981; Weinschenck, 1986; Weinschenck and Werner, 1989).

The model is an explanation model. The present state of arts is far from being able to specify the dynamic relations except with respect to soil erosion in some regions.

The concept is realistic as well as misleading. It is realistic because close interdependencies exist between the production system and the natural elements. It is misleading because the system consists of two partial systems, producing two sorts of goods:
-   marketable goods which have a price,
-   ecological quality which has a value but not a price consisting of the components
        soil erosion,
        ground water pollution,
        diversity of natural life,
        beauty of the landscape.

Only the marketable goods are subject to economic incentives and incentives initiated by changes of the technology.

There are interrelations between the two systems but only one direction is actually considered. The ecological quality of the landscape is influenced by the structure of the production system and by the intensity by which the landscape is used. Prices and the state of technology determine ecological quality almost exclusively neglecting objectives derived from the concept of sustainability.

The concept of sustainability introduces separate objectives with respect to the ecological quality and, as we have seen, with respect to social structures. These objectives are normally not consistent with the traditional economic objectives of profit maximization at the micro level and acceleration of economic growth at the macro level.

3. Problems of determining sustainable technical development

At present almost any new product or technology is introduced with the advertisement that its use will improve environmental quality, even a car using more than 15 litres gasoline per 100 km. This does not mean, of course, that the new products or new technologies will truly raise the sustainability of present economic and technical development. It means only that people have become aware that sustainability counts from the viewpoint of the society. Ecological quality is not longer considered a dependent variable but an objective. This will change the role of technology. Chart 2 demonstrates this change. The development of technology is considered as a quasi-independent variable in economic theory which determines ecological and social structures, subject only to the objectives profit maximization and the acceleration of economic growth.

Introducing the concept of sustainability changes, radically, the position of technology in socio-economic analysis. The objectives of sustainability put constraints on the realization of profit maximization and economic growth and influence the choice of technology. The methods to determine priorities of technological research change accordingly. Conway and Barbier have introduced the distinction between the "technological push approach" and the "needs pull aproach" (Conway and Barbier, 1988).

3.1. TECHNOLOGICAL PUSH AND NEEDS PULL APPROACH

The "technological push" approach is based on the traditional position of technology. It puts the innovations first. Possible gaps between the performance on the station and on the farm are investigated by "On Farm Research" which more or less only transfers the experiment field from the station to the farm. After the tests have been satisfying the innovations are presented to the economy usually accompanied by a little persuasion, but without consideration of social and ecological stability.

The starting point of the "needs pull" approach are the farmers' and the country's needs. An analysis of desirable development of future technology is based on farming systems and sectoral analysis which determines the needs and constraints. The needs and constraints are the basis for the assesment of research priorities.

The push approach is sustainable only if it considers the interests of future generations and the interests of the elements of natural life which live in the agricultural landscape. Carleton's story shows that the pull approach is no guarantee for sustainable development, if sustainability is not included in a farming system analysis. As we know to day, the American disaster would probably not have occurred if farmers had been taught to place Carleton's wheat varieties into sustainable crop rotations and into a sustainable use of the landscape.

The most important alternatives are not "push or pull" but "consideration or neglection of sustainability". Thus we are back to our initial question: Does the observation of sustainability belong to the responsibilities of research? This question disturbs the conscience of the research community at least since the discovery of

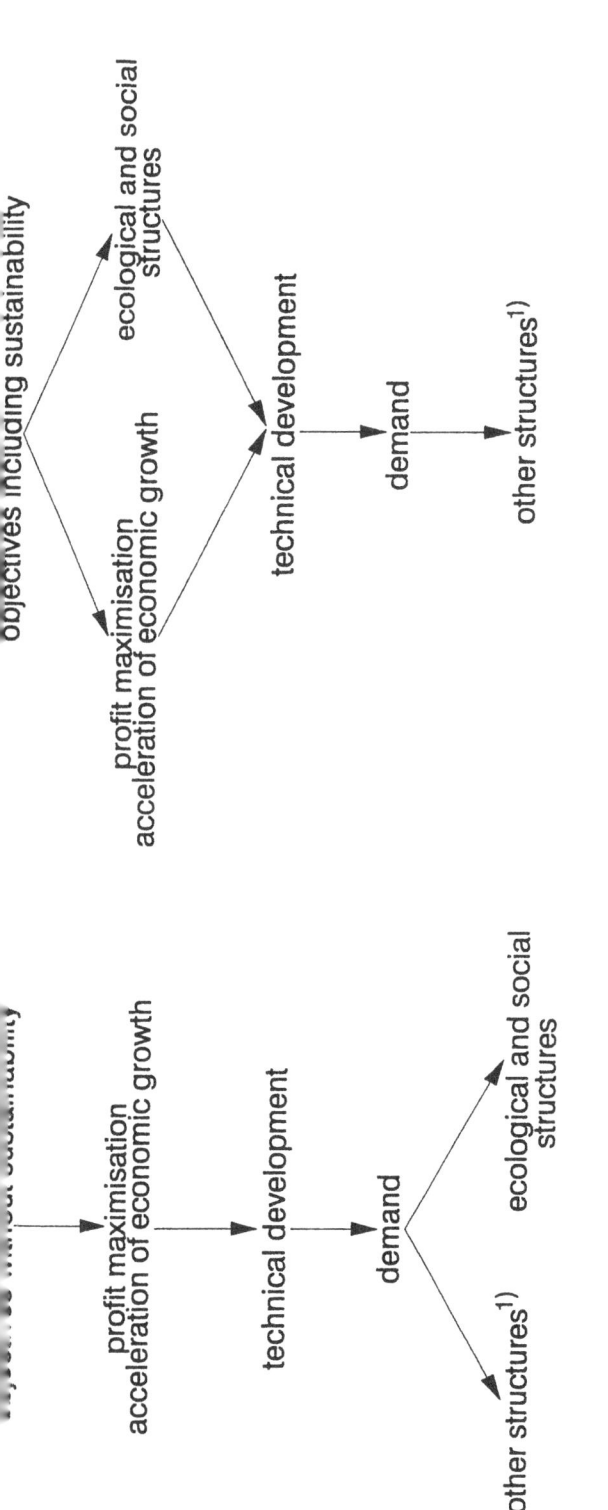

objectives including sustainability

ecological and social structures

profit maximisation
acceleration of economic growth

technical development

demand

other structures[1]

profit maximisation
acceleration of economic growth

technical development

demand

ecological and social structures

other structures[1]

[1] structures not related to ecological or social objectives

Chart 2. Change of the role of technology in the economic theory.

atomic energy. "Oppenheimer said after Hiroshima that scientists have made acquaintance with sin discovering atomic energy" (Jonas, 1985). Some years later this happened a second time. In biotechnology, especially in genetic engineering scientists met sin again. The prospects of "recombining DNA research" alarmed scientists to an extent, that they agreed on a moratorium, in order to examine the consequences and to look for safety rules. It was the first time - as far as I know - that researchers realised the principal novelty and the risk of their doing and that they revealed their concern to the public. However, the moratorium did not last long, it was released after a few years. It had been exaggerated and not necessary we were told, without explaining why the evaluation of risks and ethics has been changed. Maybe the explanation is that the scientific discoveries had been passed to private industry which is less sensible than anxious scientists.

Anyway the "devil is out of the bottle". Does it mean that research is not willing or able to care for the sustainability of its discoveries? The answer is "yes but". The classical ideal that science exists for the sake of science does not exist any longer. Science has married utility. Theory and practice (fundamental research and applied research) can hardly be distinguished any more. There is almost no discipline of natural science of which the discovery could not be used in the technical world and as a consequence there is a continuous feed back from industry resp. agriculture to research of which one must not underestimate the financial and the intellectual value. If a discovery can be transformed into a profit by developing a corresponding technology, it will be transformed regardless of the risk with respect to sustainability.

The development of atomic energy and the recent development of biotechnology show that the development of a new technology cannot be prevented by the mere risk of a catastrophe. The demand of Jonas to observe the principle in "dubio pro malo" does not find consent. One argues that the probability of a catastrophe is very low without considering that the indefinite scale of possible catastrophes makes the measurement of probabilities absurd, let alone ethical consider- ation. But it would be unfair to say science does not care for sus- tainability. Environmental research has become an almost classical example of pulled research, initiated by the care for sustainability.

In recent years a great deal of efforts of agricultural research has been concentrated on environmental problems. However, not the objectives but the disasters were first. Researchers began to concentrate on erosion problems after the American catastrophe in the thirties. Ground water pollution became a research problem after the water had been extremely polluted and books on ecological economics came out like mushrooms, after it has become obvious that economics was affected by the already existing environmental problems. The pull character of modern environmental research is undeniable but is has repair or ex post character. One tries to shut the stable door after the horse is stolen. This is certainly a necessary approach after the problems are there and the structure cannot be changed immediately. But in the long run one will have to ask how to prevent the problems.

Future oriented sustainable research has to ask for sectoral agricultural systems consisting of sustainable agro-ecosystems. What does
that mean in a dual world divided in rich and poor societies?

3.2. SUSTAINABILITY IN A DUAL WORLD

In our days, in a world with a rapidly growing population and almost 1
Milliard undernourished people, Carleton would probably, like his
colleague Borlaug, win the Nobel prize, especially if his wheat
varieties would be suitable for cultivation, more or less exclusively
in developing countries. But would he have been eligible for the Nobel
prize if his varieties would have been cultivable only in the United
States or Europe?

The present world situation is well known. I summarize the basic
facts:
1) Population grows by 92 Mill. annually. 88 Mill. of them live in
   developing countries.
2) 500 Mill. to 1 Milliard suffer from hunger.
3) Population growth and inadaequate production methods destroy the
   ecology and endanger the sustainability of the factor potential.
   6 Mill. ha of the agricultural area of the world are annually
   displaced by desert. (Brown et al, 1988).
4) In the surplus countries, farmers produce more than the markets
   can absorb and endanger the environment by employing specialized
   crop rotations, using too much fertilizer and pesticides,
   maintaining too high a density of animals and displacing any bush
   and flowers in the name of agricultural progress as John Stuart
   Mill warned, already in the 19th century.

The depressing conclusion is the need for food increases where
production decreases or stagnates, while production increases where is
no demand. This seems to be a perfect disequilibrium situation to be
compensated for by trade. However, where the need is, there is no
purchasing power and where the purchasing power is, there is no need.
Farmers in the surplus countries are not willing nor able to produce
without getting paid, and the hungry people in the developing
countries can neither pay nor find sufficical employment outside
agriculture.

The development of agriculture deserves high priority in the
developing countries and in the so called transition countries. That
means:
-   In developing countries in which the agricultural sector is not
    competitive on international markets and can meet only part of
    the needs of the population, imports resp. food aid must equalize
    the difference between needs and production, but both must
    observe the priority of domestic agriculture, giving priority to
    its maintenance and the development of its potential.
-   If the agricultural sector of developing countries or of the
    present transformation countries in eastern Europe produces more
    than what can be sold on domestic markets, they must have access
    to the markets of the industrial countries. For instance. What

else shall Poland or Hungary export to improve their balance of payments and promote economic development if not meat, grain and vegetables.

However, it does belong to the responsibilities of the scientific community to repeat again and again that there is no chance for a sustainable development without a drastical reduction of population growth. Neither biotechnology nor any other kind of technical development are able to solve the problem of sustainable development at present growth rates of population.

In the EC-countries the answer to the question what sustainable development means is fairly easy. With respect to the development of production the upper limits have been determined by the recent agreement between the EC and Washington, preceding the probable GATT agreement. The EC must reduce its production of any single product (milk, wheat, eggs etc) until production does not exceed 95 per cent of the domestic consumption.

The reduction of production can be achieved principally by three ways:
1) By displacing the production in marginal areas principally without changing the intensity of the use of the landscape in areas which remain in production.
2) By quotas and set aside programs, paying the farmers for not producing.
3) By improving the sustainability:
   - reducing the inputs of fertilizer, herbicides and pesticides,
   - reducing the specialization of farms by reintegrating livestock in the farming systems,
   - by applying methods of animal husbandry which are true to kind,
   - by improving and encouraging biological farming systems,
   - by using the set aside program to implement "biotops" not only in the marginal areas.

New technologies which increase the productivity of land, animals and labour are attractive for the single farm, if they are profitable. However, they are of little value from a sectoral point of view. Let us consider the discourse on the introduction of BST (Bovines Somatotropin) as an example. The introduction of BST would
   - increase the milk yield per cow,
   - reduce the number of cows,
   - displace grassland because the input of concentrates per kg milk will increase.

Considering the long run sustainability one must realize that imported feed used by cows at present will be needed sooner or later by direct human consumption at least in some of the exporting countries. From a global point of view the concept of long run sustainability demands therefore to emphasize the original role of cattle, to transform grass in human food. Additionally one might expect a reduction of the consumption of milk if BST is introduced because consumers might be afraid that the consumption of milk implies a health risk.

Mutatis mutandis similar arguments hold for the plant sector. Consumers are aware and afraid of the risks of modern plant production, especially associated with modern biotechniques and genetic engineering.

The increasing demand for products produced by biological farming indicates: The determination of quality is changing. Quality is not longer only determined by the product in the view of modern consumers, but also by the way it is produced.

Hence I see little room for the application of modern biotechnology in european agriculture which justifies the risk, which is associated to it. Research which feels responsible for sustainability might rather concentrate on the improvement of what is at present called biological agriculture. This does not hold for developing countries. Here, even at an optimisitc point of view the static stage of population growth will be reached only after a transition period in which the population will grow with declining growth rates ending with a population of more than 10 Milliards against about 5.3 Milliards at present.

Hence the observation of at least a minimum of social and ecological sustainability requires an increase of the productivity of labour and land.

It is beyond this paper to investigate how the necessary increase of the productivity of land and labour can be achieved. The alternatives are the development of the new methods which are in the pipeline of modern technology or the continuation of the traditional effort by classical breeding methods and improvement of cultivation methods. In determining research priorities one must realize that the countries with the largest differences between its need for food and food production are countries in which the agricultural sector 15 consists of small family farms employing 50 - 80 % of the population. The supply of food for the cities is usually their only source of cash income. The farms have to support not only the labour forces actually needed, but all family members unable to find employment in the cities. Social sustainability requires to rely on production methods which provide employment and income to the peasant farmers as long as there are no employment alternatives in the cities.

Carleton has looked for the social needs of "his" farmers. This was the starting point of his work. Therefore I think he deserves a monument. Whether this can be said from the biotechnology remains to be seen. Carleton's varieties could have been cultivated in proper and well known crop rotations without any risk to ecological sustainability. The risk of biotechnology to social and ecological sustainability is inherent in its methods and products and cannot be excluded.

4. References

Batie, Sandra S. (1989) 'Sustainable Development: Challenges to the Profession of Agricultural Economics', American Agricultural Economics Association, Vol 71/5.
Benn, G. (1968) 'Gesammelte Werke 3, Essays und Aufsätze'. Wiesbaden, Zürich.

Brinkmann, Th. (1922) 'Die Ökonomik des landwirtschaftlichen Betriebes'. Tübingen.

Brinkmann, Th. (1950) 'Fruchtfolgebilder des Deutschen Ackerbaues'. Bonn.

Brown, L. *et al*. (1988) 'State of the World'. New York, London.

Conway, G.R. and Barbier, E.B. (1988) 'After the Green Revolution. Sustainable and Equitable Agricultural Development'. FUTURES, December 1988.

De Kruif, P. (1929) 'Bezwinger des Hungers'. Leipzig, Zürich. Immler, H. (1990) 'Vom Wert der Natur'. 2. Aufl, Opladen.

Jonas, H. (1985) 'Zur Praxis des Prinzips Verantwortung'. Frankfurt am Main.

Lanzer, E.A. and Paris, Q. (1981) 'A New Analytical Framework for the Fertilizer Problem', American Journal of Agricultural Economics, February.

Sachs, W. (1989) 'A Critique of Ecology: The Virtue of Enoughness', New Perspectives Quart. 6, p. 16-19.

Weinschenck, G. (1986) 'Ethische, analytische und wirtschaftspolitische Fragen zum Thema Landwirtschaft und Landschaft', in Berichte über Landwirtschaft, Band 64, Heft 3, S. 398-407.

Weinschenck, G. (1991) 'Ethik und Ökonomik des sorgsamen Umgangs mit natürlichem Leben in der landwirtschaftlichen Produktion', in H. Rahmann and A. Kohler (Hrsg.), Tier- und Artenschutz, 23, Hohenheimer Umwelttagung, Weikersheim.

Weinschenck, G. and Werner, R. (1989) 'Methoden und Modelle zur Optimierung der Landschaftsnutzung durch Landwirtschaft', in Neure Forschungskonzepte und -methoden in den Wirtschafts- und Sozialwissenschaften des Landbaues. Schriften der Gesellschaft für Wirtschafts- und Sozialwissenschaften des Landbaues e.V., Band 25, S. 91-100, Münster-Hiltrup.

FARM ECONOMIC MODELLING OF INTEGRATED PLANT PRODUCTION

G.A.A. WOSSINK AND J.A. RENKEMA
Wageningen Agricultural University
Department of Farm Management
Hollandseweg 1
6706 KN Wageningen
The Netherlands

ABSTRACT. In analysis of future agricultural change the trade offs between environmental and economic goals of agricultural production is an important issue. The present paper focuses on the use of linear programming for interdisciplinary research by economists, agronomists and soil scientists. Application of the environmental economic model for scenario analysis allows the effects of technical developments, environmental policy and price and market policy on future farm organization in arable farming to be assessed. The approach is evaluated and priorities for further development are discussed.

1. Introduction

Increasing environmental problems in agriculture urge policy makers to develop instruments to reduce and control the pollution caused by current intensive farming practices. These measures should be both effective from the ecological point of view, which is a public goal, as well as acceptable at the farm level with regard to private goals such as income and continuity of the farm. Therefore, information is required concerning the complex interaction of production intensity, environmental aspects and farm income.
Environmental economic modelling aims to improve the insights in the complex relations between the economic, technical and ecological aspects of plant production. The results from these analyses can be used to support (a) farm management decision making (planning), (b) agricultural policy development and the selection of policy instruments (conditional forecasting) and (c) drawing up the future research agenda by indicating relations insufficiently investigated and their relative importance in farm organization.
  In the present study the individual farm is chosen as the starting point. Within this context the paper focuses on (a) structure and data requirements of environmental economic models, (b) the inter-disciplinary collaboration with crop scientists and environmental experts, (c) application of the environmental economic farm model for

23

P. C. Struik et al. (eds.), Plant Production on the Threshold of a New Century, 23–36.
© 1994 *Kluwer Academic Publishers.*

a scenario study of Dutch arable farming and (d) priorities for further research. The study was part of a larger project (see Wossink, 1993).

2. Environmental criteria and policy instruments

Environmental pollution in arable farming is caused by the use of minerals, in particular nitrogen, and biocides. Measuring the ecological damage as such is very difficult. Instead, usually, criteria are chosen as indicators of the factual or expected damage. In fact these criteria reflect an acceptable level of environmental quality. With regard to nitrogen the concentration of nitrate in the groundwater is used as the standard measure. For biocides the concentration in the groundwater, the emission to the surface water and the toxicity for aquatic organisms are well-known criteria.

For efficient policy instruments detailed knowledge of input/damage relationships is needed, both of the current production practices and with regard to new techniques. However, in agriculture, identification of this relationship is complicated by the role played by natural conditions in determining pollutant emissions, even when the production practices are the same. Note further, that the standards of environmental quality are formulated in relation to the intensity of production (units of emission to the groundwater per ha) and not as another efficiency measurement (i.e. input/(negative) output relationships, so units of emission per kg agricultural product) because of the in *natura use* of environmental sources and sinks. Hence, a clear distinction is to be made between the economic goals of efficient allocation and sustainable scale[1]. This point is particularly worth noticing in the context of the dialogue between agronomists and economists on resource use "efficiency" in agriculture (see e.g. De Wit, 1992; NRLO, 1993).

The alternative policy instruments to influence pollutor's behaviour can be classified into three categories (Baumol and Oates, 1979): (a) moral persuasion by publicity or social pressure, (b) direct controls and (c) methods that rely on market processes.

At the moment direct controlling is the main instrument used in Dutch environmental policy. Examples are the so-called Manure Law and the Admittance Law for biocides. The Manure Law indicates maximum input levels for P_2O_5 specified for types of soil and the crops grown. For N the same approach is in preparation.

The Long-term Crop Protection Plan (Min. LNV, 1990) presents quantitative objectives. The major strategic headline is to reduce the

---

[1] Daly (1992): "In loading a boat we also have the problems of allocation and scale: allocating or balancing the boat is one problem, and not overloading even a well-balanced boat is another problem. Economists who are obsessed with the allocation to the exclusion of scale really deserve the environmentalists' criticism that they are busy rearranging deck chairs on the Titanic".

input of biocides (in terms of the weight of active ingredient compared to the average use over 1984-88) by 35 % in 1995 and 50 % in 2000. Each sector of agricultural production has been given its detailed goals. For arable farming the percentages in Table 1 are the guideline. The Dutch farmers' union signed an agreement in May 1993 that commits them to achieve these goals. A levy on biocides (imposed by weight of active ingredient) was suggested at the political presentation of the Long-term Plan in June 1991 in case of defaulting on the 1995 goals.

TABLE 1. Reduction goals for biocide use in arable farming in the Netherlands.

Category	Percentage reduction total kg active ingredient[a]	
	by 1995	by 2000
Nematicides	46	70
Herbicides	30	45
Insecticides	15	25
Fungicides	15	25
Others	42	68
Total	39	60

[a] Compared to the average use over 1984-1988.
Source: Min. LNV, 1990

Further future goals for nitrogen and biocides relate to the maximum concentration of biocides and nitrate in groundwater. These restrictions are: maximal 0.5 microgram/l groundwater for biocide leaching per crop and 0.1 microgram/l per individual biocide per crop and maximal 50 milligram NO_3^-/l groundwater for nitrogen (Min. VROM, 1989).

3. Model specification and data requirements

3.1. METHOD

For simulating the economic decision making process at farm level linear programming (lp) and its extensions are frequently employed. These methods present the collection of relevant technical opportunities offered to the farm by separate activities in a process matrix. Programming methods are well suited for environmental economic research because: (a) many activities and restrictions can be considered at the same time, (b) an explicit and efficient optimum seeking-procedure is provided, (c) results from changing variables can be calculated easily and (d) new production techniques can be incorporated easily by means of additional activities in the model.

An environmental-economic farm model based on this technique covers -- besides the regular items of production such as cropping pattern, cultivation operations, labour supply and requirements and investments -- an additional component, which incorporates the environmental parameters selected for the cropping activities.

3.2. MODEL STRUCTURE

The general form of the lp model is shown in Figure 1. The activities out of which the optimal combination is to be chosen by the solution procedure, are shown across the top in Figure 1 under five headings: production activities representing different crops and cropping variants per crop, variable operations (choices among using own mechanization or contract work and among methods of control), casual labour, 0/1 activities representing new machinery for chemical and mechanical crop care and activities for the use of the whole range of biocides.

The rows of the matrix indicate the type and form of the constraints included: total land, rotation restrictions, supply of fixed and of casual labour, several coupling restrictions and the discharges of biocides and nitrogen to groundwater. Thus, each unit (hectare) of a production activity requires inputs represented by its specific vector in the lp matrix. Among the inputs the type and quantity of biocides (in kg active ingredients) related to a production activity is specified.

The leaching figures for biocides and nitrate into the groundwater are added to every cropping activity as quasi-external data. The assessment of the emission figures is described in section 3.3.

The model distinghuishes a maximum of six different operations per crop, namely: land preparation, ploughing, seedbed preparation, planting/sowing, fertilization, crop care and harvesting. In addition there are different methods for several operations. As the intention is to indicate the effects of environmental policy regulations considering technical developments such a detailed specification is a requirement.

As indicated in Figure 1 the linear programming model covers the option of several investments in new machinery, namely: (a) new spraying device, (b) new mechanical technique for weed control in potato and (c) mechanical technique for haulm killing in potato. The innovations considered provide for a reduction in the input quantities and costs of biocides. On the other hand labour and tractor hours increase when adopting the new machinery. Both effects are accounted for by linking the investments to specific cropping variants using the machinery. The capacity of the new machinery is assumed to be never limiting in the situation considered, as expressed by the -999 figures.

The linear programming model optimizes the farm labour income, being the difference between the total of the gross margins of the crops in the optimal plan and the costs for biocides, contract work, additional investments, casual labour and fined charges.

For a specific policy instrument the optimal farm organization is assessed, indicating changes regarding: income, cropping pattern and

ACTIVITIES	Production activities 1,........n	Seasonal labour 1,.........	Var.operation: methods of control and own mechanization or contract work 1,.........m	New machinery for crop care 0/1 a b	Biocides c 1,........q	Right hand Side
CONSTRAINTS						
max. hectares	+1					Total land
Rotation restrictions	+1					Max. ha of each crop or group of crops
Fixed labour in period of 14 days	+a$_{1j}$	-1	+a$_{1j}$			Available fixed labour in hours
Seasonal labour in periods of 14 days		+1				Available var. labour in hours
Coupling production activities and var. operations	+1		-1			<= 0
Coupling production activities and new mach.	+1 +1			-999 -999		<= 0 <= 0 <= 0
Coupling production activities and biocides	+a$_{1j}$		+a$_{1j}$	999	-1	<= 0
Discharges of biocides and nitrogen to groundwater	-a$_{1j}$		-a$_{1j}$			
OBJECTIVE FUNCTION	Gross margins excl. Costs of biocides	Costs per hour	Costs per hectare	Annual costs a b	Costs per kg a.i. c	

FIGURE 1. Structure of the linear programming model.

cropping variants, variable operations, labour and tractor hours used and input and emission of biocides and nitrate.

The matrix described, contains about 200 activities and circa 210 constraints. The initial farm situation is specified by circa 70 non-zero right hand side values, depending on the number of crops in the rotation scheme.

3.3. EMISSION ASSESSMENT

Each cropping activity in the model has ascribed quantities of biocides and a certain N-dressing. In future regulations maximum concentrations in groundwater will be important. In model constructing both the input figures and the leaching figures are regarded to be able to analyse the implications of present and of future environmental policy.

Emission models were used to translate a quantity of discharge of nitrogen or biocides into concentrations level in groundwater. To quantify the emission of biocides by percolation into the upper groundwater the model developed by Van der Linden and Boesten (1989) was used. The input variables of this model are the persistence of a biocide (DT_{50}) and the absorption coefficient per weight unit of organic matter (K_{om}). The DT_{50} and K_{om} values were derived from a report of "De Werkgroep Bestrijdingsmiddelen in grondwater naar aanleiding van de notitie Milieucriteria" (The Workgroup Biocides in the ground-water actuated by the note Environmental Criteria) (Van den Berg, 1990; Brouwer, 1990). In line with the tenets of environmental policy, the emission value for a sandy, wet nonhumous soil, is used, even though on clay soil, for instance, the real percolation emission will be lower (De Koeijer and Wossink, 1990).

Analysis of the emission of nitrate to the groundwater started with a balance approach for the different crops. Of the N surplus assessed in this way the larger part denitrificates[1]. The remainder is emitted to the groundwater. The resulting nitrate concentration in the groundwater is added to the different production activities in the linear programming model as another environmental parameter (De Koeijer and Wossink, 1990).

3.4. CROPPING VARIANTS

In the present study for every crop several so-called cropping variants are defined which differ in economic and environmental values. The variety of these cropping variants with regard to environmental parameters is represented in the technical coefficients in the lp matrix. Financial differences are given by means of the gross margin figures in the objective function (see Figure 1).

---

[1] For the region considered in the model computations (section 4), a denitrification rate of 80 % is assumed (Breeuwsma *et al.*, 1987).

In defining the cropping variants, which differ in process variables, such as fertilization and crop protection practices, the authors were advised by crop scientists and soil scientists. For potato, for instance, seven process variables are considered (Schans, 1991; De Koeijer, 1991):

Process variables for potato and number of alternatives

-rotation	4*
-variety	2*
-nematode control	5*
-N-dressing method	3*
-late blight control	2*
-haulm killing	2*
-weed control	3*

With these process variables it is possible to build 4 * 2 * 5 * 3 * 2* 2 * 3 = 1440 cropping variants for potatoes only. Putting all those variants in the model would make the model much too big, the more so because the other crops also have lots of variants. To reduce this number, a selected number of combinations were chosen of the first three process variables. Some of the possible combinations were not logical, for instance soil fumigation and a rotation of 5 years, so the amount of combinations could be limited further. Further potato cropping without any biocide use (ecological variant) was excluded as being too risky. Eventually 23 cropping variants for potato were integrated in the model[1].

The alternatives of the process variables haulm killing, late blight control and weed control were supposed to have no influence on the yield of potato and were built in the model as apart activities (see methods of control in Figure 1). So, the optimal combination can be chosen by the model.

In the extended model each crop has cropping variants ranging from the intensive to the ecological production system, representing a discrete set of production alternatives per crop which cover both: (a) successive points on a non-linear production function, and (b) points on different production functions using different technology. With this range of variants it is possible to investigate the effects of technical innovation and of several policy instruments on farm organization.

---

[1] Note that two potato varieties were considered, namely Bintje and a PSR variety (resistant to potato nematodes pathotype A). For the environmentally friendlier cropping variants a fixed combination of 50 % Bintje and 50 % PSR variety was chosen, as it is expected that growing more of the PSR variety will stimulate the development of new pathotypes from the present nematode population in the soil.

4. Application for scenario analysis of arable farming

4.1. ASSESSING A REPRESENTATIVE FARM

The region of the North Eastern Polder in the Netherlands was selected
for model demonstration because of its intensive cropping pattern and
for its distinct natural geographic boundaries and data availability.
For the calculations a model farm was selected, representing 239 of
the 864 specialized arable farms situated in the North East Polder[1].
In the assessment of the basic situation, i.e. the amount of biocides
currently used and the present farm organization and income level, the
model was allowed to choose the current production practices only. The
resulting intensive cropping pattern for the representative farm of 30
hectare of heavy loam soil, includes one third potato, a quarter
sugarbeet and a quarter wheat, and some hectares of chicory and onion.
This crop mix fits the statistical information on the factual cropping
pattern.

Table 2 (first column) indicates current biocide use of the
representative farm. Total use of biocides, and of nematicides in
particular, is very high as compared with the estimated average use of
biocides in Dutch arable farming which is 18.6 kg a.i./ha per year, of
which 12.7 kg nematicides (Min. LNV, 1990). At the model farm no less
than 85 percent of the total amount of biocides is used for soil
fumigation.

4.2. COMPOSITION OF THE SCENARIOS

The computations performed with the lp model are of a comparative
static nature and enable the changes in farm organization resulting
from six scenarios (see Figure 2) to be ascertained.
These scenarios represent combinations of (1) technical developments[2],

---

[1] The total population of 864 specialized crop production farms was
reduced to 8 representative farms by means of cluster analysis. Each
model farm represents a specific share of the population. The
analysis is restricted to one of these model farms.

[2] The mechanical innovations considered in the present study were: (1)
improvements of spraying techniques in crop protection, and (2)
introduction or re-introduction of mechanical crop protection
techniques. Biological innovations are represented by: (3) increase
in yield. Innovations which combine mechanical, biological and
organizational elements are: (4) application of integrated cropping
techniques and (5) introduction of ecological cultivation
techniques. New crops considered were: (6) hemp (*Cannabis sativa*),
(7) oil flax (*Linum usitassimum*), (8) Corn Cob Mix (*Zea mays*) and
(9) chicory (*Cichorium intybus* for liquid sweeteners). In collecting
the information a Delphi procedure was followed to retrieve a
consistent and unambiguous data file.

TABLE 2. Results of the computations.

	Basic situation	Optimization with all cropping variants included	Optimization with all cropping variants and yield increase	Scenario I	Scenario II Levy Dfl 60	Scenario V Levy Dfl 40
	t = 1989	t = 1989	t = 2000	t = 2000	t = 2000	t = 2000
Cropping pattern (ha)						
Wheat standard						
WT 130 kg N	7.5	7.5	3.75	3.75		
Wt 100 kg N no fungicides					4.6	3.1
WT-eco						
WT-Integrated						2.9
Potato standard Bintje	10.0	5.0[a]	5.0[b]	5.0[b]	5.0[b]	5.0[b]
variant5: Bintje + PSR variety		5.0[a]	5.0[b]	5.0[b]	5.0[b]	5.0[b]
Sugarbeet standard	7.5				7.5	7.5
SB-3 2.4 kg a.i. herb.						
SB-5 1.4 kg a.i. herb.		3.1	3.2	5.9		
SB-6 3.2 kg a.i. herb.		4.4	4.3	1.6		
Onion	3.0	3.0	3.0	3.0		
Chicory (vegetable)	2.0	2.0	2.0	2.0		
Peas					2.0	2.0
Peas-Eco						
Set-aside			3.75	3.75		
Use of pesticides (kg a.i.)						
Nematicides	1740.00	0.00	0.00	0.00	0.00	0.00
Herbicides	95.84	63.49	56.95	68.04	21.64	26.12
Insecticide	12.20	12.20	14.54	14.54	9.01	9.01
Fungicides	167.51	159.00	152.16	142.54	110.82	110.85
Other	13.50	13.50	10.12	6.75	0.00	0.00
Total	2029.05	248.00	233.77	232.21	141.50	145.99
Nitrogen use (kg N)	6010	4189	3852	3740	2930	2336
Farm Labour income (Dfl/year)	1270	17077	43859	- 29630	- 31990	- 16250

a Haulm killing mechanical, weed control mechanical, split N fertilization + petiole analysis and reduced fungicide application for late blight control.
b Haulm killing chemical, other optional operations see a.
c Obliged.

(2) environmental regulations, and (3) different forms of price and market policy. Scenario I was used to indicate the effects of technical developments compared with the basic situation. Comparing the results of Scenarios I and II shows the impacts of standard environmental policy regulations[1], whereas the effect of differences in price policy[2] can be assessed from comparing Scenarios I and IV and Scenarios II and V, respectively. The computations for Scenario III and VI indicate the effects of a compulsory switch to ecological farming. The scenarios I, II and V were considered as the combinations with the greatest practical relevance. The time horizon in the scenario computations was set at the year 2000.

External determinants   Scenario tree
of farm organization

* Technical developments   decor variant
* Institutional developments:

  - EC agricultural policy   market   production
                              oriented policy   restricted policy

  - environmental policy   not standard strict   not standard strict

SCENARIOS   I   II   III   IV   V   VI

FIGURE 2. Construction of the scenarios for the MIMOSA system.

---

[1] In this standard variant the objective for Dutch agriculture (Min. LNV, 1990) is to reduce the total amount of biocide use (in terms of the weight of active ingredients compared to the average use over 1984-88) to 50 % by the year 2000. The detailed goals for arable farming are given in Table 1. A levy on biocide use (in proportion to the weight of active ingredients) was suggested as a policy measure (see Oskam *et al.*, 1992); calculations were made for this specific incentive.

[2] The "market oriented" variant assumes sharp price reductions (based on minus 4.5 percent anually for wheat), supplemented by voluntary set-aside regulations. The other variant reflects a policy of "production restrictions". In this case price decreases are moderate (based on minus 2.5 percent anually for wheat) in combination with set-aside (15 % of the acreage). Price developments for fixed and variable inputs are also a part of the variants.

4.3. EFFECTS OF TECHNICAL CHANGE

Next, in addition to the assessment of the basic situation (year 1989) in which only current production practices are used, calculations were made by optimizing with all cropping variants for all the crops. Also new crops and set aside were taken into account. The comparative static comparison of the model outcomes expresses the possibilities for environmental-economic improvements if the most modern techniques were to be introduced on the representative farm. The optimal plan calculated for the extended model shows financial results which are about 16 000 guilders higher compared to the basic situation. At the same time, total use of biocides reduces to just 12 %. As shown in Table 2 this reduction depends very strongly on the elimination of nematicides for soil fumigation. The cropping pattern as such does not change. Compared with the basic situation, however, a different selection of cropping variants is made. There is a particularly important change-over regarding the potato variants. Instead of Bintje in a rotation of 1:3 with soil fumigation, a combination of 50 % Bintje and 50 % PSR variety is selected. On the representative farm more than 90 % of the possible increase of income of 16 000 guilders is due to higher returns for potato. So, according to the model outcome, much money can be made and significant reductions in biocide use can be achieved by changing-over to the most modern techniques. These techniques are however not (yet) generally adopted in practice. It has to be kept in mind, however, that the final reduction goals of Table 1 are formulated for the year 2000. Diffusion of the innovations as reflected by the cropping variants can be assumed to be complete by then.

4.4. COMPARISON OF THE SCENARIO RESULTS FOR THE YEAR 2000

Considering changes in physical output from the basic year gave only minor changes in cropping pattern compared with the basic situation but the income change is significant (third versus second column of Table 2). Finally the price changes of a market-oriented price policy were added. This yields the implications of Scenario I for the model farm. The reduction in income is dramatic. The general conclusions regarding Scenario I can be summarized as follows: (1) income decreases severely compared with the basic situation, hence techno-logical change cannot compensate for the price reductions of a market-oriented price policy, (2) there are significant reductions in the use of pesticides, because of innovations in cropping techniques, (3) no new crops are selected.

   As indicated, technical innovations in cropping techniques can reduce the use of pesticides significantly. According to the normative lp procedure and the assumptions made, the reduction targets for total biocide use in kg a.i. and for nematicides, as given in table 1, can be achieved by technical change. Optimization of the extended model for t = 2000, with an increasing levy in guilders per k.g. a.i. to realize the reduction goals for the remaining categories of biocides induced a change-over to other cropping variants and to other crops, accompanied by reductions in income. In summary: the share of sugarbeet and potato remains constant; sugarbeet changes over to a variant with less input of herbicides by row spraying, (2) set-aside becomes relevant in the case of a levy and replaces wheat, (3) a levy of Dfl 60 must be imposed to achieve the reduction targets, the income reduction after restitution of the levies paid is Dfl 2 360, (4) insecticides appear to be the category determining the appropriate levy of Dfl 60.

To consider the restricted production price policy in the computations, set-aside was minimized at 4.5 hectares without a premium for Scenario V. Next, the same computations were made as for Scenario II. A levy of 40 guilders by weight of active ingredients was required to achieve the quantitative reduction targets for biocide use in the case of Scenario V. In conclusion: (1) the main difference in the two price policies, given technical developments and standard environmental policy is in their implications to income, (2) the composition of the cropping patterns in both instances is very similar and (3) because of the set-aside obligation according to price policy 2, quantitative input restrictions for pesticides are less limiting in case of price policy 2 and the income losses smaller.

In conclusion: Scenario V gives the least worse prospects. Even in this case the income reduction compared with the basic situation is more than Dfl 17 000. The price policies appear to be most important for the scenario results. From the viewpoint of the farmer a production restriction price policy, as in Scenario V, is preferable. The targets formulated for the reduction of pesticide use and the emission contraints for pesticides and for nitrogen are relatively easily met, *i.e.* with small losses in income. In the case of a regulatory levy on biocide use the income loss totals Dfl 2 360 for Scenario II and Dfl 1 930 for Scenario V. In the case of Scenario I no environmental restrictions are imposed. Nevertheless pesticide use and nitrogen use decrease significantly because of the application of environmentally-friendlier techniques with relative economic advantages.

Note that future induced technical innovations (and additional on-farm and off-farm activities) were not included in the scenarios. From the lp results, however, follow the bottlenecks in future farming and these bottlenecks can be expected to have major attention in research and extension, resulting in lower income losses then computed here. An example of such an item resulting from our study is the need of alternatives for chemical insect control.

5. Discussion and outlook

Based on the experiences so far, the environmental-economic model, presented in this article, is considered to be a useful instrument to gain insights into the interactions of production intensity, environmental aspects and farm income and to compare the implications of different policy instruments.

It must be pointed at, that relationships and trends are more important than the absolute figures as the presented model application covers only a specific group of farms. Moreover, the model contains, for instance, several extensive cropping variants which are based on assumptions and experiments that have never been tested in practice.

The detailed information needed for the assessment of the cropping variants was obtained by consulting crop scientists and environmental experts. In this manner variables and data were verified before integrating them in the model (see De Koeijer, 1991). The cooperation with these experts is given attention not only during model development and implementation but also in the evaluation of model functioning and of test results.

Further refining of the lp model for normative research should focus on four items: (1) including risks, particularly those associated with crop care, (2) integration of an organic matter balance, (3) continual updating of the lp model to take account of new developments in technology and planned regulations and (4) adjustment of the organization of model input so that farm specific constraints and price and yield figures can be modified in a more user-friendly way. In the present study risks are not taken into account. In practice farmers try to reduce risk. Hence, unlike the model they may prefer a crop variant with less profit above a crop with a higher profit but also with more risks.

The environmental-economic model can be used for many purposes. The approach described is intended to give insights both with regard to decision making at farm level (planning) and for policy assessment (conditional forecasting). Regarding the first item the role of the model is in particular in the elaboration of (precise) technical specifications in the areas of environmental pollution and technical innovations. For further applications, the study of these relations, will require a continuous dialogue with experts of several disciplines for updating the model for technical change and for more restrictive or other environmental criteria.

With regard to the second orientation the method presented is an instrument in encouraging and structuring the debate by enabling an evaluation of alternative environmental policy instruments on their efficiency by comparing the costs involved and the reduction in pollution achieved. For assessing the actual changes in farming to expect, however, additional elements must be considered, namely behavioural and family-related factors. Hence, before aggregating lp model outcomes to the regional or sector level, fine tuning the lp results for these factors is preferable as indicated by Wossink (1993).

6. References

Baumol, W.J. and Oates, W.E. (1979) 'Economics, environmental policy and the quality of life', Englewood Cliffs: Prentice Hall.

36

Berg, R. van den (1990) 'Verdunning en omzetting van bestrijdings-
    middelen in grondwater', RIVM rapport no. 72 58 01 00 2,
    Bilthoven: Rijksinstituut Volksgezondheid en Milieuhygiëne.
Breeuwsma, A., Schoumans, O.F., de Vries, W. and Kraft, J.F. (1987)
    'Bodemkundige informatie voor een globaal vermestingsmodel',
    Stichting voor Bodemkartering, report nr. 2007, Wageningen.
Brouwer, W.W.M. (1990) 'Personal communication', Plantenziektenkundige
    Dienst, Wageningen.
Daly, H.E. (1992) 'Allocation, distribution and scale: towards an
    economics that is efficient, just, and sustainable', Ecological
    Economics 6, pp. 185-193.
Koeijer, T.J. de and Wossink, G.A.A. (1990) 'Emissie van meststoffen
    en bestrijdingsmiddelen in de akkerbouw', Vakgroep Agrarische
    Bedrijfseconomie/Werkgroep Landbouwpolitiek, Landbouwuniversiteit
    Wageningen.
Koeijer, T.J. de (1991) 'Integratie van milieu-aspecten in bedrijfs-
    modellen', Vakgroep Agrarische Bedrijfseconomie, Landbouw-
    universiteit Wageningen.
Linden, A.M.A. van der, and Boesten, J.J.T.I. (1989) 'Berekening van
    de mate van uitspoeling en accumulatie van bestrijdingsmiddelen
    als functie van hun sorptiecoëfficiënt en omzettingssnelheid in
    bouwvoormateriaal', RIVM rapport no. 72 88 00 00 3, Bilthoven:
    Rijksinstituut Volksgezondheid en Milieuhygiëne.
Min. LNV (1990) 'Meerjarenplan gewasbescherming: beleidsplan', Den
    Haag: Ministerie van Landbouw, Natuurbeheer en Visserij.
Min. VROM (1989) 'Milieucriteria ten aanzien van stoffen ter bescher-
    ming van bodem en grondwater', Tweede Kamer 1988-89, doc. 21 01 2
    nr. 1-2.
NRLO (1993) 'Verkennende studie over input-output relaties', rapport
    no. 93/9, Den Haag: Nederlandse Raad voor Landbouwkundig
    Onderzoek.
Oskam, A.J., van Zeijts, H., Thijssen, G.J., Wossink, G.A.A. and
    Vijftigschild, R. (1992) 'Pesticide use and pesticide policy in
    the Netherlands: an economic analysis of regulatory levies in
    agriculture', Wageningen Economic studies no. 26, Wageningen,
    Pudoc.
Schans, J. (1991) 'Optimal potato production systems with respect to
    economic and ecological goals', Agricultural systems, 37,
    pp. 349-348.
Wit, C.T. de (1992) 'Resource efficiency in agriculture', Agricultural
    Systems 40, pp. 125-151.
Wossink, G.A.A. (1993) 'Analysis of future agricultural change: a farm
    economics approach applied to Dutch arable farming', PhD Thesis,
    Department of Farm Management, Series Wageningen Economic Studies
    no. 27, Wageningen: Pudoc.

INNOVATIVE RESEARCH WITH ECOLOGICAL PILOT FARMERS

P. VEREIJKEN AND H. KLOEN
DLO-Centre for Agrobiological Research (CABO-DLO)
Postbox 14
6700 AA Wageningen
The Netherlands

ABSTRACT. The paper describes a research project aiming at the design, development and evaluation of ecologically advanced arable farming systems. An experimental layout is chosen with a pilot group of ecological farmers. Three new components, viz. multifunctional crop rotation, ecological nutrient management and a farm-specific ecological infrastructure form the basis for the prototype farming systems. Prototype systems are evaluated and improved with a set of 4 criteria regarding each component, viz. its readiness for use, its manageability by farmers, its acceptability for farmers and its effectiveness.

1. Introduction

Integrated farming systems should be considered as a feasible first step to alleviate the consequences of the ongoing agricultural crisis in EC. However, it cannot change the fact that agrotechnology is clearly beyond its optimum, causing degradation of nature and landscape, pollution of the environment and overproduction of food. The latter is a major cause of decreasing incomes and employment in rural areas. The dumping of growing EC surpluses on the world market is frustrating agriculture in other industrial countries but also in developing countries, in a way that is no longer tolerated by the EC trade partners. Therefore, the only long term solution of the current crisis would be advanced ecosystem-oriented farming systems principally carried by a strong home-market with quality labels and premium prices to ensure sufficient management achievements and economic margins (see Vereijken, 1992 for more details on this vision).

The CABO-DLO contribution to bring this solution nearer is the design, development and evaluation of ecologically advanced arable farming systems, which may be considered as most consistent variants of integrated arable farming systems. To combine prototyping and evaluation of farming systems, we have chosen for an experimental layout with a pilot group of ecological farmers, including the

P C Struik et al (eds), Plant Production on the Threshold of a New Century, 37–56
© 1994 *Kluwer Academic Publishers*

ecological prototype system of the experimental farm at Nagele (Wijnands and Vereijken, 1992). It may also permit to determine to what extent the success of integrated systems depends on skills and attitudes of farmers. The first experimental year 1991 could be started with a well-prepared and well-selected group of nine arable farmers. All of them fulfilled the demand of being authorized to carry an organic trade label, which would support a consistent approach of integrated agriculture meeting strict environmental norms, such as no emission of pesticides and maximum emission of 25 mg/l NO3 to ground- and shallow waters. In conformity with the contract with EC as major financer (CAMAR-program), appropriate variants of an initial prototype system have been designed, laid out and evaluated in 1991. Based on the results, the prototypes have been improved, again laid out and evaluated in the second experimental year 1992.

2. Materials and methods

2.1. DESIGN AND LAY-OUT OF PROTOTYPE FARMING SYSTEMS

Current organic farming offers a promising model for sustainable development of agriculture, since it is based on shared responsibility of urban and rural communities for the vital functions of the rural area as expressed in premium prices for quality production. However, some major shortcomings of organic farming have to be solved before it can be called "consistently integrated" or "ecologically advanced". It concerns the insufficient replacement of chemical inputs notably pesticides, at the detriment of yield and product quality, the environmentally hazardous and sometimes unacceptable use of organic fertilizers and the insufficient care of nature and landscape (Vereijken, 1993). To overcome these shortcomings, three new components will be developed as a basis for the prototype farming systems.

The first is a farm-specific multifunctional crop rotation, aimed at preserving soil fertility in physical, chemical and biological terms and achieved by proper temporal and spatial distribution of crops across the farm. It implies limited frequencies of single crops (1:6) and related crops (1:3), maximum separation of successive crops (to escape from semi-soil borne pests) and alternation of crops with positive and negative effects on soil structure, N reserves and weeds. These demands can be met by a rotation model, built of crop blocks with corresponding needs of and effects on soil fertility. Eventually, such a multifunctional crop rotation should provide for the base of crop vitality and quality production.

The second component is an ecological nutrient management, aimed at agronomically wanted and environmentally acceptable soil reserves and achieved by proper tuning of inputs and outputs of nutrients. It implies regular monitoring of available soil reserves, quantification of nutrient outputs, estimation of needed inputs, choice of an appropriate combination of organic manure, biological N fixation and single fertilizers, and appropriate timing and application of organic fertilization to minimize nutrient losses, notably nitrate.

The third component is a farm-specific ecological infrastructure, covering at least 5% of the farm surface and primarily consisting of field margins, ditches and other line elements, supported by buffer strips. It will take quite some years to achieve an optimum state of this infrastructure, following an appropriate plan and a consistent management. These three components are described in more detail in (Vereijken, 1993).

A major demand at the design and lay-out of prototype systems is to insert the three new components in a non-conflictive way. In practice the multifunctional rotation and the ecological nutrient management appear to support each other in such a way, it can be stated they are mutually indispensable. The ecological infrastructure has only a few direct links with the other two components. At first, it is dependent on a careful nutrient management, without which it would risk for eutrophication. Therefore, buffer strips are designed to separate fields from the main elements of the ecological infra-structure, such as ditch banks and hedges. Secondly, the ecological infrastructure may support the rotation in suppressing pests, diseases and weeds. However its major function is to provide for an extra dimension of the products to make them more attractive for the part of the population which is more interested in the quality of environment, nature and landscape than in chemical-free food production. In general, all three components should sustain the economic viability of the prototype systems.

2.2. EVALUATION AND IMPROVEMENT OF THE PROTOTYPE SYSTEMS

2.2.1. Technological innovation

Prototype systems are evaluated and improved with a set of four criteria regarding each new component.
1. is it ready for use?
2. is it manageable by the farmers?
3. is it acceptable for the farmers?
4. is it effective?
In case of the latter criterion a representative set of indicators is to be monitored and quantified for each new component.

Wheat, potato and onion are prevalent crops at the pilot farms. Therefore, the quantitative and qualitative performance of these crops are considered the best indicators for the functioning of the rotation. The annual balance sheets of PKN-inputs and outputs, the soil reserves of PKN and the leaching of nitrate are quantified as indicators for the functioning of the nutrient management. For the ecological infrastructure to be developed along the field margins, a set of indicators is monitored such as percentages of bare soil, coverage of grasses, annual and perennial dicots and abundance of plant and animal species.

By comparing observed and targeted signals from the indicators, the prototype systems can be improved. Besides, farmers themselves may suggest improvements or may want adaptations. For example, the composition of the rotation should keep following the changes of the market. With the multifunctional crop rotation model it is always possible as far as the replacing crops comply with the character of the crop blocks in question and the limitations on crop frequency and crop succession.

2.2.2. Economic innovation

Considering the economic evaluation and improvement of the prototype systems, we have abandoned our initial plan of conventional economic bookkeeping. Parameters such as net margin of the current enterprise, based on current yields, prices, fixed and variable costs are hardly influenced by the technological innovation, during the first years of a pilot project. By the years conventional bookkeeping data of the participating farms will probably be more influenced by the techno-logical innovation, but they will never reflect its full potential, since these data will always be dominated by non-specific factors such as farm size and related levels of costs. Consequently, technological innovation should be followed by economic innovation focusing on the question: "What should be the size of a farm to obtain an appropriate income with an ecological advanced system and management, taking into account perceived levels of yields, prices and inputs including labour?" It requires farm structure optimization, which will be done in cooperation with the Agricultural Economics Research Institute (LEI-DLO) in the follow-up project (EC-AIR program) from 1993-1997.

2.3. INTERACTION WITH THE PILOT GROUP

Regularly, meetings are arranged with the pilot group to evaluate results and to improve components and management of the prototype systems. Subject matter for these meetings are the data from both sides, assembled chronologically and systematically in spreadsheets and derived graphs (Microsoft Excel). Each farmer registers his management data on a spreadsheet based form, listing his major and minor fields/crops in the rows and the management items in the columns. Behind these we add the field/crop specific data of evaluative research, listing the various indicators in the columns.

Before each meeting the spreadsheets forms are updated, in a printed version by the farmers and on the computer by us. In this way the interaction is based on a commonly made and used database, to support a systematic evaluation and steady improvement of systems and management. Besides these large farm databases, three small crop databases are made each year, to support the evaluation and improve-ment of the cropping systems of potato, onion and wheat. In each of these three databases the total group of farms is listed, each row representing the crop data of a certain farm. To assure the continuity of the project, an (updated) agreement of cooperation is signed each year, specifying the inputs of both parties involved. In this agreement also the demands are specified for each innovative target which the farmers have to meet in order to obtain an annual compensation of costs of 1100 ECU per target.

3. Results

3.1. DESIGN AND LAY-OUT OF THE PROTOTYPE FARMING SYSTEMS

In 1991 the pilot farms converted to the prototype systems. It implied a drastic break from routine practices such as crop plans not based on a steady rotation, routine (over) dosage of organic manure and management of ditch slopes as a breeding place of weeds.

In 1992 the agreed multifunctionate crop rotations were laid out for a second time, in most cases with minor adaptations to market changes whilst maintaining the general model.

On average, some 80% of the farm area was in line with the agreed models in terms of frequency and succession of crops (Figure 1). Farms 1, 6 and 9 clearly trespassed the limitations in crop frequencies of wheat and/or the cereal group. As a reason it was called lack of an alternative cash crop. Farms 3, 6 and 7 clearly trespassed the limitation in succession of lifted crops ('no lifted crop after a lifted crop'). In most cases it was an inevitable incident as a result of conversion to the new rotation model. The layout of the spatial component of the model in terms of maximum displacement of crops from one year to another was somewhat retarded, due to adjustment of the models of some farms.

Except for farms 8, 1 and 6, the pilot group sufficiently succeeded in tuning the P input by organic manure to the P need of the crop plan, considering the P soil reserves and the P output by the crops (Figure 2). It is a first criterion in ecological nutrient management that already could largely be met in 1991 by changing from poultry manure high in P to cattle manure low in P. Except for farm 8, the group also succeeded in K management. The management of N required a complicated and accurate estimate, to avoid overdosage causing yield depression through diseases and lodging, followed by N leaching, but also to avoid underdosage causing yield depression through undernourishment of crops. Six of the nine farms had a net N input, deviating more than 10 % of the N output by the crops. Three of these six (8, 1 and 10) restricted the use of organic manure because of P soil reserves exceeding the optimum range and did not grow enough legumes to compensate for the restricted N input by organic manure. Nevertheless, none of the mentioned side effects of N underdosage was observed in these farms.

Compared to the first year, the group showed an increasing willingness to adopt an ecological infrastructure. As a result, ditch management improved and most farmers no longer cultivated the soil up to the ditch slope, to avoid its destabilisation. Besides, they switched from chopping to mowing of the ditch vegetation and removing the hay, in order to encourage the settlement of new species. Because it takes too long before species with heavy (so not-airborne) seeds

42

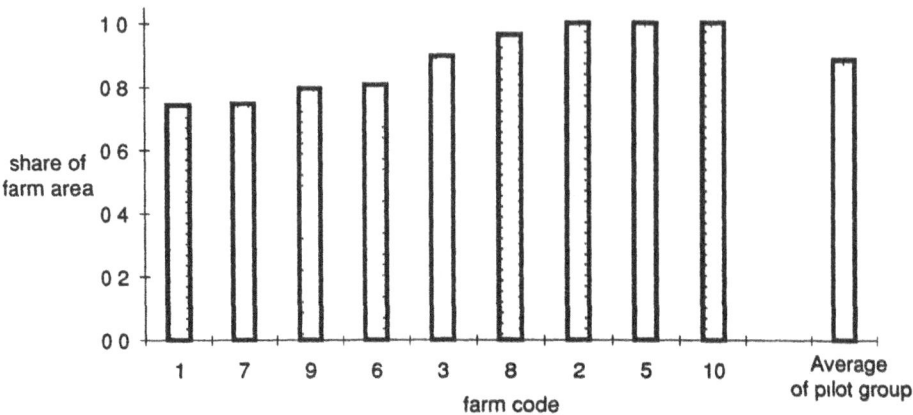

a Area in line with crop rotation model

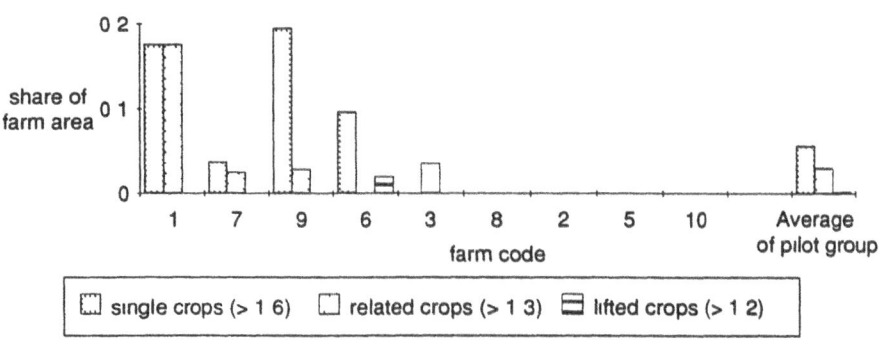

b Area too high in crop frequency

single crops (> 1 6)   related crops (> 1 3)   lifted crops (> 1 2)

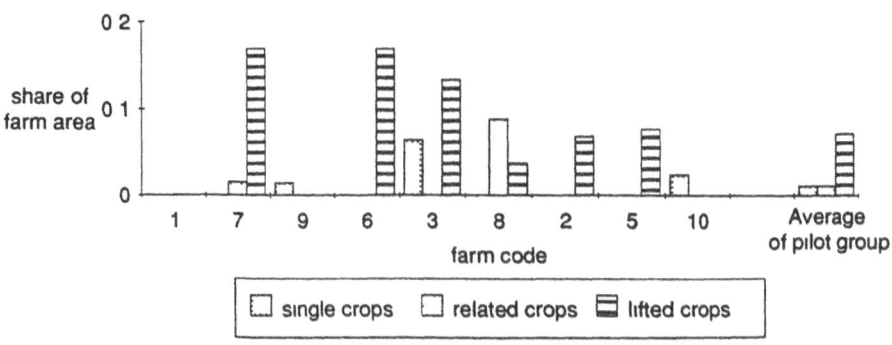

c Area too fast in crop succession

single crops   related crops   lifted crops

Figure 1   Layout of a multifunctional crop rotation by the pilot farms in 1992.
(share of farm area with crops whether or not correct in frequency and succession, spatial
component not considered).

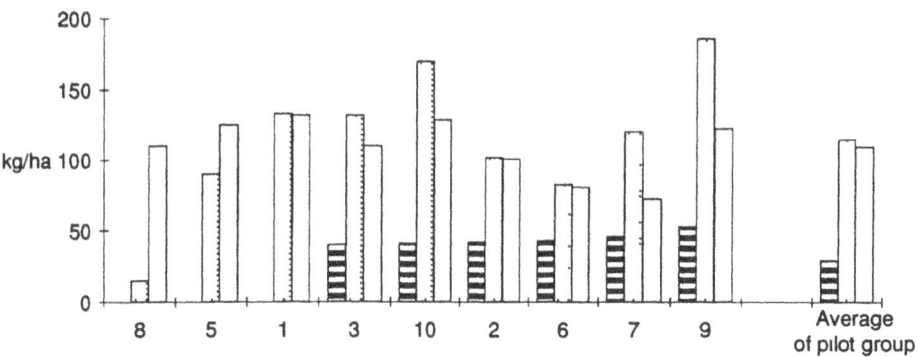

a Estimated needs

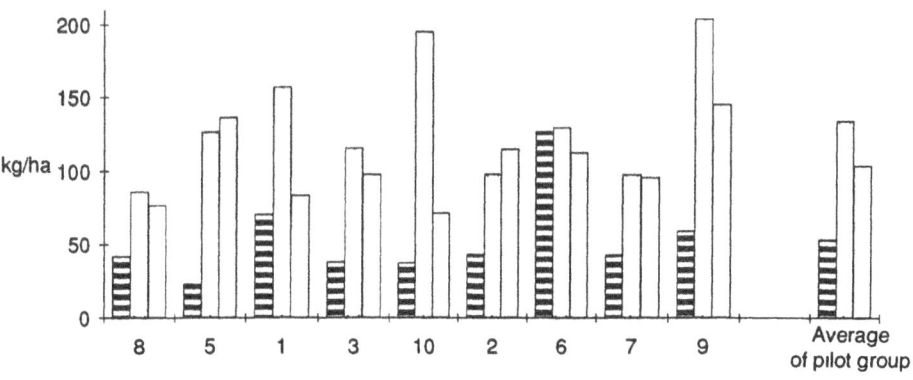

b Achieved inputs

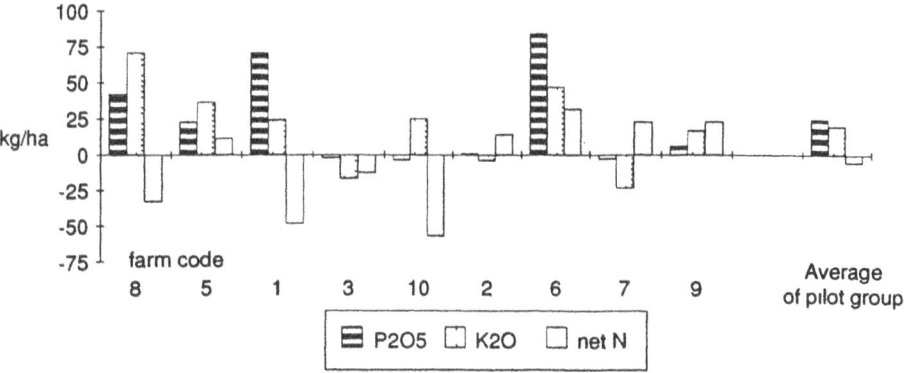

c Inputs minus needs

P2O5   K2O   net N

Figure 2   Layout of ecological nutrient management by the pilot farms in 1992
(ranged according to increasing P-need)
a Estimated needs based on available soil reserves and outputs by products
b Achieved inputs based on organic manure, deposition and biological N-fixation
c Inputs minus needs (ideally = 0)

may settle, we decided to collect seeds and sow a selection of attractively flowering species of this kind, fitting into this habitat. The farmers also started to lay out buffer strips alongside the ditches, though only farm 3 can as yet meet the criterion of 5% of farm area devoted to the ecological infrastructure (Figure 3).

3.2. EVALUATION AND IMPROVEMENT OF THE PROTOTYPE SYSTEMS

3.2.1. *Technological innovation*

After the conversion year 1991 progress in 1992 was satisfactory (Table 1). Subsequently, progress in rotation based quality production, nutrient management and ecological infrastructure will be highlighted.

3.2.1.1. Quality production

To evaluate and improve the performance of the pilot farms towards quality production, the most prevalent crops potato, wheat and onion were studied again in their yield formation. As in 1991, it appeared a considerable variation in yields: seed potato 30-45, onion 30-60 and winter wheat 5-8 tons/ha. In seed potato, the yield variation was clearly related to variation in planting density and subsequently in tuber density (Figure 4). In onion, planting density also was a major limiting factor in yield formation, though length of growing period as expressed in bulb weight seemed of importance, too (Figure 5). The cause of the longer growing period of the onion crops in the three farms with the highest yield could not be established (no relation with Nsoil reserves, maybe diseases?). In winter wheat the yield variation was far more related to grain density than to grain size. Yield and density of grains were not related to sowing density or plant density, though strongly related to N soil reserves, especially at the end of the growing period (Figure 6). However, the N soil reserves could not be related to N input as organic manure in the preceding autumn. Apparently, other N sources such as biological N fixation, crop residues, green manure and mineralisation of organic matter contributed more to the available N reserves. The current results indicate yields of potato and onion can be improved by higher planting densities.

Quality of the products is gaining importance the more offer is exceeding demand! However, in technological terms there is such a disencouraging number of parameters that quality is only overall to quantify by comparing the achieved price per kg with the price per kg of top quality products. Therefore we will overall quantify quality production in the future by a Quality Production Index (QPI).

$$QPI = \frac{\text{achieved price kg-1}}{\text{top quality price kg-1}} * \frac{\text{marketed kg ha-1}}{\text{field grown kg ha-1}}$$

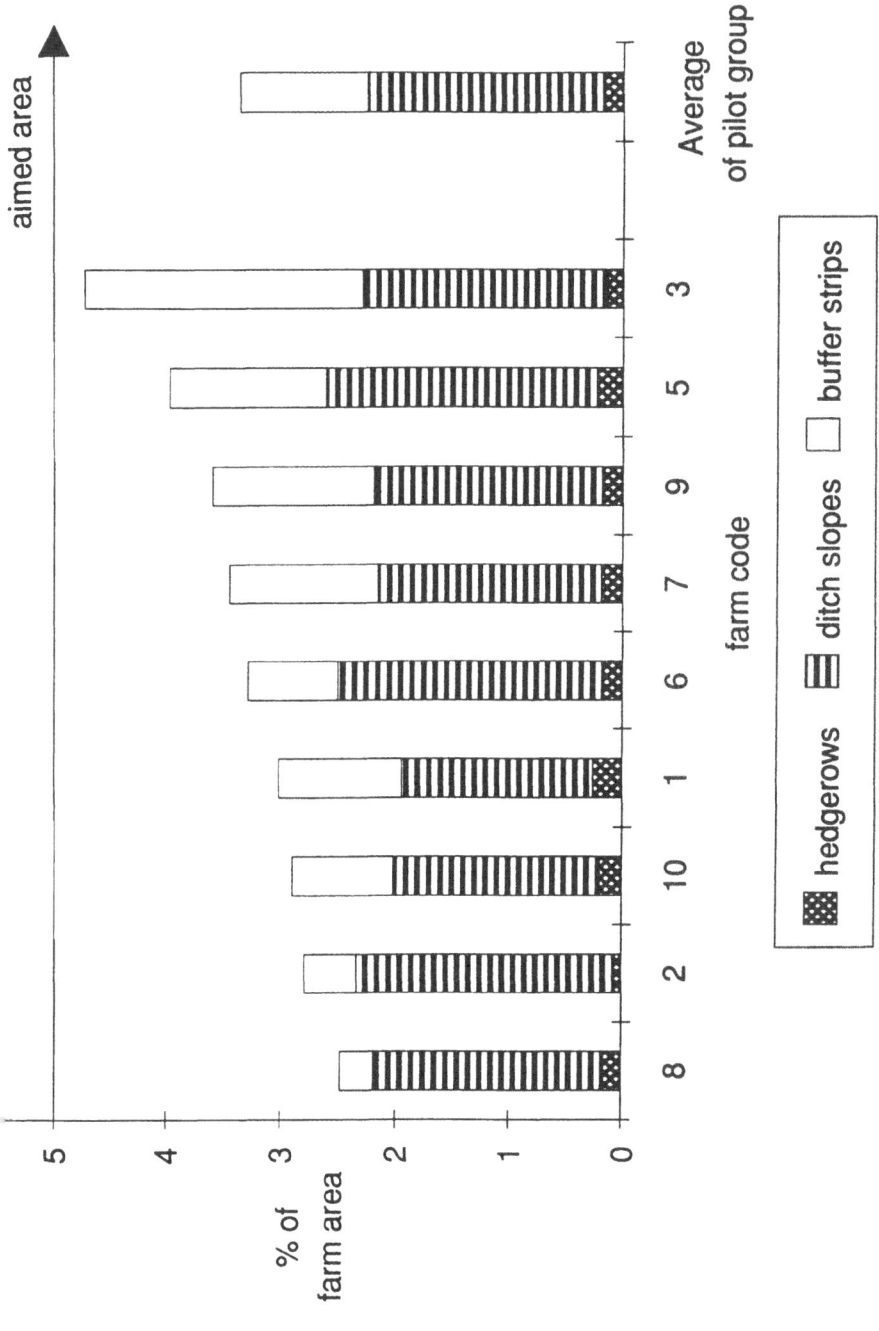

Figure 3. Layout of an ecological infrastructure by the pilot farms in 1992.

TABLE 1. Progress in 1992 of the prototype systems based on three new components.

evaluation criteria	components of the prototype systems		
	multifunct. models of crop rotation	ecol. nutrient management	ecol. infrastructure
ready for use?	ready and unchanged	improved on N management	design and management of ditch banks improved, of yards to be elaborated
manageable?	not easy, especially the spatial component due to turbulent market	not easy, especially of N, due to complicated estimates and actions requiring accuracy and good timing	not easy, due to limited insight and experience
acceptable?	some doubt about use of spatial dimension and wheat 1:6 at max	some doubt about N strategy	growing willingness to adopt
effective?	only to be judged on the long term, preferably with a reference group of organic farms without this component	for P and K only to be judged on the long term, effects of N should become visible in a few years	to be judged on medium and long term, regarding effects on flora and fauna as well as on competitiveness of the prototype
indicators of effectivity	weeds, diseases and pests, yield and quality of crops and products	available soil reserves, input/output balances of P, K, N and N leaching	useful or attractive species (biodiversity), stabilisation/improvement of product prices and profits

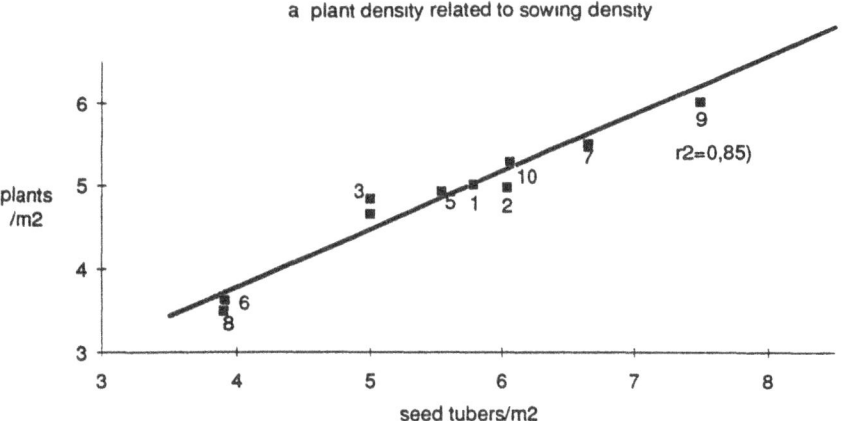

a plant density related to sowing density

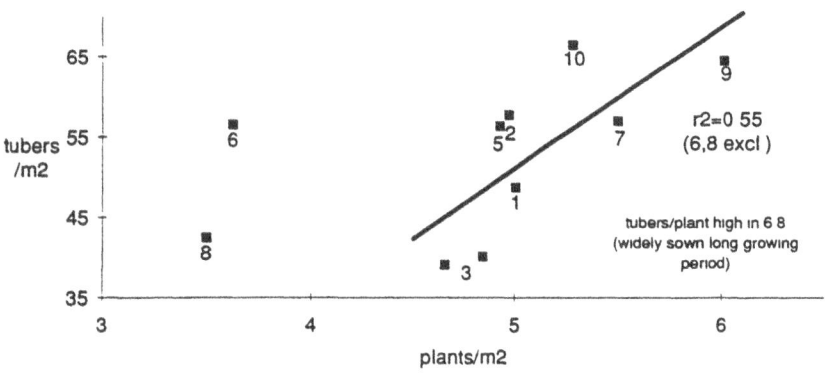

b tuber density related to plant density

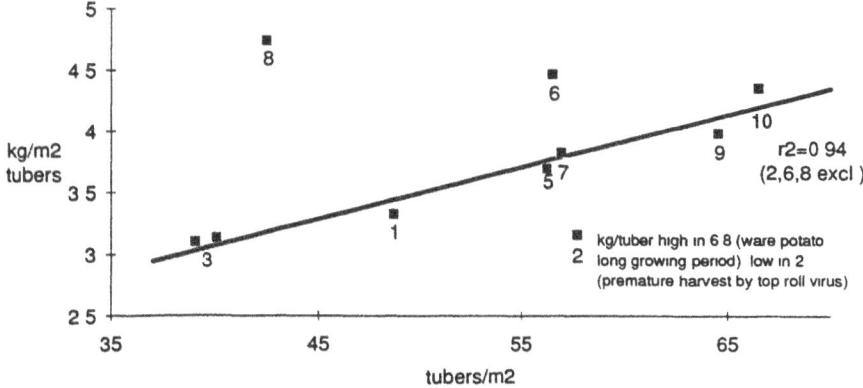

c tuber yield related to tuber density

Figure 4   Evaluation of seed potato production in pilot farms 1992
(averages of 12 monitoring plots of 10 plants row length per farm)

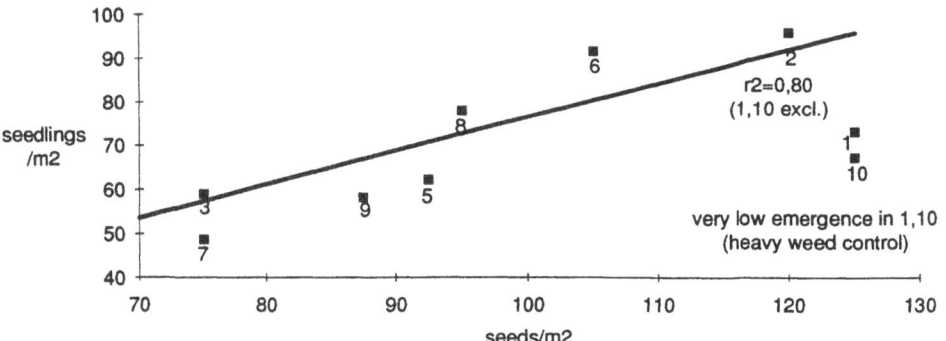

a. plant density related to sowing density

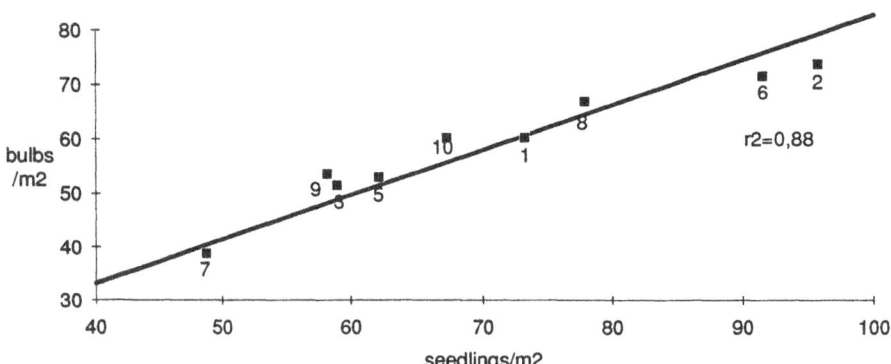

b. bulb density related to initial plant density

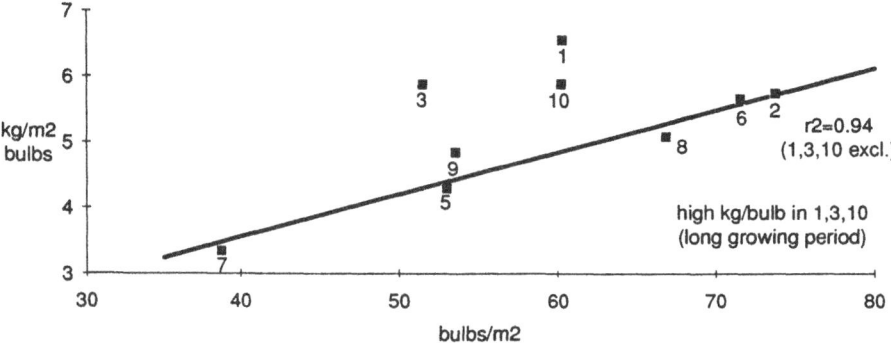

c. bulb yield related to bulb density

Figure 5. Evaluation of onion production in pilot farms 1992
(averages of 12 monitoring plots of 1.5 m2 per farm).

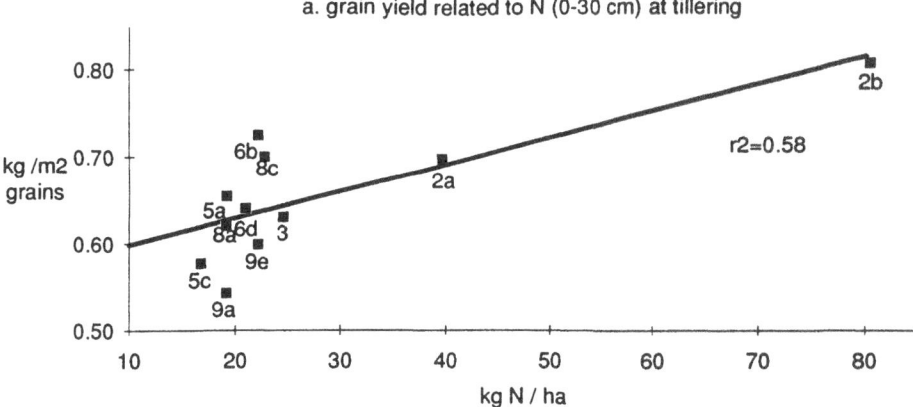

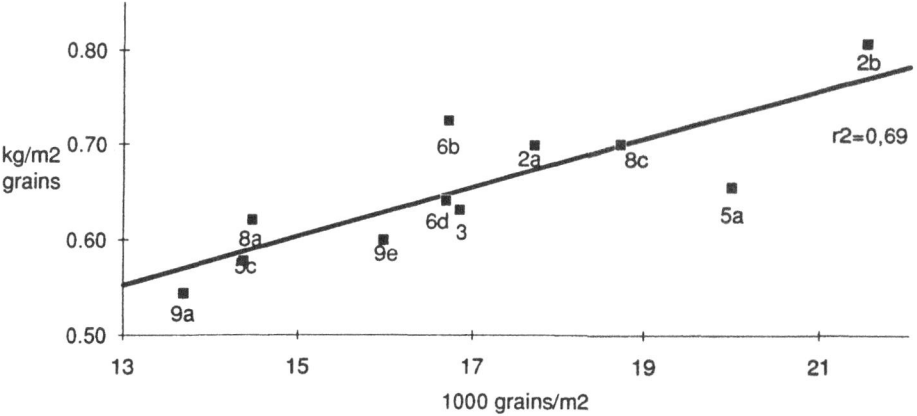

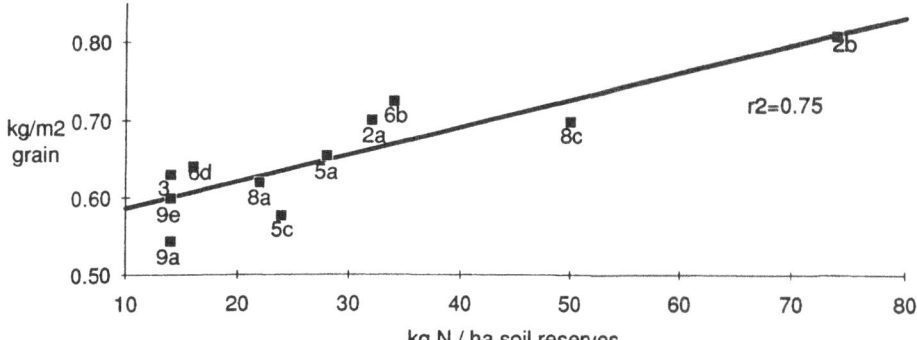

Figure 6. Evaluation of winter wheat production in pilot farms 1992
(averages of 12 monitoring plots of 0.25 m2).

For each product a wanted signal can be defined in the range $0 \leq QPI \leq 1$.

3.2.1.2. Ecological nutrient management

To evaluate and improve the performance of the pilot farms toward ecological nutrient management, soil reserves of P, K and post harvest soil reserves/leaching of N were monitored, as indicators of the effectivity of this new component for the long and short term, respectively. However, as stated in Table 1, the pilot group is still in the stage of learning and accepting. Therefore, the current P, K reserves should rather be considered as a base line. Most of the pilot farms have built up considerable soil reserves of P in particular, beyond the optimum range (Figure 7). In terms of ecological nutrient management these surplus reserves are to be reduced gradually by a consistent policy of less input than output. It implies restrictive use of organic manure and strong reliance on biological N fixation.

Considering N reserves post harvest, only farms 8 and 9 clearly exceeded the scheduled Netherlands norm, derived from the EC-norm of NO_3-N in drinking water (50 mg/l NO_3 = 11.2 mg/l NO_3-N, Figure 8). Subsequently, farm 9 seriously exceeded the EC-norm in N leaching. Besides, farms 1 and 3 clearly exceeded the EC-norm. It concerns farms with rather light soils (sandy clay) and a rather low N absorption capacity.

3.2.1.3. Ecological infrastructure

To evaluate and improve the performance of the pilot farms towards the ecological infrastructure, we monitored the development of vegetation and abundance of wild plant species in the ditch slopes, as the main element of the ecological infrastructure. Within two years, the avoidance of soil cultivation up to the ditch slopes and the lay-out of buffer strips alongside the ditches resulted into a fargoing succession of the initial pioneer vegetation (arable type) by stable vegetation (coarse and grassland type) (Figure 9). In 1992 12 coarse and 7 grassland species of the ditches slope vegetation were identified as target species, because of their attraction for man and animal. Primarily, it was considered if the species had conspicuous flowers and did not act as a weed or as a host of major pests. Besides, it was considered if the species provided for nectar and pollen for attractive or beneficial insects or for shelter to mammals and birds. In this initial period, farms 1, 2 and 5 had the highest frequencies of these species.

3.2.2. Economic innovation

Before entering the stage of economic innovation the prototype should technologically be optimized. Considering the large variation in crop

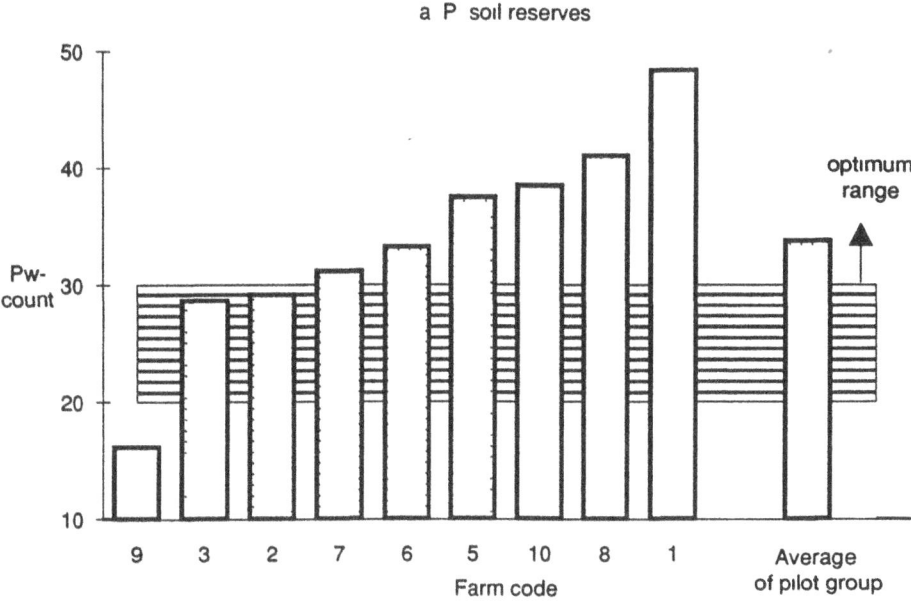

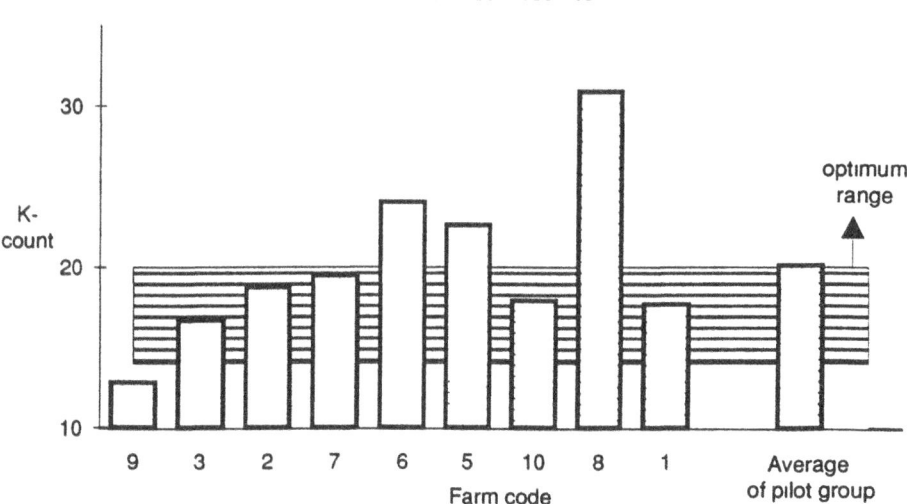

Figure 7   P and K soil reserves of the pilot farms 1991-1992
(averages of all 6 fields per farm, sampled (0-30 cm) in spring and early summer,
optimum range according to extension service, for K it concerns soils >10% clay)

a   P reserves = mg P2O5/l soil, 1 60 extraction with water (Pw-count)
b   K reserves = mg K2O/100 g of air dry soil, 1 10 extraction with 0 1 n HCl (K-count)
(farms arranged according to increasing P reserves to demonstrate the possible need of
additional K fertilizers besides P dosaged organic manure)

a N soil reserves

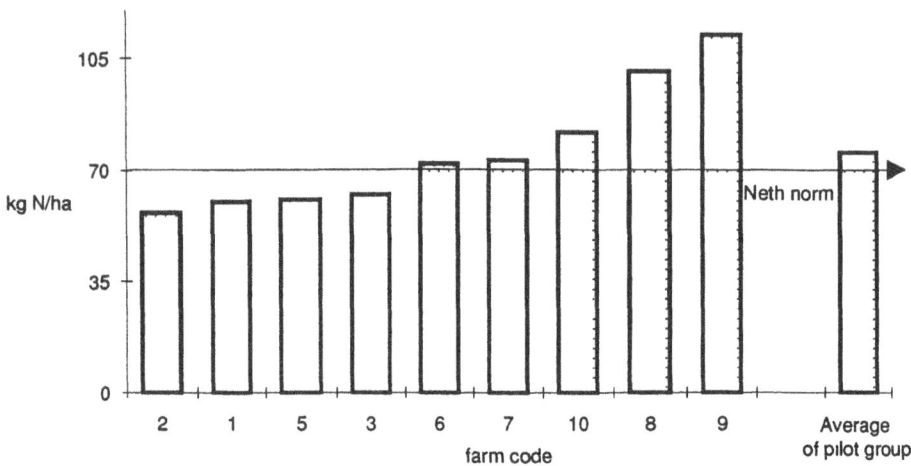

b N leaching

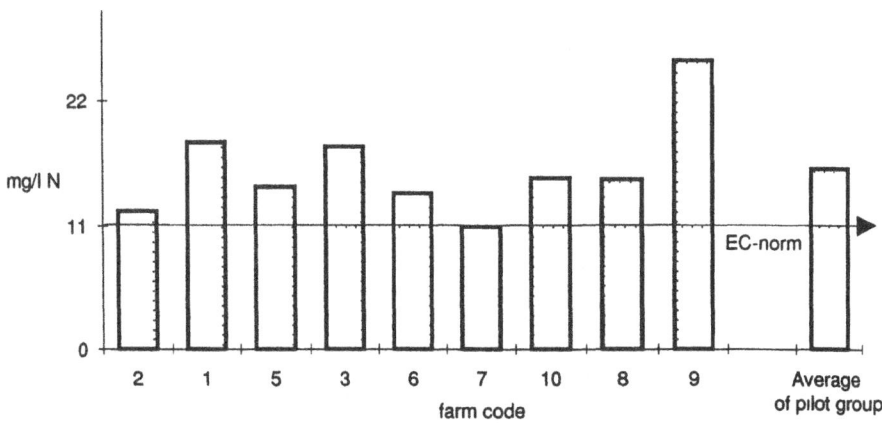

Figure 8   N soil reserves and N leaching of the pilot farms in 1992
              (weighed averages of the crops according to the crop rotation model per farm)

a   N reserves = content of mineral N in the soil layer 0-100 cm at start leaching period,
       samples of 20 probes of each crop, which covers at least half a field
       Neth norm scheduled norm for 1995, derived from EC-norm for drinking water of 11 2 mg
       N-NO3/l (assuming 300 mm precipitation surplus and 50% leaching)
b   N leaching = mg N/l (NH4 and NO3) in drainage water, samples of 3 drainpipes per crop,
       average of 2 samplings in november 1992 and january 1993

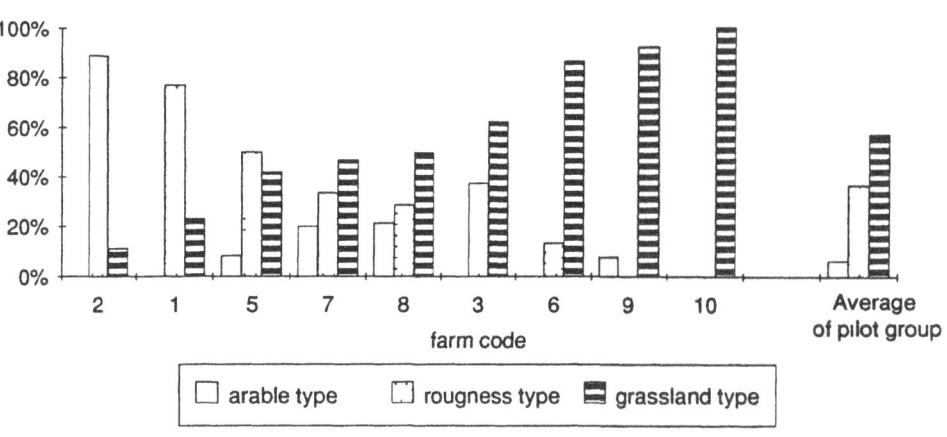

a Relative frequency of vegetation types

arable type   rougness type   grassland type

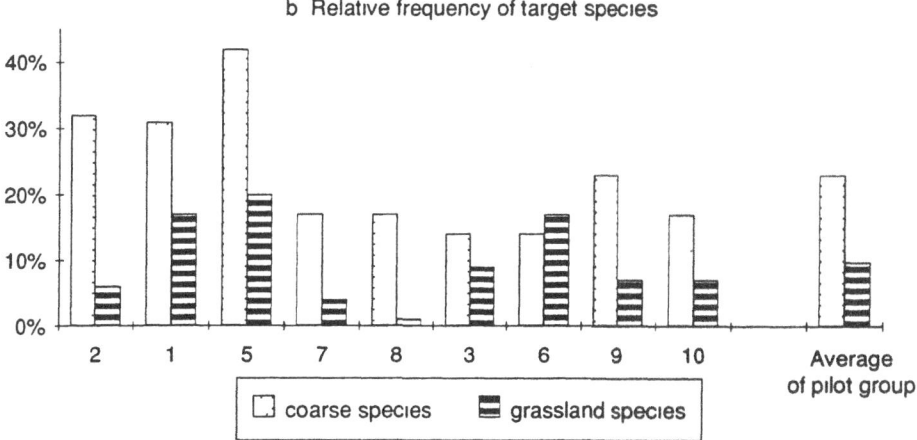

b Relative frequency of target species

coarse species   grassland species

Figure 9   Vegetation types and species diversity of ditch slopes in the pilot farms in 1992

a   Relative frequency of vegetation types present in 13-18 monitoring sections of 100 m length
      per farm
      arable type = at least 50% bare soil or dominated by annual arable weed species
      grassland type = at least 50% of perennial grassland species (mainly grasses)
      coarse type = other vegetation of biennial/perennial species
b   Relative frequency of target coarse and grassland species in the monitoring sections

Target species present in1992

coarse species(12)

Anthriscus sylvestris	Phragmites australis	Cardamine pratensis
Epilobium hirsutum	Ranunculus repens	Cerastium fontanum
Eupatorium cannabinum	Symphytum officinale	Lotus corniculatus
Glechoma hederacea	Taraxacum officinale	Pastinaca sativa
Heracleum spondylium	Urtica dioica	Plantago lanceolata
Phalaris arundinacea	Valeriana officinalis	Ranunculus acris
		Rumex acetosa

grassland species (7)

yields, current cropping systems and management do not provide for a reliable basis. (Figures 4-6). Besides soil productivity, labour productivity needs further improvement, considering the huge labour requirement in weed control, which in most farms may hardly or not be covered by the farmer himself (Figure 10). As a result, labour is to be hired which is often hard to get, expensive and doubtful in quality. Most of the labour appears to be required by a small group of lifted crops with little weed competitiveness, such as onion, carrot and chicory. Appropriate drilling and mechanical weed control systems may reduce the requirement of hand labour. However, prevention by a weed suppressive rotation requires less inputs of labour, machines and support energy and should therefore be preferred. The multifunctional crop rotation is aimed at weed prevention. To what extent it is successful will be assessed by further monitoring the input of hand labour in weed control (Figure 10).

4. Follow-up and dissemination

The first and second year of the project brought satisfactory progress, especially in methodology, which is of major importance considering the lack of insight and experience in innovative research on pilot farms. In case of field experiments on the crop rotation level aimed at development and testing of farming systems, a period covering at least one rotation is minimally required. Therefore, the research will be continued for another four years, to enable reliable conclusions on the effectivity of the three components and subsequently on the feasibility of the prototype systems. It will be done in a new EC-project (AIR program) with partners from Louvain-la-Neuve (Belgium) and Wexford (Ireland).

The project has met increasing interest of policy makers, researchers, extensionists and farmers, especially as a result of various lectures at national symposia, congresses and courses. In this sense, we think we are successful in profiling organic farming in an avantgarde role at the development of Netherlands agriculture towards more sustainability. If, as is stated in agricultural policy papers, our agriculture should be fully converted to integrated systems by the year 2000, such a group of avantgarde farms is needed to show the way. The potential role of organic farming in the sustainable development of agriculture, the shortcomings to be overcome before it can play this role and the targeted contribution of the project in this were the basic elements of the lectures (Vereijken, 1993).

Besides, a special issue of the Netherlands Journal of Agricultural Science was edited on Dutch research in integrated farming in the major plant and animal production sectors (Vereijken and Van Beusichem, 1992). Based on 14 years of experience in the arable farming sector, a methodic way was described towards more sustainable farming systems (Vereijken, 1992). Subsequently design, lay out and improvement of prototype systems for arable farming were described, amongst others (Wijnands & Vereijken, 1992).

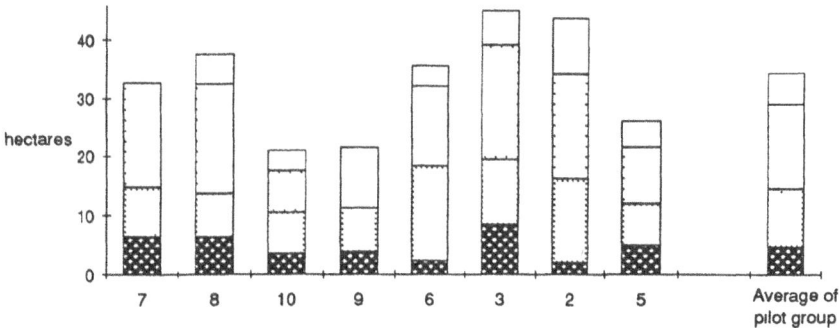

a area per crop group and overall the farm

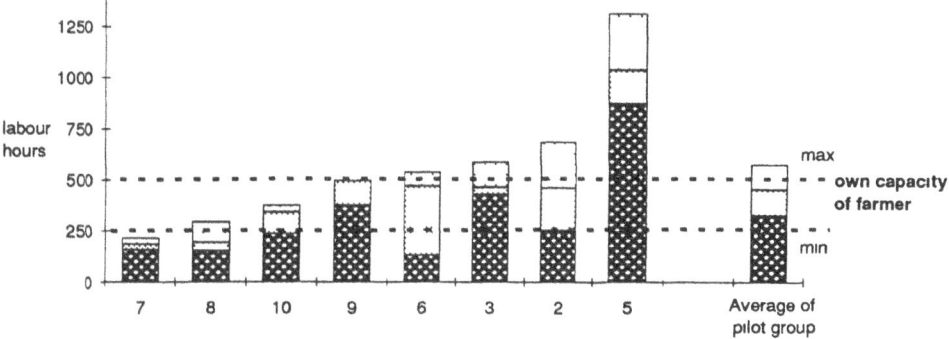

b hand labour per crop group and overall the farm

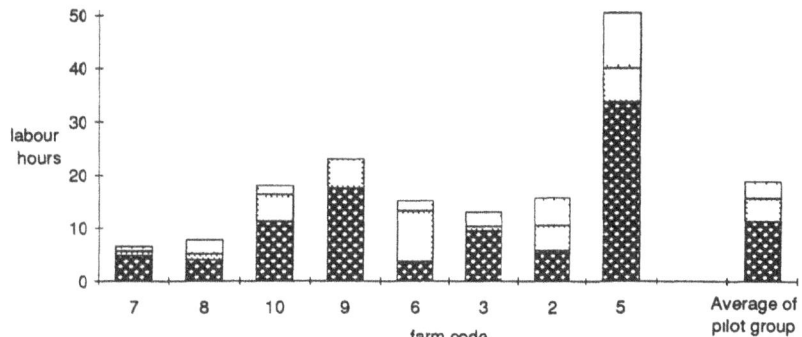

c hand labour per hectare (weighed for areas of crop groups)

group A lifted crops, little weed competitive (onion carrot and chicory)

group B lifted crops, moderately weed competitive (potato, beet, cabbage, celery)

group C mowed crops, moderately weed competitive (cereals, maize, pulses, herbs)

group D mowed crops, highly weed competitive (grass and lucerne)

Figure 10   Input of hand labour in weed control in pilot farms 1992

5. References

Vereijken, P. (1992) 'A methodic way to more sustainable farming systems'. Neth. J. of Agric. Sci., 40: 209-224.

Vereijken, P. (1993) 'Targeted innovation of technology for sustainable development of agriculture'. Proceedings EC-workshop "Potential and limits of organic farming" Sept. 17-18, 1992. Louvain-la-Neuve, in press.

Vereijken, P. and Beusichem, M.L. van (1992) (eds) 'Research on integrated farming systems in the Netherlands'. Neth. Journal of Agriculture Science 40: 207-337.

Wijnands, F.G. & P. Vereijken, 1992 'Region-wise development of prototypes of integrated arable farming and outdoors horticulture'. Neth. J. of Agric. Science 40: 225 - 238.

PLANTS OF THE FUTURE. CAN PAST HISTORY TEACH ANY LESSONS?

JOAN THIRSK
1 Hadlow Castle
Hadlow
Tonbridge
Kent, TN11 OEG
England

ABSTRACT. The present-day search for alternative crops is not a new experience in western Europe. Using the English evidence, this essay examines the strategies sought and found on three previous occasions, in the fourteenth century after the Black Death, between 1650 and 1750, and in the late nineteenth century. The recurring patterns and parallels are assembled and discussed.

1. Introduction

As a historian, I much regret the lack of interest among farmers and agriculturists, at least in my country, in the history of agriculture, as though it had nothing to teach them. But visiting Wageningen some months ago, I was pleased to learn from a young agricultural historian that when he gives lectures in country areas in the Netherlands nowadays, some farmers do actually ask him if there has ever been a similar experience in the past to the one which they are undergoing now. Does this mean that a new age is dawning, that farmers are ready to listen to the historians? I hope so, for I believe that historians can give a perspective on the present which is more illuminating than anything that the journalists can offer. To substantiate that assertion I shall in due course offer a book; here I offer a very brief summary.

My short answer to the Dutch farmers' question, whether any past experience resembles the one that they suffer at present would be, "most definitely yes". We have been through this experience three times before in our history since the twelfth century, and further back we cannot go for lack of sufficient documents. In fact, our history, in the long term, is made up of alternating phases of what may be called mainstream agriculture, when the principal objective has been to produce the basic necessities, grain and meat, followed by an alternative phase when a surplus of cereals and meat has forced farmers to look for other crops and other ways of using the land.

P. C. Struik et al. (eds.), Plant Production on the Threshold of a New Century, 57–70.
© 1994 *Kluwer Academic Publishers.*

I speak of 'our farming history', but my viewpoint should be explained more precisely. My knowledge of the sequence of events is drawn from the English evidence, but the broad trends of agricultural history in each changing phase have always taken much the same direction throughout western Europe. The differences, of course, have also been significant: they have affected the precise dating of the phases; and the different social structure of our farming populations has greatly affected the degree of commitment shown by farmers to different solutions at different times. So I emphasise at the outset that I offer only the English view of the story. But I know that you will recognise strong parallels in the history of the Netherlands at every stage.

We have lived through a long period of mainstream agriculture, and we are now plunged into the middle of a different phase, demanding what is called an alternative agriculture. In south-eastern England, where I live, the most conspicuous alternative at the moment seems to be golf courses. But my county is Kent, the garden of England, and we are also seeing a more diversified horticulture. In that respect, we are actually repeating very closely our seventeenth-century experience, which was then taken up again in the late nineteenth century, continuing from where it had left off around 1750. The similarities between the second occasion in the seventeenth century, the third from about 1880 onwards, and now the fourth, raises some interesting questions about the first experience, which occurred at a time when our documents are far less abundant. Certainly, I now see it in a new light.

2. Three phases of alternative agriculture

2.1. 1350-1500: FIRST PHASE OF ALTERNATIVE AGRICULTURE

In order to set the whole problem in its full historical context, the circumstances on all three past occasions requiring an alternative agriculture need to be described. The first occurred in the fourteenth century as a result of the Black Death in 1348-50, when it is generally reckoned that one third the population of England, and indeed of Europe too, died of plague. As a result, the demand for grain collapsed; so too did the supply of labour to cultivate the arable land. The principal solution to subsequent farming problems was to put a large amount of land under grass. As the wool textile industry was expanding at the time, the most appropriate and successful measure was to put arable to pasture and keep sheep; their numbers greatly increased. Much later on, Sir Thomas More was to talk of "sheep eating men".

But we also see in some areas at this time the hesitant beginnings of commercial dairying. The evidence comes from the monastic houses, which were already selling wool in bulk, and now sold ewes' milk cheese as well. They were not involved in cow keeping for the dairy on a commercial scale; that was the affair of small men only, and while small cow keepers sold butter and cheese, theirs was a local business, and certainly not conducted as a far-reaching national trade.

We also see at this time some experimental trials with alternative arable crops, such as rapeseed, madder, flax, buckwheat, and teasel. The monastic houses in East Anglia produce the evidence for this, and that is significant because the monks were well placed to keep in touch with the latest farming practices on the Continent, and in Flanders and the lower Rhineland rapeseed, buckwheat, madder, and so on were flourishing and would persist. But these crops did not take firm root in England at this time. Interest quickly died out.

Nevertheless, this phase of alternative agriculture lasted for about 150 years, from 1350 to around 1500. Putting arable land to pasture seems to have been the principal solution, but commercial dairying and trials with some new crops were modest initiatives here and there. Then population began to rise again, from the early decades of the sixteenth century; the need for grain increased gradually, then rapidly, and a great deal of land that had been under grass for a century or more had to be ploughed up again.

2.2. 1500-1650: MAINSTREAM AGRICULTURE

In this next phase of mainstream agriculture, throughout the sixteenth century and beyond, farmers put great efforts into increasing grain production: they ploughed up old pasture, they increased the use of fertilizers, they intensified rotations. After two generations, their remarkable achievements were seen around 1590-1600 in faint signs of an actual surplus of cereals; it showed itself in sharply falling grain prices. A great craze for growing woad, the blue dye crop, took hold, and farmers justified this by saying that only the profits of woad could pay their rents. Certainly, cereals did not pay. But that alarm passed when bad harvests brought recurring periods of food scarcity all through the first half of the seventeenth century; then in 1642 came the outbreak of our civil war, and the farming routine was seriously disturbed in some parts of the country. Yet food for the army made heavy demands, and in some places output in consequence was further increased.

2.3. 1650-1750: SECOND PHASE OF ALTERNATIVE AGRICULTURE

Finally the peace came, and 1650 seemed like the beginning of a new era. Yet by 1653 food surpluses were so great and grain prices were so alarmingly low that Parliament took the remarkable step, in 1656, of allowing grain to be exported (hitherto, all home-produced grain was guarded for domestic use, except for modest exports which were sometimes allowed under special licence). Just under twenty years later the government actually agreed to pay bounties (cash sums) to farmers who managed to export grain abroad, in order to ensure that they kept on growing it. Thus the second experience of an alternative agriculture had begun. Grain prices continued to be low for about one hundred years from the mid-1650s to about 1750, and throughout that time farmers had to cast around for alternatives.

On this second occasion, we have enough documents to show more exactly the expedients that were adopted. All sorts of social changes had been under way in the period of mainstream agriculture which now suggested practical solutions to the problems of finding alternatives. You will recognise here marked similarities with our own day. Already in the sixteenth century Englishmen had worried about the rising prices of certain industrial raw materials which they had to import from abroad. Olive oil, used in cloth finishing, making soap, and so on, was becoming ever more expensive; so were dyestuffs (woad for the blue dye, madder for the red, and weld for the yellow); so were linen and canvas from France and the Low Countries. So efforts began to find an oil to replace olive oil, and that is how rapeseed returned in the 1560s. It came into its own as an alternative crop when the fenlands of eastern England began to be drained in the seventeenth century, and rapeseed proved to be entirely suitable; it was securely established there during the 1650s and it subsequently spread all down the eastern half of England. It remained a commonplace sight until the second half of the nineteenth century when foreign oilseeds began to be imported and mineral oils began to be used. When rapeseed returned in the late 1970s the journalists fulminated against this alien crop, which they associated with Van Gogh and France and thought totally unsuitable in the English landscape of Turner. All memory of the past had gone!

Other industrial crops which were developed as import substitutes in the sixteenth century, and which settled in firmly in the seventeenth, included woad, weld (the yellow dye crop which was localized in Kent), hemp and flax; hops which became concentrated in the south-east, most of all in Kent, but which then began to spread successfully in another area altogether, in the West Midlands, and tobacco, which was first grown in 1618 in Gloucestershire, and although banned by the government in 1619, quickly spread as a poor man's crop, and in 1670 was growing in 22 English and Welsh counties. It was not stamped out till 1690. Another less durable crop was madder. In certain years, when shortages in Zealand sent prices rocketing, this red dye crop flourished, but it faded away when Dutch production and prices resumed their normal level. Safflower, used as a pink dye, was tried but, pricewise, it did not manage to compete successfully with the safflower grown in Alsace.

These ideas for alternative industrial crops had all filtered into England as merchants traded abroad. But more ideas still were born as English gentlemen travelled on the Continent and were introduced to many new foodstuffs: vegetables like asparagus, globe artichokes, and Jerusalem artichokes, that were new to England: greatly improved varieties of the familiar vegetables, like onions cabbages, and carrots; and fruits, including better varieties of the familiar apples and pears, plus new fruits like cherries and apricots.

Noblemen, gentry and merchants travelling abroad dined with foreigners who served all these delicious foods at their tables. They developed a taste for them and brought back foreign gardeners to their country houses to cultivate them in England. Then refugees escaping religious persecution in the Netherlands and France came to England, bringing their cooking and eating habits with them. They became close neighbours of ordinary Englishmen, especially in the towns, and this was important from 1560 onwards in spreading new ideas about diet. Some of the refugees from the Netherlands and France themselves set up market gardens in and around London; they were supremely successful at Sandwich in Kent, whence they sent their produce by coastal vessel to London, and were notable gardeners on the outskirts of the City of London and Westminster.

Interest in growing more varied vegetables and fruit had thus begun to build up before any urgent economic necessity pointed to them as alternative crops. But when it did, they were among the obvious choices. That is when the potato established itself, having made a poor beginning when it was first introduced around 1600. It spread successfully after the 1670s, and then took its place as a mainstream crop, alongside cereals.

In England we recognise in all this some marked similarities with our present-day experience for new vegetables, fruits, and horti-cultural systems generally, have again found favour as our English travellers have observed them abroad in the 1960s and 1970s, and brought them home; or else they have been introduced by immigrants, by West Indians and Indians settling in our midst.

Another strong similarity between the two experiences shows in a pronounced surge of interest, in the sixteenth and seventeenth centuries and now, in herbal medicines. It is associated in the sixteenth century with the wider circulation of information on distilling the essences of plants. This became a routine activity in gentlemen's houses, where before it had been mostly reserved for monastic houses and apothecaries. A great many books on distilling began to appear, which are found in gentlemen's libraries. Along with this went a growing prejudice in favour of home-grown herbs, now considered preferable to the drugs brought at great expense from the east. Large acreages, of course, were not involved in this case, but sometimes the profits from small pieces of land were remarkable, and could greatly help a farmer to diversify his sources of income. Saffron is a good example: as a medicine it had many uses, including the general one of lifting depression; it had a culinary use; and it was used as a dye. In the 1660s one acre was expected to bring a profit of £20 above all charges (even though the labour costs were £14). The grass mown off at midsummer was a bonus, and you finished at the end of the year with three times more saffron bulbs than when you started. It was a most successful crop in eastern England, and continued especially in Cambridgeshire well into the eighteenth century.

So vegetables, fruits, and herbs commended themselves as alternative crops on economic grounds. But they had yet another virtue. The kingdom was plagued with unemployed poor. These crops required little land but large amounts of finicky labour. So poor men seized their opportunity. The most professional of them gave enormous care to manuring beds, hand sowing, weeding, protecting their plants against wind and cold with cloches and mats, and bringing the produce to market without damage. People positively welcomed the work which these garden crops created, and quite a discussion developed about the merits of spade cultivation. It was recognised that, in comparison with cultivation by the plough, production per acre was higher, and so enabled three or four times the number of people to be fed on the same amount of land.

Dairying also began at this time a marked phase of expansion, having been stimulated most of all by the civil war, when huge supplies of cheese were ordered to feed the armies. For the first time ever legislation regulated the activities of the cheese and butter merchants, a sure sign that their activities had undergone a qualitative change. Some large dairy farmers now appeared in certain districts of England, developing unwontedly large enterprises.

Poultry keeping was another alternative activity, which expanded notably in eastern England. The influence of foreigners from France and the Netherlands was significant here, for when unusually large numbers of turkeys, geese, ducks and hens feature in farmers' inventories from Norfolk and Suffolk, it often turns out that the farmer has a foreign surname. By the time of Defoe, geese walking to London from East Anglia were a familiar sight.

When attention is turned to the special problems of the large landowners at this period, their strategies show a distinct interest in the profits of rabbit warrens, a tendency to revive and rejuvenate deerparks, for the enjoyment of hunting, and for the venison as well, a notable drive to clean out old fishponds or to make new ones (the profits from a fishpond were reckoned to be much higher than from the same acreage of meadow), and, towards the end of the seventeenth century, a renewed concern for the professional management of wood-lands because timber had greatly risen in price. At the same time, landowners in certain parts of the country, especially north of London, favoured putting down arable land to grass. Their chief objective now was cattle fattening, rather than sheep keeping, for store beasts coming from the north could pause on their way to the London meat market.

Thus, an alternative agriculture was assembled which had something to offer all classes. The choices, it should be noted, include not a few which have come to the fore again in recent years.

2.4. 1750-1879: MAINSTREAM AGRICULTURE AGAIN

Then the wheel turned again, round about 1750. Population had reached a plateau after about 1650 and stuck there until 1750. Then it began to rise again, and at once a fresh demand for mainstream foodstuffs made itself felt. Farmers turned away from alternative crops to concentrate on grain, and now potatoes, and meat. England was launched at the same time on a strong phase of industrial development, which made heavy demands on farmers to increase basic food production using less labour. Thus was inaugurated the agricultural revolution, which greatly changed systems of arable farming, through the introduction of row cultivation, wheeled implements, then machines, and new methods of scientific livestock breeding. The alternatives which had been developed in the interval, horticulture, dairying, and poultry keeping, did not disappear, but they entered a quiescent phase, deprived of any strong stimulus to expand or innovate.

The momentum which had driven alternative agriculture forward for a century was lost only gradually, so that people hardly appreciated the decline. Things wound down slowly and the deterioration in produce at the market and the opportunities lost were only appreciated when the wheel turned yet again, the phase of mainstream farming came to an end, and people looked around with a more critical eye; then they realised what had happened. Vegetable supplies to the towns were thoroughly impoverished; ordinary families ate green vegetables no more than once a week; greengrocers did not find it profitable to transport cheap greens like cabbages; they preferred to offer asparagus to the better off. In Lancashire it was usual to sell greens in a chopped up mixture of shabby cabbage, leeks, and parsley to make pottage. Orchards had not been regularly renewed with fresh plantings, and many had deteriorated so badly, and the trees were so old, that they had been grubbed up, and the land used for other things. If the poor ate fruit at all, it was said, they only ate the poorest quality or in periods of glut. The quality of dairy produce was much complained of, there being no uniformity in the butter and cheese bought from the same supplier between one order and the next. The low standard showed up badly against the far superior quality of the dairy produce coming from Holland and Denmark; foreign produce was already filtering into England in increasing quantity. Likewise, the supply of poultry was pitiful, and well-to-do customers in the north of England routinely sent to London for their supplies.

2.5. 1879-1914: THIRD PHASE OF ALTERNATIVE AGRICULTURE

Mainstream agriculture had reigned supreme from 1750 for 130 years. When it collapsed, it happened comparatively suddenly in 1879. A series of bad seasons ruined the harvests one after another, but farmers were not compensated by the high prices of scarcity; instead grain poured into the country from north America, while meat arrived in increasing quantities from Australia and New Zealand. It was just as traumatic a time for farmers as is the present.

Their first recourse was the one most favoured after the Black Death, of putting large acreages down to grass. In only ten years, between 1877 and 1887, permanent grassland increased by 14 per cent, and in one county, Dorset, the increase was 61 per cent! This time, we have a most instructive discussion about the difficulties and the cost of laying arable to grass. We do not have any such literature from the fourteenth century, but the late nineteenth-century discussion makes clear that the problems were exactly the same. Land could be left to fall down to grass, and for lack of money some farmers did exactly that; alternatively, much money was spent on good seed, only to be seen to be wasted when the good grasses were swamped by poor, and pasture quality rapidly deteriorated. This experience was discussed at length in the farming journals, and sheds a wholly new light on the fourteenth and fifteenth centuries. Historians studying that period have observed the spread of ley farming beginning at that time. To put arable to grass for a short period of years was then a novel procedure in the central regions of common-field farming, for the common field system relied on keeping arable permanently cultivated, pasture that was also permanent, and permanent meadow, always mown for hay. Ley farming went against strong traditional practice. But just as many farmers in the late-nineteenth century found it the most effective method of management, rather than aspiring to create fine quality permanent pastures on land that had been under the plough for decades, so did the medieval farmer. Both had learned exactly the same lesson, but their experience was too far separated in time for the late nineteenth-century farmer to profit from it.

Trials to find the best way of laying land to grass were legion at this time, and J.B. Lawes at Rothamsted was one of the scientists who was especially active in this search. Another batch of experiments centred on the hope that the cultivation of cereals could still prove profitable, despite foreign competition, by growing them at less cost. This meant using chemical fertilizers more scientifically. But, in the event, the heroic achievements of chemical fertilizers came in the next phase when mainstream agriculture returned to dominate the scene.

For the owners of great estates, having large acreages of land to deal with, and having many farms thrown on their hands, the familiar solutions from the past served their turn again. Much of the poorest heath, moorland, and thinnest chalk downland were turned over to rabbit warrens, reviving a choice which had featured on the two previous occasions; alternatively, moorland was used for winged game - partridges and pheasants - reviving an interest that had been modestly evident in the seventeenth century. But now it was so successful that it seems to have filled the place that might have been taken by a revival of deer parks.

The greatest success stories of the late nineteenth century, however, picked up seventeenth-century projects and carried them forward in a remarkably innovative way: these were dairying, horti-culture, and poultry. The dairying story has a special twist to it for although some improvements in butter and cheese quality were achieved by factory production in these years, they did not amount to a massive transformation, nor was there any sign of the great diversification of dairy products, which we are seeing nowadays. Such possibilities were frustrated by the fact that liquid milk sales expanded so rapidly and profitably that, except where farmers had no accessible towns, milk was their least troublesome option. Moreover, farmers in the Netherlands already had a firm hold on the English market for butter and cheese. So the English dairy farmer concentrated on liquid milk and prospered. Indeed, not a few arable farmers turned to dairying as soon as the railways arrived and gave them access to large towns.

Poultry production for eggs and meat also made great strides, notably in Sussex, around Heathfield, and in the Vale of Aylesbury. It was a speciality of small farmers, and when bred for the meat market, it involved the cramming of chickens. We have detailed descriptions of the Heathfield procedures, which were especially successful because of a cooperative system of collecting and despatching the poultry to market, devised by small men among themselves. But we also see the beginnings of larger-scale poultry keeping here and there, when individual arable farmers, hard hit by the depression, somewhat reduced their acreages and specialized in poultry, giving their birds the chance to range over large areas of grass. But mostly the small men were kings in this kingdom, people recognising that the risks of disease were much reduced when poultry keeping was carried on in small, rather than large, units.

Horticulture, in all its aspects, was the most astonishing success story of all. In the seventeenth century fruit and vegetable growing had been an interest first of gentlemen, then of some middling farmers, and it showed distinct regional concentrations. Now it spread greatly, creating many more specialized areas. The great achievement of these years was much higher production along with much improved quality, which brought hard, and soft fruit, and vegetables within the reach of ordinary people on a more generous basis. Illustrating the former disregard of soft fruit, it is noteworthy that before 1887 acreages were never counted, although agricultural statistics generally had been collected for twenty years. Similarly, while fruit and vegetable acreages in market gardens were called for, they were not collected on farms. The resulting figures were totally misleading.

The transformation of horticulture in these years is vividly described by Mr Assbee, who was superintendent at the Covent Garden Market, watching the ever larger space needed to deal with all the produce that flowed in there. The producers grew improved varieties of everything, many introduced from the Continent (though some critics were already saying that the so-called improved varieties lacked the flavour of the old), and greatly increased the use of glasshouses. They then broadened out from vegetables and fruit into plants and cut flowers. The whole story amounted to a horticultural revolution, and yet as I use that term you will realise that it left plenty of room for yet another leap forward, which we are seeing in our own day.

All this innovation involved the education of the public in a more diversified diet, and the efforts are seen in the literature, urging the English, among other things, to dress their salads with oil and vinegar as the French did, a practice which the English did not heed then, and are only learning now. Watercress, however, entered the diet with a great flourish, being well received when eaten plain and unadorned. The prejudice against tomatoes, because they were thought to be poisonous, was whittled down, though one greengrocer said that most of his tomatoes were bought by his customers to be eaten as a fruit or in salad. Attempts were made to make shoppers familiar with celeriac, sugar (mange tout) peas, chicory, endive, cardoons, and aubergines, and one writer, gathering news of new foodstuffs eaten in Japan, recommended their radishes, turnips, and soybeans. Such propositions fell on deaf ears, and have waited another hundred years for a better chance.

Horticulture in the Middle Ages and in the seventeenth century had been an activity carried on by individuals on small parcels of land. And whenever middling farmers pursued it, it was as a branch activity, alongside arable farming; such men, in consequence, were often called farmer gardeners. Now horticulture began to emerge as a commercial activity, involving large acreages - as much as 200 acres - in the hands of one enterprise, even while, at the same time, small men were still making a satisfactory living on 15 acres, or a little more. Opinion was still firmly in favour of their survival. In fact, a lively debate was carried on concurrently, defending smallholdings against those who argued that the future lay only with large farms. It is noticeable, moreover, that all the idealist colonies set up at this time by anarchists, socialists, or religious groups, relied on horticulture for their livelihood, along with the other thriving alternatives of dairying and poultry.

2.6. 1914-1970: MAINSTREAM AGRICULTURE

The third phase of alternative agriculture was abruptly halted when war broke out in 1914, and the kingdom had to produce as much of its own grain, potatoes, and meat as possible. Mainstream agriculture returned, therefore, and though it yielded far from happy experiences through the 1920s and 1930s, prosperity for the farmer returned during the 1939-45 war, and for about twenty-five years after. Yields of arable crops were then dramatically raised, and farms became larger and larger.

3. Current situation: in search for alternative systems once more

We have now entered another phase, when the overproduction of grain has created a fresh crisis, and we seek alternative systems once more. Many present-day parallels have emerged in this survey of the three previous phases of alternative agriculture. The resemblance with the seventeenth century is particularly strong in certain respects. The problem has stolen up gradually as it did then, and not suddenly as in 1349 and in 1879. The decades of the 1960s and 1970s saw a great many people travelling abroad, as did the gentry and merchants in the sixteenth century. Our modern travellers eat different foods, and show a desire to see them in the shops at home. We have also given a home to many immigrants. These two developments have both stimulated ideas for alternative farm produce. Singular individuals experimented in the 1960s and 1970s with crops and systems which are now established, and are proving their economic worth. We are plainly living through another horticultural revolution, enjoying more fruit, vegetables, and flowers, and seeing trials with more varieties of everything. Among the latest of many suggestions being offered to farmers as alternative crops are sorrel, rocket, and okra.

Dairying too is experiencing another innovative phase, with the development of yoghurts, soft cheeses, fromage frais, and new ice creams; we have surely not yet seen the end of the novel uses for milk. Poultry farming has also undergone a revolution since the 1960s, moving in and out of the battery system. Many new vineyards have been set up in recent years, as in the seventeenth century (space has prevented a description of those earlier efforts, but they featured as expected, in tune with their time), and English wines are winning awards in competition with other European countries. The wine enterprise may be killed by common market regulations if production increases above a certain level; the axe is poised over the grapegrowers' heads at this moment. Owners of large acreages have started to produce venison for the table from deer parks. Coppicing is being carried on at a more professional level than for many decades, and some charcoal-burning is producing charcoal for barbecues. Industrial crops have returned, as in the seventeenth century: thus rapeseed is grown for a vegetable oil, and serious consideration is now being given to its use as an industrial oil, thus repeating the solution of the seventeenth century. Flax is being cultivated for linseed oil - but why not for linen too, since the fashion for linen is reviving? Nor should we altogether exclude the possibility of a return of vegetable dyes like woad and madder. An advertisement appeared recently recommending indigo-dyed jerseys because they had the merit of fading unevenly! The Japanese, it should be noticed, have deliberately fostered a revived fashion for indigo-dyed fabric, in order to save their dyeing arts from extinction. Why not do the same for woad, since it is the same dye in a less concentrated (faded?) form?

As one speculates further on what alternative agricultures from the past might prove serviceable in the next century, it is also instructive to survey some of the background circumstances that have loomed prominently every time a return has been made to alternative agriculture. We are nowadays seeing a revived interest in the use of plants in medicine, and much more sympathy for homeopathy. This trend matches the strong interest in medicinal plants and the distillation of plant essences shown in the sixteenth to seventeenth centuries. We are also seeing a growth of vegetarianism, reproducing exactly the surge of interest which occurred in the late nineteenth century. It also occurred in a more subdued way, for it was then eccentric, in the later seventeenth century. Perhaps it even found followers in the fifteenth century, though our documents on the subject fail us.

Yet other repetitions prompt reflection. It is a characteristic of alternative agriculture that it offers a variety of solutions that are specific to certain localities. It does not offer one remedy that will fit the situation more or less everywhere as does mainstream agriculture. I notice that whenever I discuss alternative agricultures with economists, they always dismiss every single suggested alternative from the past because it is not a mass solution. The economist's mindset seems to be determined by the conventions and assumptions of mainstream agriculture, and it is plainly uncomfortable with the notion of multitudes of choices, different for different localities, which alternative agriculture offers. But from the past that is one of the most instructive lessons of the phase: the salvation may be rhubarb in one place, liquorice in another, teasel in another, celery in another; and the list of possibilities is very long.

Another characteristic feature of successful alternative agriculture is the mixing of several small pursuits on one farm. That lesson is embedded in a most discerning sentence in <u>Maison Rustique</u>, a French book of husbandry of 1600 by Estienne and Liebault, emphasising the farmer's duty to concern himself with many small things: "A good farmer will make profit from everything, and there is not (as we say) so much as the garlic and onion which he will not raise gain of by selling them at fairs most fitting for their time and season, and so help himself thereof and fill his purse with money". Again and again, both in the seventeenth and in the late nineteenth centuries, commentators said the same thing in different words: the farmer must show attention to a number of small things, and attention to detail that was the secret of success in those years.

There is yet another rhythm in all this past experience which concerns the differing contributions of men and women, as one phase of dominant agriculture has moved into another. It is a generally recognised characteristic of women that they show attention to small things and attention to detail. So, in the event, it was the women who showed the way forward whenever a new phase of alternative agriculture returned. They had nurtured the small things, which then came to the fore as alternative pursuits. They were originally in charge of the dairy; they were the horticulturists, intent on gathering vegetables and herbs for the kitchen and for medicinal use; they were the keepers of the poultry. But when an alternative agriculture was needed, the menfolk recognised the commercial possibilities of the women's occupations, one after another. They adopted them, but they also developed them in fresh directions. Thus they took over dairying from the women and set up dairy factories; they set up glasshouse horticulture; they set up large poultry enterprises. A most discerning observer in the late nineteenth-century saw one of these developments at a critical moment in its progress; knowing the potential for growth in poultry-keeping, he noted the current obstacles preventing its quicker expansion. It was due, he said, to the prejudices of the menfolk. Farmers held poultry-keeping in contempt. It was the concern of their wives, and they did not even admit that it made a profit. Often they positively hindered their wives' efforts. But in the end, the prejudice was broken down and they entered the field.

The past history of alternative agriculture suggests that the present phase in farming fortunes could well last for two generations or more, but it will not last for ever. In England many leisure projects are being launched, which occupy the land with golf courses, cart racing, mountain bike riding, and horse riding. These are the present-day equivalent of the deer parks and grouse moors of the past, though now they serve other classes besides the gentry. But the new circumstances also prompt reflection on our high levels of unemployment. They may not fade away in a year or two as the politicians expect. Should we not foster alternative agricultures as a way to employ many people? It was an outstandingly successful feature of the seventeenth-century phase, when unemployment was a major concern. So why not learn a lesson from the past, why not grow labour-intensive crops again? Why not saffron which once grew most successfully in England? Nor should we forget alternatives that can be assembled into a mixture of activities. That is another successful strategy; in fact, we are actually seeing something of this sort on smallholdings which are combined with teleworking. Two different occupations are combined to make a full living.

The trend in favour of smallholdings leads on to reflections on the size of farms. In the most recent phase of mainstream agriculture the drive to create ever larger units was strong, and seemed destined to continue. Now it creates difficulties for anyone wishing to engage in some experimental branch of alternative agriculture, for the new venture is likely to be labour intensive, and require small units of land. But if anyone wants a small slice of land, and a large farm is broken up, the smaller units need farmhouses, and planning regulations greatly hamper such developments. The merits of family farms and smallholdings were urged in the late nineteenth, and in the seventeenth century. From the fifteenth century no debate survives, but plenty of evidence demonstrates how the large home farms of the gentry were broken up into small farms; it happened, whether it was discussed or not. Shall we see a more sympathetic discussion developing over the next decades in favour of family farming, moderating the adverse criticism that is routinely directed, in England at least, at family farms?

4. Conclusion

This survey of the past does not produce certain answers concerning the alternative agriculture which we now seek for the future. But it establishes a framework in which certain truths are embedded. These have proved valid on three occasions in the past, and they embody experiences which could guide farmers now. The scene, of course, is viewed from a very different angle from that taken by the report to your government of the Netherlands Scientific Council, entitled <u>Ground for Choices. Four Perspectives for the Rural Areas in the European Community</u>. That is a survey of the whole European continent and considers four alternative strategies for the same vast territory. It is a view of the scene from on high, by those who can influence the framing of policy. The account given here represents a worm's eye view from closer to the ground, seeing things from the viewpoint of much smaller groups of countrymen, who in the past found their way along many separate paths. They were exploiting their distinctive local resources, and developing economic insights that pertained to their own small regions. They devised solutions that fitted quite specific local circumstances. Their experience is offered here as a different but, nevertheless, illuminating perspective on the possible shape of things to come.

5. References

The documentation supporting the arguments in this paper is too voluminous to be given here. It will be fully set out in my book, <u>Alternative Agriculture, Past and Present</u>.

FARMING SYSTEMS CHARACTERIZATION THROUGH CROPPING PATTERNS; experiences from lowland vegetable production in Indonesia

J.S. BUURMA
Agricultural Economics Research Institute LEI-DLO
P.O. Box 29703
2502 LS   The Hague
The Netherlands

ABSTRACT. This paper aims at translating the farming systems concept & in technical terms in order to improve the communication between social and technical scientists. It illustrates how vegetable growers in Java adjust their cropping patterns to the prevailing natural and socio-economic circumstances. Altitude and distance to big cities are the decisive factors for the location of vegetable production in Java. These factors imply a matrix of ecological and economic zones, resulting in production regions which are contrasting in vegetable crop choice. Within the production regions differences in water availability are decisive for the composition of the cropping pattern. A formal survey in the Brebes' production region for shallots revealed five distinct cropping patterns which are contrasting in production season. Within these cropping patterns the farmers' socio-economic positions are still different and depend on land size. This means, that a cropping pattern is characteristic for a farming system so far as a limited range of farm sizes is considered.

1. Introduction

Farming systems are the cumulative result of farmers' decisions during a long period of time. They represent the farmers' adaptation to the prevailing natural and socio-economic circumstances. This definition makes it plausible, that there are many different farming systems, but the way how to classify them remains unclear.

It is the objective of this paper to make the definition of "farming system" operational and clear for the different disciplines involved in farming systems research. Subject of research is the vegetable production sector in Java, the main island of Indonesia, containing wide ranges of natural and socio-economic circumstances.

The research questions for this study are: where is the production of vegetables located, which factors explain the location of vegetable production centres, how are the vegetables organized in cropping patterns, which factors explain the composition of these

71

patterns, are these cropping patterns homogeneous from the socio-economic point of view?

CIMMYT (1980) developed a very useful package of concepts and procedures for the planning of research and development. One of the quintessences of this package is to arrive at homogeneous regions or groups of farmers. Such a group of farmers is called a recommendation domain and may be defined in terms of both natural and socio-economic factors. The drawback of this way of defining is again, that it does not provide an operational picture of the farm. For that purpose the cropping pattern is much more informative.

The purpose of distinguishing homogeneous regions and/or groups of farmers is providing a tool for the planning of research and development, especially for the selection of target regions and target groups. As such the results of this study were applied in a Dutch-Indonesian development project (ATA-395) at the Lembang Horticultural Research Institute (LEHRI). More detailed information is given by Buurma and Sinung Basuki (1991) and Buurma (1992).

After this introduction the attention is firstly focused on the distinction of ecological zones and economic zones, resulting in a matrix of production regions. Next the distinction of cropping patterns in the main production centre for shallots is described. Finally the results are positioned with regard to limiting conditions and decision making at farmers' level.

2. Production regions

A computerized databank was established to answer the research questions on the location of vegetable production. This databank contained the over 1987 planted areas of 18 vegetable crops and 3 ecological characteristics for all 1650 subdistricts in Java. The ecological characteristics were altitude, soil type and rainfall zone.

Plotting the areas of the different vegetable crops in the subdistricts' map of Java showed a significant pattern of production concentrations. The few typical highland subdistricts showed very high production concentrations, largely resulting from highland vegetable crops. In lowland the production package was changing with the distance to big cities: leafy vegetables nearby; fruit-bearing vegetables further away and bulb vegetables far away from the big cities.

In order to arrive at a division of Java into roughly homogeneous production regions, boundaries have been drawn through the transition zones between (1) lowland and highland; and (2) the production centres for leafy, fruit-bearing and bulb vegetables. This exercise resulted in a matrix of 23 production regions: 16 in lowland and 7 in highland. This classification in production regions proved to be very useful for the selection of target areas for research and development.

The boundary between lowland and highland was defined by computing the cumulative area percentages at increasing altitudes for the different vegetable crops. Two of the resulting curves are presented in Figure 1. Typical lowland vegetable crops (like yard-long

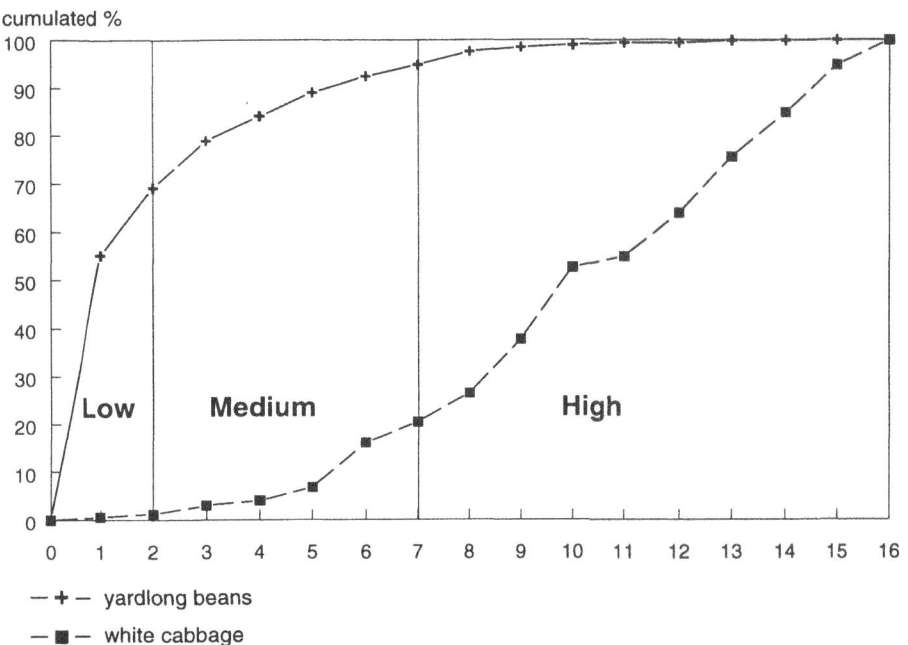

cumulated %

Low Medium High

— + — yardlong beans

— ■ — white cabbage

Figure 1. Cumulated area-% by altitude .

Source: Databank Vegetables Java 1987

beans) showed a convex course and typical highland vegetable crops (like white cabbage) a concave course. On the basis of the above curves the following ecological zones for vegetable production were defined: typical lowland = below 200 m a.s.l.; typical highland = above 700 m a.s.l.; and a transition zone = medium land = between 200 and 700 m a.s.l.

3. Cropping patterns

As target area for research and development in shallots, the lowland production region around Brebes was selected. This region represents about 45 % of the shallot area in Java. Shallots and hot peppers are the dominant vegetable crops in the region (72 % resp 16 % of total area). A formal survey among 140 shallot growers in the region revealed the following cropping patterns.

```
Code    Jan Feb Mar Apr May Jun Jul Aug Sep Oct Nov Dec
-----------------------------------------------------------------
SP--    shallots/p e p p e r s
RS-S    r  i  c  e       shallots          shallots
RSSS    r  i  c  e       shallots   shallots   shallots
RSPS    r  i  c  e       shallots/p e p p e r s shallots
RSPP    r   i   c   e      shallots/p  e  p  p  e  r  s
-----------------------------------------------------------------
```

Discussions with farmers made clear, that water availability played an important role in the composition of the cropping pattern. Table 1 confirms this proposition and shows, that water shortage at most "SP--" parcels already starts in or before June. Parcels with an "RS-S" cropping pattern get water shortage around July and "RSSS" around August. The cropping patterns "RSPS" and "RSPP" mostly occur on parcels with late (=> September) or no water shortage.

TABLE 1. Number of parcels at beginning of water shortage period (month) and cropping pattern.

Month\Pattern	SP--	RS-S	RSSS	RSPS	RSPP	Total
<= June	24	21	2	2	4	53
July	7	22	11	7	6	53
August	4	17	17	7	14	59
=> September	6	3	10	10	9	38
No shortage	0	3	7	9	10	29
Total	41	66	47	35	43	232

An indicator for the farmers' financial position is the status (owned/
rented) of his land. Ownership indicates that the farmer was not
(financially) forced to sell his land in the past and simultaneously
it means that the farmer can save the expenses for land rent in the
future. Analysis of the land status revealed high land rent
percentages below 0.6 ha and above 3.0 ha. The group below 0.6 ha
probably included many farmers who cannot afford to buy land. The
group above 3.0 ha probably included "gentleman farmers" who rent
large areas of land for large scale shallot production. This finding
makes clear, that farmers with a similar cropping pattern may operate
from different economic positions. Such positions within cropping
patterns roughly correspond to farming systems. The classification in
cropping patterns and farm sizes proved to be very useful for the
selection of target groups for research and development.

4. Discussion

As stated earlier, farming systems are the result of the farmers'
decision making. Decision-making is usually sub-divided into three
levels: strategic, tactical and operational decisions. These three
levels can be roughly associated with the limitations and the matching
adaptations as reported in the previous sections. The associations
concerned are depicted below.

Limitation	Adaptation	Decision
Climate -------> Transport ----->	Ecological zone Economic zone	Strategic
Demand --------> Policy --------> Irrigation ---->	Potential crops Compulsory crops Production period	Tactical
	Cropping Pattern	
Technology ----> Technology ---->	Crop management Pest management	Operational

The picture shows the translation of environmental (climate) and
economic (transport) limitations into ecological and economic zones.
These zones imply a strategic decision of the farmer resulting in a
limited number of potential vegetable crops. Dependent on market
perspectives (demand), production regulations (policy) and water
availability (irrigation) the farmer makes a cropping pattern, which
is a tactical decision. The cropping pattern is the starting-point for

operational decisions with regard to crop management, pest management, etc. The picture illustrates, that the cropping pattern is the meeting point of the economic and technical disciplines. As such it should be the starting-point for the planning of research and development.

The advantages of defining cropping patterns in stead of recommendation domains can be substantiated with the results of the previous sections. The cropping pattern does not only reveal differences in water availability, but also the consequences for crop choice and timing of production. The latter information is not included in the description of recommendation domains. CIMMYT admits, that the boundaries of recommendation domains are arbitrary in case of gradual variations of circumstances (like water availability). The cropping patterns concept removes this problem, results therefore in more homogeneous groups, provides a better grip for the identification and selection of target groups and facilitates the organization of on-farm research.

A cropping pattern may cover more than one farming system. Farms with a similar cropping pattern may differ in labour organization, land owner-ship, financial resources, etc. This difficulty can be largely removed by a sub-classification on a socio-economic factor like land size. Consequently a cropping pattern is characteristic for a farming system so far as a limited range of land sizes is considered.

5. Conclusions

Altitude and distance to big cities are the decisive factors for the location of vegetable production in Java. Combination of these factors leads to a classification in production regions which are contrasting in vegetable crop choice.

Within the Brebes' production region for shallots water availability is decisive for the composition of the cropping pattern. The farmers' economic positions are still different within the cropping patterns and depend on land size. This means, that a cropping pattern is characteristic for a farming system so far as a limited range of land sizes is considered.

In comparison to recommendation domains, the cropping pattern also provides information on crop choice and timing of production. Thus it promotes the communication between social and technical scientists and consequently the planning of research and development.

6. References

Buurma, J.S. and Sinung Basuki, R. (1991) 'From statistical data to research regions', Acta Horticulturae 270, 147-152.

Buurma, J.S. (1992) 'Target groups for research and development in shallots in Brebes', Internal Communication LEHRI/ATA-395 No. 47.

CIMMYT, (1980) 'Planning technologies appropriate to farmers; concepts and procedures', CIMMYT, Mexico, Manual for research planning.

ECONOMIC AND ECOLOGICAL POTENTIALS OF SUSTAINABLE AGRICULTURE IN SOUTH-INDIA

A. DE JAGER
Agricultural Economics
Research Institute (LEI-DLO)
P.O. Box 29703
2502 LS Den Haag
The Netherlands

E. VAN DER WERF
ETC-Foundation
P.O.Box 64
3830 AB Leusden
The Netherlands

ABSTRACT. The comparative performance of seven farm pairs in South-India, consisting of one ecological and one conventional reference farm, is analyzed in relation to agronomic and economic performance. The first interim results of the study indicate that both for farmers and society the short-term needs for food and cash income can be successfully combined with many of the long-term needs for sustainability.

1. Introduction

The development of sustainable agricultural systems increasingly draws the attention of farmers, researchers and policy makers in both the Northern and the Southern hemisphere. However complete different socio-economic circumstances have resulted in this common interest in sustainability. In the South the lack of capital and access to technologies and inputs led to an increased use of marginal lands, adverse environmental effects and an insecure food security in the long run. In the North the excessive and inappropriate use of inputs and technologies have resulted in environmental pollution and uneconomic production in the absence of subsidies. In this paper, which focuses on the situation in the Southern hemisphere, a sustainable agricultural production system is defined as one that maintains or enhances environmental quality, satisfies future demands of society for food and fibres and assures the economic and social well-being of producers (Hailu and Runge-Metzger, 1993). However a general consensus on the conceptual research framework, criteria, parameters on sustainability and strategy is lacking. Some researchers stress the importance of optimal combinations of inputs at the lowest cost of input per unit of output (De Wit, 1992), others give priority to reliance on internal and renewable resources as much as possible and an optimization of biological balances and relationships within the farming system (Francis *et al.*, 1986).

P. C. Struik et al. (eds.), Plant Production on the Threshold of a New Century, 77–83.
© 1994 *Kluwer Academic Publishers.*

In South-India ecological agriculture is practised by a small number of farmers for a number of years. In this context ecological agriculture is defined as a production system without application of any chemical fertilizers and pesticides, optimizing use of local resources and deliberately including ecological farming principles like stimulation of diversity and complexity. Since this production system appears to meet the definitions of sustainability, an agronomic and economic evaluation of the performance of these farms may give indications on the potentials of the practised technologies for the development of sustainable production systems in the region. Therefore, in 1989 the existing Agriculture, Man and Ecology (AME) programme, implemented by ETC-Foundation, started monitoring existing ecological practices with the following objectives:
- Identifying, qualitatively and quantitatively, the viability of ecological agriculture by itself and in comparison with conventional agriculture.
- Examining the long-term prospects of ecological agriculture.

The research was executed by the AME-team and The Institute for Command Studies and Irrigation Management (ICSIM), Bangalore with assistance from LEI-DLO.

2. Methodology

Because an intensive study is required to gain insight in diversity and complexity of production techniques, to identify problem areas and since only a limited number of established ecological farms (EFs) were available, a comparative case study approach was selected (Maxwell, 1984, Werf and De Jager, 1992). Limitations of the approach are representativeness, possibilities for extrapolations to regional and national level and elimination of effects of factors not determined by the studied production system, especially farmers' management skill.

The study covers seven farm pairs in the states of Karnataka and Tamil Nadu, each consisting of an ecological and one comparable reference farm practising conventional agriculture. The EFs were selected according to the criteria in the above mentioned definition of ecological agriculture with the restriction that at EFs ecological management was practised for at least three years. Every EF was linked to a conventional reference farm (CF) close to the EF with a similar main crop trying to eliminate non-system factors like soil type, climate, topography, etc.). In addition a comparison was made with available regional secondary data.

An extensive descriptive picture of the case-study farms was made including detailed physical and socio-economic information. Every two months data were collected by AME and ICSIM staffs using a structured schedule covering all crop and livestock input-output flows in actual quantities and money value, total labour requirements and cash flow. Existing secondary data were collected from the relevant government departments. The farmers were actively involved in the data collection and results were verified and discussed extensively with the participating farmers in individual and group discussions. Considering the huge yearly variations in yields and economic results on farms and in order to be able to study stability of the results the farms are monitored over a period of five years. This paper describes the results of two years of field work (1990-1991). Although the research finds its strength in the detailed analysis at farm level, emphasis in this paper is on the general results of the studied production systems.

3. Results

3.1. FARM CHARACTERISTICS

The EFs appear to have special characteristics and entirely different cropping patterns than the studied CFs. The average holding size of EFs is considerably higher than the selected CFs (Table 1) and the average holding size of 1.0 ha in the state of Tamil Nadu. Ecological farmers are much more involved in off-farm activities than their conventional colleagues. On average the irrigated area and market orientation is slightly lower on EFs. The total assets, expressed per ha, are comparable for the two farming systems and consist mainly of farm implements and livestock.

TABLE 1. Averages of farm characteristics for ecological and conventional farms (in 1990 1 US$ = 16.5 Rs).

	Ecological	Conventional
Holding size (ha)	4.5 (1.8-12.1)*	2.6 (1.6-4.0)
Irrigated area (% of total)	54 (0 - 96)	63 (19 - 92)
Total assets/ha (x 1000 Rs)	85 (24 -136)	91 (46 -138)
Off-farm income (x 1000 Rs)	14 (0 - 42)	1 (0 - 5)

* between brackets the minimum and maximum value

3.2. AGRONOMIC ASPECTS

A more diversified cropping system on EFs is observed, expressed in a higher number of different crops cultivated per farm (Table 2). On EFs pulses, fodder crops and trees have a higher share and vegetables a lower share in the cropping system than on the CFs. Tree products in EFs include fruit, fodder, green manure, fixing of atmospheric nitrogen, fuel and construction wood species, mostly in combinations. Furthermore, trees are used as windbreaks, for the provision of permanent soil cover and leaf litter and to recycle nutrients leached to deeper soil layers. Both farming systems make use of techniques like deep-rooting crops, green-leaf and farm yard manure for soil fertility management. Composting, biogas production and mulching are additional techniques applied on EFs. Concerning plant diversity EFs apply a wide range of specific practices like multi-story cropping, selective weeding, cover crops. Mixed cropping and intercropping are practises widely applied in both farming systems. In the ecological and conventional farm management in the region, livestock forms an integrated component of the farming system. In both systems mainly cows, goats and poultry are present and in almost all cases manure is produced and applied. However on EFs the efficiency of manure production appears to be greater. The total nutrients input at farm level (including nutrients in animal feed) and the dependency on external nutrients for crop production is significantly lower on the EFs. Dependency on external nutrients on EFs was 33 % for N, 40 % for P and 34 % for K. For the reference group 57 %, 69 % and 56 % was observed respectively. Of the external N-P-K used on the CFs, 62 %, 60 % and 48 % respectively were in the form of chemical fertilizers. The use of pesticides in conventional agriculture is rather low (2-3 kg/ha), but varies strongly among the crops cultivated and different farms.

TABLE 2. Averages of a number of agronomic aspects on ecological and conventional farms.

	Ecological		Conventional	
Number of different crops/farm	10.6	(6 - 14)*	7.7	(3 - 12)
Number of trees/ha	192	(17 -519)	37	(10 - 83)
Total LWU**/ha	1.9		1.7	
Total external nutrients kg/ha	66	(23 -100)	145	(40 -465)
External nutrient dependency of crops (%)	36	(17 - 58)	58	(32 - 79)
Number of soil fertility improvement techniques practised	5.2	(3 - 8)	3.5	(2 - 6)
Number of plant diversity techniques practised	4.0	(3 - 7)	1.6	(1 - 2)

* between brackets the minimum and maximum value
** LWU (Life Weight Units) = 250 kg

At EFs similar yield levels were observed on major crops like sorghum, finger millet, groundnut and sesame compared to the CFs and statistical state averages. For irrigated rice a higher yield level was realised on the EFs in both years (Table 3).

When N-fixation of estimated leguminous crops is taken into account, in both systems a similar input of N-P-K and a negative nutrient balance for N and K is observed. On ecological farms however the deficiency levels were considerably higher.

TABLE 3. Average yield (kg/ha), average inputs of N, P and K (kg/ha) and average nutrient balance at field border (kg/ha) for irrigated rice on ecological and conventional farms.

	Ecological		Conventional	
	Input	Balance	Input	Balance
N*	87	-27	90	-4
P	17	3	16	4
K	51	-42	55	-29
Yield	4370		3290	

* including estimated N-fixation leguminous crops

3.3. ECONOMIC ASPECTS

The economic results differ enormously among farms in the same group and among the two years. No significant differences are observed between the average gross margin per ha of EFs and CFs. Similar product prices are realised on EFs and CFs.

TABLE 4. Averages of a number of economic key figures on ecological and conventional farms.

	Ecological	Conventional
Gross income/ha (x 1000 Rs)	14.7 (4.6-53.4)*	14.2 (3.6-30.3)
- share from crop activities (%)	75 (25 - 98)	88 (77 - 99)
- share sold (%)	48 (32 - 66)	53 (9 - 79)
Variable costs/ha (x 1000 Rs)	7.4 (3.5-23.8)	5.6 (2.8-11.8)
Gross margin/ha (x 1000 Rs)	7.3 (-0.3-29.6)	8.6 (0.2-22.9)
Labour days/ha	229 (79 - 531)	260 (140 - 435)
Returns/labour day (Rs)	28 (6 - 65)	30 (8 - 68)

* between brackets the minimum and maximum value

Surprisingly no difference was observed between the average labour days per ha and the returns per labour day between the two studied systems. In both systems approximately 50 % of the labour is supplied by women, on EFs slightly more use is made of hired labour (72 % compared to 55 % of the total labour input). Gross margins of crops on CFs tend to be higher than on EFs, while similar gross margins are realised for livestock activities. However on CFs livestock is mainly kept for traction purposes, on EFs also production of manure, milk, eggs and meat are of significant importance (higher gross income). Livestock activities had considerably lower gross margins than crop activities.

TABLE 5 Average economic results of crop and livestock activities on ecological and conventional farms

	Ecological		Conventional	
	Crops	Livestock	Crops	Livestock
Gross income/h	11.0	3.7	12.5	1.7
Variable costs/ha	4.6	2.8	4.8	0.8
Gross margin/ha	6.4	0.9	7.7	0.9
Gross margin/LWU	-	0.5	-	0.5

4. Conclusions

The studied farms in both groups can be classified as a well-above average group of farmers in terms of skills and resources considering the average holding size and return per labour day in comparison with state averages. The importance of off-farm income on EFs indicates that currently only farmers with a sufficient alternative source of income have been willing to bear the risks involved in the process of transition and experimentation towards a complete ecological farming system.

On EFs the use of fertilizers and pesticides is replaced by a wide variation of practices mainly geared towards creating crop diversity and maintaining soil fertility. This results in a more diversified cropping system on EFs apparently requiring a larger total arable area. Soil fertility management practices result in a considerable lower dependency on external nutrients. The use of own produced and often composted manure and fixation of nitrogen by leguminous crops are essential elements in the supply of nutrients for crop activities. Due to a different cropping pattern the composition of the output on EFs differ from conventional agriculture: more pulses, tree products, fodder crops and livestock products. Crop yields and the overall economic results show no significant differences between the EFs and the CFs.

The results indicate that both for farmers and society the short-term needs for food and cash income can successfully be combined with many of the long-term needs for sustainability. The absence of pesticides, the use of renewable and internal sources, nutrient recycling are essential positive aspects towards sustainability on EFs. However, like in conventional agriculture negative nutrient balances at crop level occur and may also negatively influence yields on EFs in the long run.

5. Discussion

The long-standing experiences in ecological farming are an essential source of information which should be used to increase the sustainability of conventional farming. This is particularly true for internal nutrient recycling, improved manure management, use of nitrogen-fixing species and aspects of crop rotation. Eventually this data set covering five years of agronomic and economic data at activity and farm level, combined with results of on-station and on-farm experiments, will be an important input for modelling and simulation activities to define optimal land use systems using criteria of sustainability, food security and economic performance.

6. References

Francis, C.A., Harwood, R.R. and Parr, J.F. (1986) 'The potential for regenerative agriculture in the developing world', American Journal of Alternative Agriculture 1(2), 65-74.

Hailu, Z. and Runge-Metzger, A. (1993) 'Sustainability of land use systems', Margraf, Weikersheim, Germany.

Maxwell, S. (1984) 'The role of case studies in farming systems research', IDS, Sussex.

Werf, E. van der and Jager, A. de (1992) 'Ecological agriculture in South-India, an agro-economic comparison and study of transition', ETC-Foundation/LEI-DLO, Leusden/Den Haag.

Wit, C.T. de (1992) 'Resource use efficiency in agriculture', Agricultural Systems 40, 125-151.

THE USE OF FARM ECONOMIC MODELLING TO DETERMINE THE INFLUENCE OF MACSHARRY ON THE DEVELOPMENT TOWARDS SUSTAINABILITY

C. VAN DER HOUWEN
Rijkswaterstaat Directorate Flevoland
P.O. Box 600
8200 AP Lelystad
The Netherlands

ABSTRACT. In the development of sustainable farming systems, both environmental and economic aspects need to be studied. An interesting question is whether the restructuring of the Common Agricultural Policy by EC commissioner MacSharry, which aimed at reducing production surpluses, will be an incentive for extensification.

A farm economic Linear Programming model was made for arable farms in a newly reclaimed area of the Netherlands, called South Flevoland. The model was extended with environmental parameters like use of nitrogen fertilizer and pesticides. Different cropping methods were included in the model. This way the optimal farm organization under different circumstances and the effects on farm income and environment could be studied.

Model calculations showed that on the farms in question no use was made of set aside and that the MacSharry measures were hardly an incentive for extensification. The limitations of the model are discussed and it is indicated how the model can be improved.

1. Introduction

The Common Agricultural Policy (CAP) has used a system of protectionism and price support, in order to secure a sufficient income for the farmers. Indeed for many years farmers have benefitted from this system. Because of this market regulation together with technological developments, agricultural production has increased enormously. This has caused overproduction and environmental problems.

Because of production surpluses the European Community is facing budgetary problems. Therefore EC-commissioner MacSharry induced a restructuring of the CAP. During the years 1993, '94 and '95, intervention prices will be lowered stepwise, while direct income support is given by premiums per hectare. The amount of income support depends on whether or not land is withdrawn from production (set aside).

It is interesting whether the reform of the CAP will be an incentive for extensification. Wossink *et al.* (1992) and Wossink (1993)

P. C. Struik et al. (eds.), Plant Production on the Threshold of a New Century, 85–91.
© 1994 *Kluwer Academic Publishers.*

introduced a method for modelling economic and environmental relationships at farm level. The aim of this paper is to demonstrate how an
environmental economic linear programming-model was used to investigate the effects of the MacSharry measures on farm income and
environment and to what extent the model can be used.

2. Method

A model was made of specialized arable farms in South Flevoland. South
Flevoland is Holland's youngest polder. This polder with a surface of
43,000 hectare was reclaimed in 1968. Since 1978 farms have been
rented out to individual farmers. In 1991 there were 153 arable farms
with an average surface of 55 hectare. In the polder the soil type is
clay which is very fertile and therefore yields are very high. Farm
structure is rather homogeneous in this region.

The created model is a normative optimization model. With this
comparative static method optimal farm organization was assessed under
different internal and external conditions. The matrix for the mixed
integer linear programming was made in Lotus 1-2-3. It was tried to
make optimal use of the spreadsheet functions to link input data to
the matrix. The model optimizes net farm result. The activities or
variables that contribute to the objective function are for example
cropping activities, hired labour, investment in machinery and use of
nitrogen fertilizer and pesticides. The restrictions are for example
available land, available labour and crop-rotation requirements.

The crops that were available in the model were: ware potato,
sugar beet, winter wheat, spring barley, onion, pea (dry or fresh
harvested) beans, seed grass and set-aside (perennial rye grass). For
the four first mentioned crops different cropping methods were
defined, which differed in use of nitrogen fertilizer, pesticides and
input of labour and machinery. Table 1 shows the number of variants
for these crops. The four variations in use of nitrogen fertilizer
were 25 %, 50 %, 75 % and 100 % of the recommended dose. The effects
of nitrogen fertilizer input on production output was determined by
means of region-specific production functions. The use of organic
manure was left out of consideration. For winter wheat and spring
barley a current and a reduced level of pesticide input were
distinguished. Together with the four possibilities of fertilization
this resulted in eight cropping variants. It was assumed that the
reduced pesticide input results in a 10 % lower yield. Thus a certain
interaction was assumed between fertilizer and pesticide input. For
sugar beet three methods of weed control were distinguished, which
were assumed to have no effect on the yield of sugar beet. Ware potato
is the most important crop both as it concerns the environment and for
farm income. Therefore variants were defined for several cropping
activities. Two potato varieties were considered. "Bintje", the main
potato cultivar in Holland which is rather liable to diseases, and a
variety that is resistant to nematodes. The latter was assumed to have
a 5 % higher yield. For rotation a rotation of three years and a
rotation of four years were possible. The rotation of three years was

assumed to have a yield reduction of 8 %. The variants for weed control, haulm killing and late blight control were assumed to have no effect on the yield of potatoes.

TABLE 1. Numbers of cropping variants.

	Nitrogen	Pesticides	
Winter wheat	4*	2*	
Spring barley	4*	2*	
Sugar beet	4*	3*	weed control
Ware potato	4*	2*	variety
		2*	rotation
		2*	weed control
		2*	haulm killing
		2*	late blight control

For the base-year 1992 and the years 1993, '94 and '95 when the MacSharry measures are implemented, different product prices were used. The MacSharry measures were put to the model. In the model there are two possibilities. The premium for cereals is obtained for a maximum of 12.9 hectare of MacSharry crops and no land is set aside. Or 15 % of the surface for which support is given is set aside and premiums are obtained for all the MacSharry crops.

When the model is run, the optimal farm organization is determined, that is the cropping pattern and cropping variants that generate the maximum net farm result. Whereas for each cropping variant that is used the amounts of nitrogen fertilizer and pesticides used are known, the environmental effects can be assessed.

3. Results

The model was run for a farm of 55 hectare for the year 1992. This is the year before the MacSharry measures are implemented. At first only the standard cropping variants were adjusted to the model in order to simulate the actual situation in arable farming in this region. In Table 2 it is shown that the maximum net farm result is obtained with a cropping pattern of winter wheat, sugar beet and potato each on a quarter of the surface and onions and beans in a proportion of two to one on the last quarter. The net farm result is almost Dfl 10 000 in deficit while the use of pesticides is on average 9.32 kg active ingredient per hectare and 158 kg nitrogen is on average applied per hectare.

Next the alternative cropping variants were added to the model. This way it could be studied whether under the current circumstances environment-friendly cropping methods would be used in the optimal farm organization. In Table 2 we can see that by using alternative cropping methods net farm result can be raised by Dfl 1700. For sugar beet an alternative way of weed control is used and in potato a mechanical weed control is applied. A different way of haulm killing is introduced also. The use of pesticides is decreased with 13 %. Nitrogen fertilization was maintained at the recommended dose. Because of the appearance of seed grass in the cropping pattern, which was caused by a shortage of labour due to the extra labour needed by the cropping variants, the use of nitrogen was even increased to 163 kg N/ha.

TABLE 2. Model outcomes for a farm of 55 ha in year 1992 with standard cropping methods and with cropping variants.

	cropping methods	
	standard	variants
cropping pattern (ha)		
winter wheat	13.75	8.25
ware potato	13.75	13.75*
sugar beet	13.75	13.75*
onions	9.17	9.17
seed grass		5.5
beans	4.58	4.58
net farm result (Dfl)	-9972	-8290
input of pesticides (kg a.i./ha)	9.32	8.07
nitrogen fertilizer (kg N/ha)	158	163

* alternative cropping method used

Subsequently computations were made for the years 1993, '94 and '95. The effects were calculated both for a farm which made use of the standard cropping methods and for a farm which had all cropping variants available. Therefore it was possible to investigate whether the MacSharry measures would direct arable farming towards a more sustainable production. The effects on net farm result are depicted in Figure 1.

In both situations net farm result decreased in the year 1993 with a good Dfl 1500. For the years 1994 and 1995 the net farm result remained almost the same. The difference in net farm result between the farm with standard cropping methods and the farm with cropping variants, remained constant over the years at about Dfl 1800.

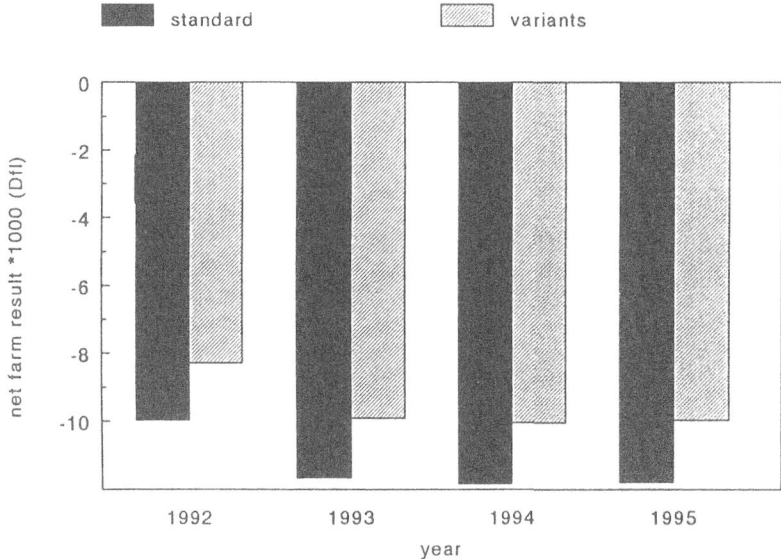

Figure 1. Effects of MacSharry measures on net farm result for an arable farm of 55 ha.

The effects on the environment are shown in Table 3. In the calculations with the standard cropping methods the change in use of fertilizer and pesticides is a consequence of a change in the cropping pattern. Five and a half hectare winter wheat was replaced by seed grass. No use was made of set-aside. If the cropping variants were available, in winter wheat the use of nitrogen fertilizer was reduced to 75 %. This also gives a little reduction in use of pesticides, because the growth regulator chloormequat is not needed when N-fertilization is reduced. In 1995 winter wheat is replaced by spring barley and therefore nitrogen and pesticide use were further diminished.

TABLE 3. Effect of MacSharry measures on nitrogen and pesticide use for a farm of 55 ha with standard and alternative cropping variants; a.i. = active ingredient.

| year | standard | | variant | |
	nitrogen kg N/ha	pesticides kg a.i./ha	nitrogen kg N/ha	pesticides kg a.i./ha
1992	158	9.32	163	8.07
1993	163	8.93	159	7.93
1994	163	8.93	159	7.93
1995	163	8.93	155	7.76

4. Discussion

The environmental-economic linear programming model is very detailed (see also Van der Houwen, 1993). The crop production is included in the model in such a way that each cultivation activity is described separately. To determine the exact amount of pesticide used, it is necessary to describe the crop protection activities very precisely. In this the main restrictions of the model are found. In practical farming there is a large variation in use of pesticides. There is no statistical material available from which the cropping variants can be obtained. The aid of crop scientists is necessary in assessing the cropping variants. In the model cropping variants were defined only for the most important crops. If alternative cropping methods would have been included in the model for every crop, the model outcomes would probably indicate further reduction in pesticide use.

Another difficulty in model building is the lack of precise knowledge of the relationships between inputs, cropping activities and physical production output. For example production functions in dependence of nitrogen fertilizer are difficult to obtain and exact figures about the interaction between use of fertilizer and liability for pests and diseases are not available at all. The technical relationships are often very complex and therefore difficult to describe in a model. In the present study the application of organic manure was left out of consideration, because no figures were available about the effects on physical output.

Another problem is that in order to investigate the effects of different cropping methods and farm organization on the environment, the damage to the environment has to be quantified, but no reliable data are available. In the present study the environmental effects are expressed in terms of kilograms active ingredient. Of course this is a very rough manner, because it is obvious that one pesticide is more harmful than another.

A restriction of the modelling technique is that non-linear relations like production functions cannot be modelled. In order to model the relation between nitrogen fertilizer and physical output, successive points on the production function were considered as separate cropping variants. Because only four points on the production function were used, a lot of information was neglected.

Because the environmental-economic linear programming model is a model at farm level, the outcomes of the model calculations are valid for that specific farm. Therefore the model calculations cannot be generalized for a larger region where many different farm types are found.

The level of farm income calculated by the model is strongly dependent on the product prices that are used. Therefore the changes in farm income for different model calculations are more interesting.

5. Conclusion

A model was made for arable farms in a Dutch polder with which the interrelationships between cropping methods, farm income and environment could be studied. It was shown that by using environment-friendly cropping methods, farm income could be increased. The effect on net farm result of MacSharry was confined to Dfl 1500 and no use was made of set-aside. The MacSharry measures were hardly an incentive for extensification.

Because the outcomes of the model are totally dependent on the underlying assumptions, precise information is needed about technical relationships. Co-operation with crop scientists is very important.

The type of modelling discussed above is very labour intensive. The method could be improved if the input data could be decoupled from the matrix. This would make it easier to update the model and to use it for different situations.

6. References

Houwen, C. van der (1993) 'Economische modellering van akkerbouw-bedrijven in Zuidelijk Flevoland in relatie tot milieu en EG-beleid', Directie Flevoland Rijkswaterstaat. (in preparation).

Wossink, A., Koeijer, T. de and Renkema, J. (1992) 'Incorporating the environment in economic modelling of farm management', Sociologia Ruralis Volume XXXII (1), 115-126.

Wossink, A. (1993) 'Analysis of future agricultural change: a farm economics approach applied to Dutch arable farming'. Dissertation Agricultural University Wageningen.

REVIEW AND APPLICATION OF THE FARMING STYLES CONCEPT: THE CASE OF DUTCH ARABLE FARMING

J.H. VAN NIEJENHUIS AND G.A.A. WOSSINK
Wageningen Agricultural University
Department of Farm Management
Hollandseweg 1
6706 KN Wageningen
The Netherlands

ABSTRACT. In farm economics it is realized that among farmers there is a large variation in capacity and willingness to reach the economic optimal organization of production. This behavioral aspect influences farm structure by the production technology chosen from current options (static situation) and also by differences in the time taken to the adoption of innovations (dynamic situation). In sociology, the resulting differences in farm organization under similar natural, social-economic and technical conditions are referred to as "styles of farming". The present paper focuses on the assessment of different styles of farming among arable farmers in the Netherlands and their viability perspectives considering two scenarios of technical and institutional changes.

1. Introduction

The research described in this paper aims at contributing to investigating and modelling the adaptation processes of family farms in response to changing external conditions. An impression of the reactions of farmers to alternative policy regulations and instruments can be very informative in the process of policy development. In farm economics linear programming (lp) for representative farms is the common procedure for this kind of analysis (see for instance Wossink and Renkema, 1993). In the present study the sociological concept of farming styles was used to derive the representative models. Being a new item in farm economics, the concept of farming styles is discussed first and the differences with the traditional farm economic classifications are highlighted. Next, for specific farming styles the effects are analysed of different forms of technical innovation, agricultural price policies and environmental regulations on farm organization, farm income and the environmental quality of production. The population of specialized arable farms in the Frisian clay soil region in the Netherlands served as a case study.

P. C. Struik et al. (eds.), Plant Production on the Threshold of a New Century, 93–100.

2. The concept of farming styles

Farms vary in many aspects: they produce under different natural, economic and social conditions and, moreover, farmers have different goals. Even in one region the performance of farms differs greatly. In farm economic research concerning the adaptation to changing external variables, the units in the population considered must be classified into categories as simultaneous treatment of all the units is normally impossible. For such a classification several multivariate techniques are available (see Hair *et al.*, 1987). Irrespective of the technique used, the classification is normally limited to aspects of the farm business and the subsequent economic analysis focused on the optimal farm organization. Lacking in this approach is an integration of the farmer's strategy which means that the interactions between the decision making process and the technical and economic/financial status of the farm remains obscured. In farm economics it is recognized, however, that among farmers there is a large variation in capacity and willingness to reach the potential production level, for instance. In sociology, these aspects resulting in differences in farm organization under similar natural, social-economic and technical conditions are referred to as "styles of farming" (Van der Ploeg, 1990). This concept pretends to enable both the farm and the farmer's strategies to be taken into account which means that diverging responses to market changes and policy interventions can be assessed in a more adequate way.

   In farming styles analysis the population concerned is reduced to a smaller number of groups by means of statistical methods like the principal components method, applied to bookkeeping data. Principal components analysis, like any multivariate technique, implies an exploratory investigation of a data set. Therefore discussions with regional experts and with farmers concerned are the necessary follow up of the mathematical analysis to prevent misinterpretation of the styles.

3. Application in future analysis of Dutch arable farming

The starting point in farm economic research is the concept of the individual farm manager as an "adaptive man" (see, for instance Brandes, 1985). The central issue then is the adjustment in the organization of production due to a reconsideration of farmers' strategic decisions. The research presented covered a number of parts to analyse these adaptation processes and to find possible differences among farming styles: (1) assessment of farming styles for the case study region, (2) assessment and description of technical developments and alternative policy options for Dutch arable farming by means of a number of scenarios, (3) implementation of a farm model for each of these styles, and finally (4) model computations to indicate the future perspectives of the different styles.

   According to the 1992 Farm survey there were 449 specialized arable farms in the region considered. Of 27 of these farms with bookkeeping data[1] over 1990-91 were available. These are all arable farms with a farm economic bookkeeping system looked after by a commercial bookkeeping agency. To this data principal components analysis was applied and we also applied a clustering procedure to the factor scores. The last one didn't result in a satisfactory clustering of the farms. There will almost always be more clusters than factors though a cluster may be close to a particular factor. The principal component analysis resulted in a number of well recognizable styles. These styles are to be interpreted as ideal or pure types (Stewart, 1981). Considering the numerical relevance of the remaining units these were analysed as well in the present study. The original economic data of farms of a farming style were used in constructing farm models for each style.

The lp model used in the assessment of the future perspectives of the styles optimizes the net farm result, being the difference between the total revenues and the total costs (including costs for own labour and capital). For a given situation the optimal farm organization is assessed, indicating changes regarding: net farm results, cropping pattern, variable operations, labour and tractor hours used and input and emission of chemicals and nitrate. The model covers several cropping variants for every crop which differ in economic and environmental values (see Wossink and Renkema, 1993; Wossink, 1993). The matrix described contains about 170 activities and circa 175 constraints. The farm style is specified by means of the right hand side values, the yield levels of the different crops and the initial cropping plan. The other technical coefficients are the same in all situations.

To analyse the external determinants affecting agriculture the scenario method was applied (Bright and Schoeman, 1973). To define the scenarios, the conditions of agriculture at farm level were clustered into three main external determinants: (a) technical developments; (b) EC market and price policy for agriculture; and (c) environmental policy. The innovations considered in the present study were: (1) improvements of spraying techniques in crop protection, (2) introduction or reintroduction of mechanical crop protection techniques, (3) increase in yield per hectare of the various crops, (4) application of integrated cropping techniques and (5) introduction of ecological cultivation techniques (i.e. total abstention from use of chemical fertilizers and pesticides) and (6) introduction of new crops namely Corn Cob Mix (Zea mays) for fodder and chicory (Cichorium intybus) for the extraction of liquid sweeteners.

With respect to output prices, we considered two variants of price changes. The world-market oriented variant assumes sharp price

---

[1] Out of the total data set, 127 variables were selected that were assumed to be relevant for arable farms. Note that the population only represents farms with a special farm economic report in addition to the fiscal report that is compulsory in the Netherlands.

reductions, the other variant resembles the current MacSharry proposals[1].

Regarding environmental policy for arable farming the Long-term Crop Protection Plan (Min. LNV, 1990) presents quantitative objectives. The major strategic headline is to reduce the input of biocides (in terms of the weight of active ingredients compared to the average use over 1984-88) by 35 % in 1995 and 50 % in 2000. Each sector of agricultural production has been given its detailed goals[2].

TABLE 1. Farming styles in Frisian arable farming.

Style I	: The intensive farmer Small scale farm (40 ha), high labour input, accent on vegetable growing.
Style II	: The labour saving farmer Large farm (80 ha), highly mechanized, common cropping pattern.
Style III	: The fine tuner Larger farm (70 ha), specializes on seed potato cropping (high yields).
Style IV	: The middle scale farmer Middle scale farm (60 ha), hopes to expand the acreage further in future.
Style V	: "Outsitter" Small scale farm (40 ha), did not intensify and will sell farm in the medium term.

---

[1] The "market oriented" variant assumes sharp price reductions (based on minus 4.5 percent annually for wheat), supplemented by voluntary set-aside regulations. The other variant reflects a policy of "production restrictions". In this case price decreases are moderate (based on minus 2.5 percent annually for wheat) in combination with set-aside of agricultural land for farmers mainly producing supported products (15 % of the acreage). This implies that the second variant resembles the current MacSharry regulations. Price developments for fixed and variable inputs are also a part of the variants (see Wossink, 1993).

[2] Those for arable farming for the year 2000 are: herbicides minus 45 %, insecticides minus 25 %, fungicides minus 25 %, others minus 68 % and total minus 60 %.

4. The results

The styles are summarized in Table 1. In the region considered almost every arable farm grows seed potato, ware potato is not common. The first three styles resulted after principal component analysis. Style I is based on the first factor extracted, styles II and III cover the farms with a high score on the second factor split according to their percentage of seed potato. Styles IV and V represent the farms not scoring on factors 1 - 2 distinguished by size (ha). They comprise about 50 % of the population.

By combining the provisions of technical innovations with the variants for EC market policy and the expected environmental policy regulations two scenarios were constructed indicated as free trade scenario and policy trend scenario. The computations performed with the environmental economic model indicate the comparative implications for the farming styles I-IV of the two scenarios (Table 2). Farming style V was not included in this analysis because the type of farmer concerned does not aim at continuity of his enterprise.

The basic situation (t = 1990) of the four farming styles is quite different, especially with respect to the net farm result. The results for farming styles II and III are far better compared with the two other styles. The impact of the free trade scenario is dramatic. The reduction in net farm result for t = 2000 varies between Dfl 60 000 for farming style I to Dfl 120 000 for style III. Neither of the farming styles realizes a positive net result under these conditions. Under the policy trend scenario the reduction is less severe and varies between Dfl 25 000 for style I to Dfl 66 000 for style III. In relation to the basic situation both scenarios lead to significant reductions in pesticide use and in nitrogen use per farm. The price policies appear to be most important for the scenario results. The targets formulated for the reduction of pesticide use are relatively easily met. Under the free trade price conditions imposing constraints on the different categories of pesticides implies an additional reduction in net farm results of Dfl 1770 for farming style I to Dfl 5800 for farming style II, for instance. For a detailed description of the scenario results see Wossink et al. (1993).

5. Discussion and conclusion

The farming styles offered a good opportunity to study the effects of scenarios on the farming business. The normative approach indicates that some farmers will be able to adapt their management to the changing circumstances. At t = 2000 their labour income will be low, but high enough to stay in business. The analysis of the differences in the actual farming situation highlighted choices made in practice and indicated that there is more than one way to survive. About one half of the farmers is still waiting. Their labour income of today is sufficient for a living, but they will have to adapt their farm in the near future or leave business. These conclusions are drawn on the basis of a small sample of farms which pay for additional bookkeeping

TABLE 2. Summary of the scenario results for the farming styles[1].

Style	Situations		Net farm result	Return to labour (1000 Dfl)	Pesticide use (kg ai)	Nitrogen use (kg N)
I Intensive farmer	Base year	t = 1990	5	112	211.53	6 743
	Free trade	t = 2000	-57	63	84.60	6 012
	Policy trend	t = 2000	-20	99	84.60	5 883
II Labour farmer	Base year	t = 1990	73	149	464.05	14 840
	Free trade	t = 2000	-37	48	185.62	9 495
	Policy trend	t = 2000	22	107	185.62	9 495
III Fine tuner	Base year	t = 1990	68	158	391.65	12 413
	Free trade	t = 2000	-52	48	156.66	7 115
	Policy trend	t = 2000	2	100	156.66	8 165
IV Middle scale farmer	Base year	t = 1990	-24	66	291.63	9 582
	Free trade	t = 2000	-110	-10	116.65	6 523
	Policy trend	t = 2000	-72	28	116.65	6 044

1) All data are per farm
ai = active ingredient

information. In general there are more farms with an above average result in such a group. So in reality the number of farms with worse continuity perspectives might be larger.

During the use of the concept of farming styles in the present farm economic analysis some shortcomings were encountered. First, the styles are based on differences among farmers/farms. These different data are translated into the coefficients of the lp models. This implies that common characteristics in the future development of the farms are disregarded. Saving on costs, for instance, has become a general feature in arable farming due to the ongoing pressure on product prices. As the assessment of the styles focuses on hetero-geneity this aspect is not recognized whereas it may be important in forecasting future changes in farm organization. In that case the results will be better than computed here.

Further, the assessment of the future strategy of each farming style is still a difficult item. The past changes within a specific style can be assessed, of course, but to what extent can these paths be extrapolated to indicate future farm development? As follows from the scenario results, adaptation of part of the styles (style IV in particular) is required to safeguard the continuity perspectives. This aspect of "induced" adaptation is not considered as the styles found are time and location specific. Moreover, it is realized by agricul-tural sociologists that the notion of farming styles needs to be more theoretically underpinned and elaborated particulary to be useful for forecasting studies.

Despite these shortcomings it should be emphasized that the concept seems a major contribution to integrating sociological and psychological elements into normative farm economic analysis. Hence, the present study is to be seen as a first presentation of a potential fruitful field of interdisciplinary research.

References

Brandes, W. (1985) 'Ueber die Grenzen der Schreibtisch-Ökonomie', Tübingen: Mohr.

Bright, J.R. and Schoeman, M.E.F. (eds.) (1973) 'A guide to practical technological forecasting', Englewood Cliffs: Prentice Hall.

Hair, J.R., Anderson R.E. and Tatham, R.L. (1987) 'Multivariate Data Analysis', New York: Macmillan.

Ministerie LNV (1990) 'Meerjarenplan gewasbescherming: beleidsplan, Den Haag': Ministerie van Landbouw, Natuurbeheer en Visserij.

Ploeg, J.D. van der (1990) 'Labour, markets and agricultural produc-tion', Boulder etc.: Westview Press.

Stewart, D.W. (1981) 'The application and misapplication of factor analysis in marketing research', Journal of Marketing Research, vol. XVIII, pp. 51-62.

Wossink, G.A.A. (1993) 'Analysis of future agricultural change: a farm economics approach applied to Dutch arable farming', Series Wageningen Economic Studies no. 27, Wageningen, Pudoc.

Wossink, A., Niejenhuis, J.H. van en Haverkamp, H. (1993) 'Friese akkerbouw, waarheen: bedrijfseconomische toekomstverkenningen op basis van bedrijfsstijlen, Report Department of Farm management, Wageningen Agricultural University (in preparation).

Wossink, G.A.A. and Renkema, J.A. (1993) 'Farm economic modelling of integrated plant production', Invited paper International Congress Plant production on the threshold of a new century. June 28 - July 1, 1993, Wageningen.

ECONOMIC ASPECTS OF ENVIRONMENTAL POLICY IN THE MUSHROOM INDUSTRY IN THE NETHERLANDS

A.J.J. VAN ROESTEL, J.P.G. GERRITS,
AND L.J.L.D. VAN GRIENSVEN
Mushroom Experimental Station
P.O. Box 6042
5960 AA Horst
The Netherlands

ABSTRACT. Environmental policy affects the mushroom industry in the Netherlands at two levels: 1. regulations for compost production on centralized, large scale composting facilities concerning the emission of stench and ammonia; 2. regulations for mushroom farms, mainly concerning the emission of chemicals in the wastewater. Measures taken by composting companies will reduce emission of stench and ammonia before 1995 to almost zero. Emission on mushroom farms will be reduced before 2000 to almost 10 % of residues of chemicals applied in mushroom growing. This will increase average production costs (excluding harvesting labour and delivery costs) with 11 % to Dfl 2.22 per kg, mainly because of a growing compost price. In the past two years, production grew with 15 % p.a., mainly because of an increasing number of crops per m^2 resulting from changes in compost production.

1. Introduction

The cultivation of white button mushrooms, *Agaricus spp.*, in the Netherlands has become an important economic activity with about 3,500 people (including family labour) employed on mushroom farms. The growth of mushroom production has been made possible by a growing export volume of fresh and processed mushrooms (Table 1). About 55 % of mushroom production is processed. Especially the export to the German market has been important for the Dutch mushroom industry.

The driving forces behind this production growth have been the possibilities to decrease cost price by increasing output per m^2 and to benefit from economies of scale. Centralizing compost production and the use of highly standardized farming systems contribute significantly to the international competitiveness of the industry.

Environmental policy affects the industry in different ways: regulations for compost production on large scale composting facilities (par. 2), regulations for mushroom farms concerning crop protection (par. 3.1) and other environmental policy objectives (par.

P. C. Struik et al. (eds.), Plant Production on the Threshold of a New Century, 101–108.
© 1994 Kluwer Academic Publishers.

3.2). The effects on cost price per unit are summarized (par. 4). The discussion (par. 5) draws the attention to the changed competition on the compost market because of environmental policy.

TABLE 1. Development of mushroom growing (*Agaricus spp.*) in the Netherlands.

	Absolute values			Average annual growth rate (%)		
	1980	1990	1992	80-85	85-90	90-92
Production (10^6 kg)	60	147	195	11.9	7.0	15.0
Cropping area (10^3 m^2)	664	1050	1127	4.1	5.3	3.5
Yield (kg m^{-2} y^{-1})	90	140	173	7.6	1.5	11.0
Number of farms	823	852	806	0.6	0.1	-2.5
Cropping area/farm (m^2)	807	1230	1399	3.5	5.2	6.6
Export volume[a] (10^6 kg)	52	184	198	16.5	10.0	3.8
Net export value (10^6 Dfl)	176	559	628	13.8	10.5	6.0

[a] includes weight of preserved mushrooms after processing.

2. Environmental policy in relation to compost production

2.1. AIM OF THE POLICY

Phase I of compost production (Figure 1) starts with mixing and moistening the basic materials straw, horse manure, chicken broiler manure and gypsum. Then the mixture is composted in long narrow stacks between 1.5 and 2 metres in width and height (windrows). Phase I takes roughly one week during which considerable amounts of ammonia and stench are emitted into the air. In phase II compost is pasteurized and conditioned; some additional ammonia is removed. Phase II takes a week to produce a selective substrate for optimal growth of mushroom mycelium. After phase II the compost is spawned with mushroom mycelium that is grown on sterilized cereal grains (Gerrits, 1988).

Two companies produce phase I compost in the open air. They represent 66 % resp. 27 % of total production of phase I compost in the Netherlands. The largest company is owned by the Cooperative Dutch Mushroom Growers Association. Stench and ammonia emission are a nuisance for the population living near the composting facilities. This emission has to be reduced to an acceptable level. Stench is caused by a number of organic sulphur compounds (Op den Camp and Derikx, 1991). A few of them have an extremely low odour threshold. Emission of ammonia into the air is considered responsible for the acidification of land and surface water in the Netherlands. Reduction of ammonia emission is an important aim of the environmental policy in the Netherlands.

For large scale composting facilities bringing the traditional phase I composting process indoors did not seem feasible for technical reasons. Therefore it was necessary to modify the composting process. These modifications should neither be expensive nor deteriorate the productivity of the compost.

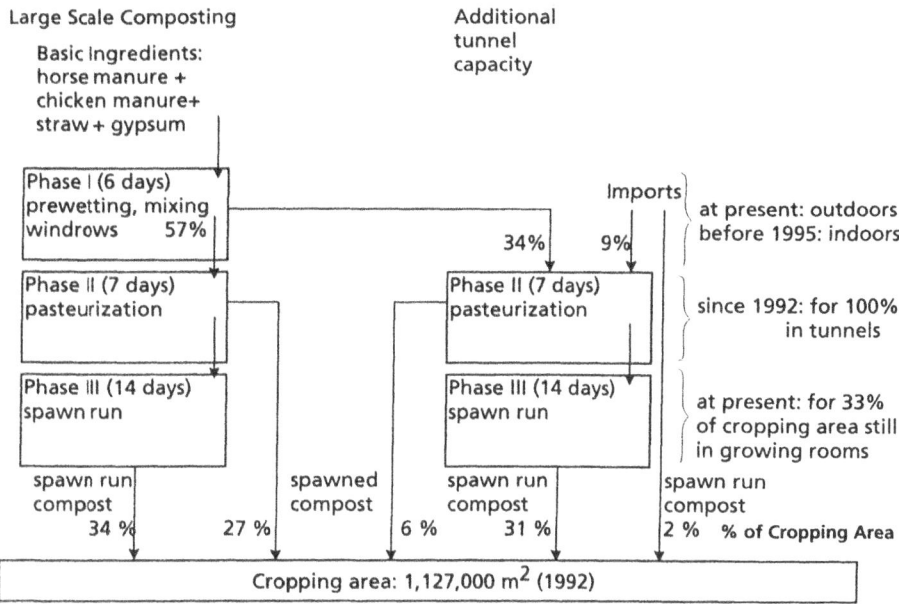

Figure 1. Compost production for mushroom growing in the Netherlands.

2.2. DEVELOPMENT TOWARDS INDOOR COMPOST

2.2.1. History

Until recently, many mushroom growers filled the shelves in their growing rooms with phase I compost. Growing rooms are provided with air handling equipment. Compost temperature during phase II and III strongly depends on the ventilation of the rooms.

Some mushroom growers (about 30 % of total cropping area) already filled their growing rooms with spawned (phase II) or spawn run (phase III) compost. Phase II resp. phase III takes place in "tunnels", specially designed climatized rooms where compost temperature is controlled by forcing air from beneath the floor through the compost.

2.2.2. Present status

Now, the compost production process is in an intermediate stage. Phase I still takes place in the open air. The most important change is that all phase II compost is produced in tunnels. In 1992, the composting companies invested about Dfl 180 million to expand their tunnel capacity. Price of spawn run compost from the largest company increased from Dfl 235 per tonne ('91/'92) to Dfl 265 per tonne ('92/'93).

2.2.3. Future

Before 1995 phase I must take place indoors. The basic material will be moistened and mixed in a closed environment. A high temperature process in specially designed rooms will substitute the fermentation process in windrows. Exhaust air will be treated. Phase II and III will not change much. The composting companies will have to make an additional investment of about Dfl 130 million. Price of spawn run compost is estimated to increase approximately to Dfl 295 per tonne compost.

2.3. CONSEQUENCES OF CHANGES

2.3.1. Environmental aspects

Ammonia emission during phase I will be reduced to 1-5 % of its present level before 1995 by washing the exhaust air. The same holds for ammonia emission during phase II. The stench caused by organic sulphur compounds requires additional air treatment. Experiments have shown that emission can be reduced to a minimum by oxidation of the air (Op den Camp and Derikx, 1991).

2.3.2. Production growth

Bringing phase II and III in tunnels causes a reduction in the growing cycle with 1 to 3 weeks. Production per year will increase because of the increasing number of crops. It can be calculated that in the last 5 years half of the average production growth of 12 % was caused by intensification of the cropping cycle. Growth of cropping area (5 %) and autonomous productivity improvement (1 %) were responsible for the other half of production growth. Since 1989 market prices for mushrooms have annually decreased by 9 %.

2.3.3. Increasing financial risk

Spawned and spawn run compost are about 2.5 resp. 4 times more expensive than phase I compost. Variable costs represent an increasing proportion of total cost price per unit. Because mushroom farms are strongly specialized, problems of liquidity will arise earlier when market price declines.

3. Environmental policy in relation to mushroom farms.

3.1. CROP PROTECTION AND REDUCTION EMISSION

3.1.1. Aim of the policy

Emission of the residues of pesticides, fungicides and disinfectants causes environmental problems. Especially the emission of these residues in the wastewater of mushroom farms has to be reduced (MLNV, 1991). It is agreed that the mushroom industry will reduce the use of pesticides and fungicides before 1995 with 40 % and before 2000 with 52 % as compared to 1988. An additional aim is to minimize the emission of chemicals into the environment. This requires additional precautions and investments.

3.1.2. Development towards 2000

Several factors contribute to the reduction in the use of chemicals on mushroom farms: Bringing composting phases II and III in tunnels (where no pesticides or fungicides are used), improvement of application techniques of pesticides directed at minimizing residuals, special attention in extension and education activities for application of pesticides, improving hygienic farm conditions and observation techniques, biological control and resistant mushroom strains.

The reduction of the emission of chemicals is expected to be in proportion to the reduction in use. Additional reduction must be accomplished by separating clean from dirty water, installation of a settling pit on the farm and connection to the mains sewerage system leading to a sewage treatment plant.

3.1.3. Consequences of changes

The costs of the proposed measures (Table 2) strongly depend on the size and design of the farm. The average farm (20 %) has 6 rooms. The larger farms (25 %) own more than 50 % of the cropping area.

Under these circumstances disinfectants will be removed completely from the wastewater. From the active ingredients supplied to a settling pit, eventually 11.5 % will remain. Based on figures of 1988, this is 16 kg of active ingredients per year (MLNV, 1991).

The disposal of the settling sediment amounts to 60 % of the increase of miscellaneous variable costs. The past four years show a decrease of emission. For 15 % of the farms, connection to the mains sewerage system is cost intensive because of the distance.

TABLE 2. Costs of emission reduction. Remaining % of farms having to invest.

	Average investment (Dfl, 1988)	Farms (%) 1988	Farms (%) 1992
1. Insolation growing rooms	12,000	25[a]	
2. Spraying tools for pest	6,000	25	
3. Separate dirty water. settling pit	25,000	50	14
4. Connect to mains a) <300 m.	15,000	35	\
sewerage system b) >300 m.	127,000	15	/ 31
5. Replace/modify growing shelves	36,000	27	< 10

Increase of fixed costs (Dfl year^{-1})	6,500 (includes 1,2,3,4a)	
Increase of var.costs: a) air filters	2,000	36 %
b) miscellaneous	3,000	100 %

[a] % of growing rooms
Source: MLNV, 1991; Van Horen and Verhoeven, 1993

3.2. CONSEQUENCES OF OTHER ENVIRONMENTAL POLICY OBJECTIVES

Because of national regulations, energy efficiency of farms has to increase considerably (estimated 50 %). A small levy on the price of natural gas is already imposed to subsidize new developments. Energy costs will probably increase.

Pumping up and discharge of ground water for air cooling will be restricted. Cooling liquids will have to meet specific conditions. Many cooling units will have to be modified. The introduction of quality control for cooling installations will also increase costs.

The annual production of 650,000 tonnes (1992) of spent compost was used as a fertilizer in horticulture and arable farming. The use will be limited in the future because of government regulations. Spent mushroom compost will meet growing competition from other composts and animal manure. This will increase costs of disposal.

Government policy is directed at prevention of waste production and recycling of waste. The costs of disposal of stems of mushrooms and plastics will strongly increase in the near future.

4. Summary of the consequences for cost price

The effects on cost price per unit are given in Table 3. Average costs of production on spawn run compost in '91/'92 are compared with average costs of production in '92/'93 and in 1995, when phase I compost will be produced indoors.

The increase of average production costs per unit will be 11 %, the rising compost price being mainly responsible for this. This calculation applies only to compost from the largest compost facility.

TABLE 3. Effects on average production costs of mushrooms (Dfl kg^{-1}). Costs of harvesting labour and delivery are not included.

	Average prod. costs (until 1992)	In-creased comp. price 1992	Emission reduction 1995	In-crease comp. price 1995	Average costs 1995	Increase (%)
Fixed costs	0.55		0.03		0.58	+ 5
Variable costs	1.45	0.09[1]	0.01	0.09[2]	1.64	+13
	----	-----	----	-----	----	---
Production costs	2.00	0.09	0.04	0.09	2.22	+11

[1] Based on spawn run compost CNC 1991/'92 and 1992/'93
[2] Estimate

The effect of emission reduction is relatively small, but will depend on farm size, because it mainly concerns fixed costs. For about 15 % of the mushroom farms, increase of costs will be higher because of the long distance to a mains sewerage system.

The effects of other environmental policy (par. 3.2) are difficult to estimate. Beginning in 1995, an increase of Dfl 0.05 per unit is realistic. The saving on the costs of pesticides will amount to Dfl 0.01 per unit.

5. Discussion

There has always been a great demand from mushroom growers for phase III compost. The large composting companies could not respond to this demand because of environmental restrictions. In the eighties, the government formulated the conditions under which the expansion of the tunnel capacity was possible. At the same time, others were permitted to expand their tunnel capacity without special effort. About 20 % of the phase I compost for these additional tunnels now is produced in Belgium (Figure 1), where conditions were less strict. Environmental policy turned out to be inadequate to prevent this situation and changed competition on the compost market significantly.

6. Conclusion

In the forthcoming years, environmental pollution by the mushroom industry in the Netherlands will be reduced significantly. Emission of stench and ammonia during the composting process and the use of pesticides and the emission of chemicals on mushroom farms will be reduced significantly. Required measures and investments are raising cost price per unit. At the same time, production per m^2 is growing because of crop intensification, resulting from changes in compost production.

7. Acknowledgement

The authors thank Lambert van Horen from the Information and Knowledge Centre Dep. Mushrooms for his help in acquiring and interpreting data.

8. References

Gerrits, J.P.G. (1988) 'Nutrition and compost', in L.J.L.D. van Griensven (ed.), The Cultivation of Mushrooms, Darlington Mushroom Laboratories, Sussex, England and Somycel S.A., Langeais, France.

Horen, L.G.J. van, and Verhoeven, R.H.M. (1993) 'Afvalwater van champignonteeltbedrijven', De Champignoncultuur, 37 (2) pp. 61-69.

MLNV (1991) 'Long-term Crop Protection Plan', Government Decision (including background document on mushroom cultivation), Ministry of Agriculture, Nature Management and Fisheries, SDU, The Hague, Netherlands.

Op den Camp, H.J.M., Pol, A., Griensven, L.J.L.D. van, and Gerrits, J.P.G (1992) 'Stankproduktie tijdens "Indoor Verse Compostbereiding" en het effect van luchtbehandeling met een luchtwasser', De Champignoncultuur, 36 (6) pp. 319-325.

Op den Camp, H.J.M. and Derikx, P.J.L. (1991) 'Stankproduktie bij buiten- en indoor-compostering', De Champignoncultuur, 35 (1), pp. 15-23.

MICROPROPAGATION, AN ESSENTIAL TOOL IN MODERN AGRICULTURE, HORTI-
CULTURE, FORESTRY AND PLANT BREEDING

R.L.M. PIERIK AND H.J. SCHOLTEN
Department of Horticulture
Agricultural University
Haagsteeg 3
6708 PM Wageningen
The Netherlands

ABSTRACT. The classical methods of in vivo cloning often fall short of
that required (too slow, too expensive, too difficult or even impossible). For that reason the technology of micropropagation (cloning
under sterile conditions) has become an important tool particularly in
horticulture. Micropropagation made it possible to produce (regenerate) viable individuals of many plant species. So micropropagation
developed into a commercial industry all over the world. In the last
15 years knowledge of micropropagation has grown quickly; this can be
illustrated by the fact that in the Netherlands 80 commercial laboratories exist with a production of 100 million plants in 1992.
    This article contains descriptions of the various cloning systems
and their application in agriculture, horticulture, forestry and plant
breeding. Particular attention will be paid to the advantages and disadvantages of the propagation systems and the possibilities to avoid
genetic variation. Micropropagation enables genetic engineering, which
would be impossible if methods for regeneration of cells and tissues
into plants were not available. Special attention will be paid to the
demands of customers, problems in tissue culture laboratories (economic, marketing and technical) and avoiding peaks in labour demand. A
global survey of the world production will be given.

1. Introduction

In vitro culture of higher plants (Pierik, 1987) is defined as the
culture on nutrient media under sterile conditions of plants, embryos,
organs, explants, tissues, cells and protoplasts. The most important
branch of in vitro culture is cloning, also called micropropagation or
vegetative propagation. The aim of micropropagation is to create safe-
ly possibilities to clone higher plants disease-free, faster and less
expensive than in vivo. In vitro cloning also offers possibilities to
clone plants in cases where it is impossible in vivo. An additional
advantage is that micropropagation enables the production of disease-

109

P. C. Struik et al. (eds.), Plant Production on the Threshold of a New Century, 109–114.
© 1994 *Kluwer Academic Publishers.*

free plants and thereby also facilitates the so-called phytosanitary transport. This paper summarizes a number of aspects of the techniques, the possibilities and its application.

2. Cloning systems

The methods summarized below have been developed to micropropagate plants, although they vary in reliability and convenience of application (Pierik, 1987).

SINGLE NODE CULTURE. This method is the simplest, most natural and safe method (no problems with mutations) with plants which elongate (e.g. potato, tomato, lilac, birch, oak, and grape), forming a stem with leaves with buds in their axils, but is impossible with rosette plants. The rate of propagation is strongly dependent on the number of nodes formed within a particular time interval. When cloning shrubs and trees, serious problems can arise with this method (dormancy of buds and failure to elongate the stem).

AXILLARY BRANCHING. Axillary buds have their dormancy broken by breaking apical dominance with cytokinin. This method has become the most important propagation method being simple and quite safe. Another advantage is that the propagation rate is relatively fast and the genetic stability is usually preserved. However, mutations can occur when adventitious buds are formed as a result of high cytokinin levels. The axillary branching method, certainly for rosette plants, has become in practice far the most important propagation method in vitro.

REGENERATION OF ADVENTITIOUS BUDS/SHOOTS. This method includes the formation of adventitious buds/shoots on explants from leaves, petioles, stems, scales and floral stems. However, the percentage of plant species that can regenerate adventitious buds is relatively small and is often restricted to herbaceous plants. The chances of obtaining mutations is much higher with this method than with the first two methods described, particularly with so-called chimaeric plants. However, it is successfully applied for *Saintpaulia ionantha*, lily, hyacinth, *Achimenes*, and *Streptocarpus*.

REGENERATION OF PLANTS FROM CALLUS, CELLS AND PROTOPLASTS. Despite claims to the contrary, cloning of higher plants through callus, cells, and protoplasts has been found to have many disadvantages. The greatest difficulty with callus cultures is their genetic instability. In the Netherlands only a few plant species such as *Anthurium andraeanum*, are cloned (partially) through a callus phase.

3. Application of micropropagation

It is especially notable in horticulture that people have quickly responded to the results obtained from research on micropropagation. A recent study (Pierik, 1993) showed that the Netherlands in 1990 had a total of 78 commercial tissue culture laboratories with a production of about 95 millions plants. From these plants 99 % are horticultural crops: pot plants (42 million), ornamental bulbs and corms (24 million), cut flowers (21 million) and miscellaneous (4 million). The three most commonly produced plants in the Netherlands are: *Nephrolepis* (a fern), lily (a bulbous plant) and *Gerbera* (a cut flower), accounting for 59 % of the total Dutch production in 1990.

An analysis has also been made of commercial micropropagation in 15 West European countries (Pierik, 1991). In 1988 Western Europe had a total of 248 commercial tissue culture laboratories with a production of 213 million plants. The most important categories of micropropagated plants (figures in millions) are: pot plants (92), cut flowers (38), fruit trees (19), and ornamental bulbs and corms (13). The most frequently cloned crops are: *Ficus*, *Syngonium*, potato, strawberry, *Spathiphyllum*, *Gerbera*, *Prunus*, *Philodendron*, and *Nephrolepis*. The largest producers of micropropagated plants are: The Netherlands, France and Germany, accounting for 62 % of the total West European production (Pierik, 1991; Riordain, 1992). An estimation in 1989 (Pierik, 1993) showed that in the world a production of 500 million of plants was reached. At present the world production will certainly be more than 1 milliard.

4. Production costs

A starting tissue culture laboratory needs an investment of about Dfl 500,000. The first five years are crucial. In this period the investment will have to be doubled, but also the break-even point must be reached. A larger part of the production must exist of cultivars in large quantities at reasonable prices. These plants will lower the unit price and take a large part in the overhead costs.

Depending on the species and cultivar the cost price of tissue culture plants in the Netherlands is Dfl 0.30-0.60. Standaert-De Metsenaere (1991) gave a general structure of the costs. The costs of labour are 60-80 % of the total costs. In smaller and specialized laboratories the cost of labour account for a relatively large part of the costs. The costs of labour (as an example 65 %) can be divided in production (32.5 %), R&D (6.5 %), and other activities like media preparation, administration, supervision (26 %). The media costs are 7 %. Equipment and laboratory costs account for 17 %. The remaining 11 % consists of costs for office, representation, electricity, etc.

In vitro plants have to compete with in vivo propagated plants and seeds. The price for extra quality is generally not paid by the customers. Therefore, in general the profits are low and consequently a low budget is available for R&D, diversification of products, automatization and development of specific features of tissue culture. The costs can be lowered by a few percent by efficient use of new equipment for culture rooms, plant density, etc. By a drastic reduction of labour costs and development of more efficient propagation techniques, the market share of micropropagated plants can expand.

5. Demands of customers

Major demands of the customers, where micropropagated plants are concerned:

PRODUCT ATTRIBUTES

1.   High quality plantlets; homogeneity and uniformity of the plants.
2.   Good plant form and not too bushy.
3.   Easy to acclimatize and to bring into production.
4.   No mutations.
5.   A low percentage of infections when tranferred to soil.
6.   Reasonable and competitive prices.

SERVICES

1.   Delivery on time.
2.   Periodical delivery of large quantities.
3.   Culture and acclimatization advice.
4.   Acquisition from more than one source to spread risks.

6. Problems in the Western European tissue culture laboratories

Problems encountered in micropropagation in Western Europe can be summarized as follows:

TECHNICAL

1.   Year-around production.
2.   Lack of acclimatization ex vitro.
3.   External and especially internal infections.
4.   Induction of dormancy in bulbous crops.
5.   Limited possibilities to automatize.
6.   Genetic instability in callus systems.
7.   Unwanted after-effects (e.g. branching).
8.   Each cultivar of a species requires different media, etc.
9.   Rejuvenation in shrubs and trees.
10. Excretion of toxic substances.
11. Problems with scaling up.
12. Problems in production management.
13. No diversification of product.

ORGANIZATIONAL, ESPECIALLY PEAKS IN LABOUR DEMAND

1. Acquisition of temporary labour.
2. Contracting out during production peaks.
3. Organization of labour shifts.
4. Staggering of labour throughout the year.
5. Choosing complementary crops.
6. Cold storage.
7. Automatization of the production processes.

ECONOMICAL

1. Financing and setting up a laboratory.
2. No intervention by third parties which increase the cost price.
3. Difficulties in lowering the costs of research, overhead and labour.
4. Profits do not come from the low-priced standard items.
5. First profits come after about one year of production.
6. No diversification of products.

MARKETING POSITION

1. Strong competition from countries with low labour costs.
2. Overproduction of a number of easy-to-clone popular crops.
3. Summer dumping.

MARKETING

1. Economic justification vis-à-vis the production costs of the same plant in vivo.
2. No access to and knowledge of foreign markets.
3. At present completely dependent on the production centres of ornamental crops, because most products are marketed in the sphere of ornamental horticulture.
4. Difficulties in finding new products and markets.
5. Lack of marketing expertise.
6. Language and communication problems.

7. Production trends in the Netherlands

The rate of increase in the number of commercial laboratories (Pierik, 1991) and also in the production of micropropagated plants is still slowing down, especially in the Netherlands, but also in other West European countries. In the period 1992-1993 a few larger Dutch tissue cultures laboratories discontinued their production. This is primarily due to competition in the West European market, where prices of a few standard crops (such as lily, *Gerbera*, *Nephrolepis* and *Saintpaulia*) have strongly decreased in recent years. Since Eastern European coun-tries (e.g. Poland, Bulgaria, etc.) and a number of the third world countries (e.g. India, Thailand, Indonesia, Singapore, etc.) have entered the West European market (Pierik, 1991), the price fall is complete which is due to low labour cost in competitive countries.

8. Conclusions

Since only 1-3 % of the total clonal propagation in the Netherlands is by micropropagation, it can be concluded that micropropagation is still in its infancy. A greater increase in micropropagation will occur when advanced production techniques (e.g. synthetic seeds, bioreactors) are developed, when mechanization is introduced (e.g. robots, etc.) and when plant material from trees can efficiently be rejuvenated. We expect that a special and large market will be opened when tree micropropagation will be possible on a large scale.

9. References

Pierik, R.L.M. (1987) 'In vitro culture of higher plants', Martinus Nijhoff Publishers, Dordrecht, The Netherlands, 344 pp.

Pierik, R.L.M. (1991) 'Commercial micropropagation in Western Europe and Israel', in P.C. Debergh and R.H. Zimmerman (eds), Micropropagation, Kluwer Academic Publishers, Dordrecht, The Netherlands, pp. 155-165.

Pierik, R.L.M. (1993) 'Micropropagation: technology and opportunities', in J. Prakash and R.L.M. Pierik (eds) Plant Biotechnology, Oxford and BDH Publishing, New Delhi, India, pp. 9-22. Riordain, F.O. (1992) 'The European plant tissue culture industry', Agronomie 21, 743-746.

Standaert-De Metsenaere, R.E.A. (1991) 'Economic considerations', in P.C. Debergh and R.H. Zimmerman (eds), Micropropagation, Kluwer Academic Publishers, Dordrecht, The Netherlands, pp. 123-140.

INTRODUCTION TO A BIO-ECONOMIC PRODUCTION MODEL FOR SUGAR BEET GROWING

A.B. SMIT[1, 2], J.H. VAN NIEJENHUIS[1] AND P.C. STRUIK[2]
[1] *Wageningen Agricultural University*
Department of Farm Economics
[2] *Wageningen Agricultural University*
Department of Agronomy
Haarweg 333
6709 RZ Wageningen
The Netherlands.

ABSTRACT. Yield prediction is a basic tool for practical purposes in sugar beet growing. Connected with an economic module, a crop model can serve as a basis for decision support at field level. PIEteR as such is a bio-economic model, mainly focusing on nitrogen fertilization, plant density and harvest date. Its main component is a crop growth model, that simulates crop responses to weather, soil factors and growers' decisions. Its input and (future) output make PIEteR a useful basis for decision support. Its potential for accurate predictions will be tested and compared with other models. If PIEteR will prove to be the most accurate and useful model, it will be developed further, thus enabling improvements in the quality of the growers' decisions.

1. Introduction

Yield prediction is a scientific challenge and a practical necessity. In science it is a way to obtain more insight in plant and crop processes. Yield prediction is also a basic tool for practical purposes in the sugar industry and sugar beet growing (Withagen, 1989), facilitating:
1. logistic planning of mainly sugar beet transportation and production, storage and sale of sugar in sugar industry (Crals and Stinglhamber, 1992);
2. planning at farm level.
Yield prediction to support planning in the sugar industry has been far more widely applied than decision support at individual farm level (Sperlingsson and Choppin de Janvry, 1992). Crop yield can be predicted by two types of models (Burke, 1992):
1. mechanistic or dynamic growth models, simulating the processes occurring in crop-weather interactions;
2. regression or black box models, based on empirical relationships

P. C. Struik et al. (eds.), Plant Production on the Threshold of a New Century, 115–122.

and using meteorological records and crop yield data from the past.

The first approach leads to a more complex model than the second. It has a wider range of application, especially as a research tool. A dynamic model may predict yields better than a regression model, but requires more information input. If this information is not available, the regression model may be the best option (Burke, 1992; Seligman, 1990).

A new field in crop modelling is the combination of crop growth and economic models. Economic factors never interact directly with plant growth, but crop models and economics can be combined to support farm management (Penning de Vries, 1990). In general, only a few crop models for farm management are successful, even when they have been especially tailored for use by farmers or extension personnel (Seligman, 1990). In the period 1987 till 1990 Biemond et al. (1989) began the development of PIEteR[1], a bio-economic model as a basis for decision support in sugar beet growing. In 1992, this research, still in its infant stage, was evaluated and re-started under the new project title 'The development of a location specific bio-economic production model for tactical decision support in sugar beet growing, with special reference to yield, quality and environment'. It focuses on tactical and semi-operational decisions in sugar beet growing at farm level.

Tactical and semi-operational decisions have to be made by the farmer before or during the growing season. Thus, repeated operational decisions, such as those on weed removal, are excluded. The most important decisions in our model are nitrogen supply, plant density and harvest date. Other tactical and semi-operational decisions are those on tillage, seed bed preparation, choice of variety, sowing time, sown area (related to permitted yield quota), resowing, and harvest method. In the following we discuss decision support in sugar beet growing (2) and describe the current status of PIEteR (3).

2. Decision support in sugar beet growing

2.1. EXPERT SYSTEMS AND DECISION SUPPORT

In the following an existing expert system for sugar beet growing is discussed and the need to develop a decision support system is explained and illustrated.

BETA is an expert system for sugar beet growing, developed by the Research Station for Arable Farming and Field Production of Vegetables (PAGV) and others (Kemp Hakkert, 1992). BETA assists farmers to make improved decisions on choice of variety, crop protection etc. It uses previous observations on root and sugar yield of varieties that were

---

[1] PIEteR means: 'Production model for sugar beet, including Interactions between Environment and growing decisions, and their influence on the quantitative, qualitative and financial Result'.

grown on different fields at the farm over several seasons in recent years. Actual information on sowing date, plant density and growing stage is necessary for advice on resowing and herbicide application to use just a few examples. BETA presents a list of recipes, such as 'If this occurs, then you can choose between the following options: ...'.

For some decisions the 'recipe method' is not sufficient. More location specific information is needed on the actual growth and production of the crop, to predict for example the influence of harvest date on yield and profit in a specific combination of weather, soil and previous treatments. This was behind the initiative in 1987 to develop PIEteR, a location specific bio-economic production model as a basis for decision support.

PIEteR is still in an early stage of development. In future it should be able to analyze a combination of different sets of input that, 'processed' by a system of processes and interactions, will offer agronomic, ecological and farm economic predictions. This output will significantly increase farmers' ability to reach optimal decisions in given circumstances.

PIEteR's inputs requirements include:
1. meteorological data (temperature, radiation, rainfall);
2. soil data, which make the model explicitly location specific (contents of silt and organic matter);
3. growers' decisions (sowing date, plant density, harvest date);
4. the payment system for sugar beet (based on harvest date, yield and quality). This system takes into account EC price policy, world market prices of sugar and the penalties that have to be paid for dirt tare.

PIEteR's outputs present the expected value of several parameters at different dates (in percentage, kg or guilders per ha):
1. root and sugar yield;
2. sugar content and extractability;
3. dirt tare;
4. remaining nitrogen in the soil, after harvest;
5. economic result.

In order to produce output from a given input, quantitative knowledge is required on:
1. the crop's response to the effects of meteorological and soil data and growers' decisions;
2. the remaining processes in the soil, that determine the losses of nitrogen, especially by leaching.

Thus, different disciplines are integrated in this project, especially crop ecology, soil science and farm economics. Moreover, different process levels are integrated: crop, field, farm and sector level. Growers' decisions increasingly have to deal with all these levels. In arable farming there is a growing need to take into account not only crop development, but to optimize farm planning as a whole and to be aware of what is happening on meso- and macro-scale in agriculture.

2.2. HARVEST TIME AS A POINT FOR ILLUSTRATION

As stated above, the most important decisions in PIEteR are nitrogen supply, plant density and harvest date. The first two will not be discussed here, but the main principle is the same as for the third: harvest and delivery time of sugar beets.

The sugar industry focuses on optimal use of processing machinery and labour during the campaign. Since processing during Christmas would induce high labour costs, the campaign should be finished before December 25th. On the other hand, the campaign should start as late as possible, since early delivery has to be met with industrial subsidies to compensate yield losses. Based on national and regional yield predictions in August the sugar industry makes a planning for daily amounts of beets to be processed.

In order to fulfil the industrial planning, the company representatives develop a time table for delivery of sugar beets from the farms in the various regions. The individual farmer has some influence on this time table and on the amounts that should be delivered at various dates. To enable him to plan in an efficient and economic way he should know at the beginning of August, which yield of root and sugar he can expect. Later on in the growing season, especially in the months of October and November, it can be helpful to harvest some weeks earlier than delivery date. This would reduce frost risks and possibly the amount of dirt tare and the risk of soil structure damage. The latter results from an increasing precipitation surplus during these months. Before harvest, internal quality may increase over time, i.e. both sugar content and extractability, thus improving the farmer's profit. After harvest, during storage, both will decrease.

The model should therefore predict the influences of a decision to harvest one week later than normal. Depending on the time of year, this could have a positive influence on yield, internal quality and storage losses. On the other hand, dirt tare, frost damage to the crop and soil structure problems could have a negative influence. The net financial effect will be spelt out to farmers using the programme.

3. PIEteR

3.1. ACTUAL STATE

The actual state of PIEteR is the result of the work of Biemond et al. (1989) and Greve (1992). They chose a black box approach, in which causal relationships are described by non-linear equations. The model is dynamic, because it simulates development and production on a daily basis. It is not mechanistic, since relationships are described only at a high level of integration. The coefficients have been determined by causal regression analysis of data from the Dutch Research Institute for Sugar Beet Growing (IRS) and the sugar industry.

PIEteR exists of a main module to which several submodules are connected. The main module describes growth and development of the crop. Five phases are distinguished. In each of the five, different factors have major influence on the biological processes that take place[1] (Biemond *et al.*, 1989; Houtman, 1992; Schiphouwer, 1992):

1. The 'emergence phase': the period between sowing and emergence; the length of this period depends mainly on soil temperature.
2. The 'phase of exponential growth': the period between emergence and the so-called 'growing point date' (GPD), the day on which an average root contains 4 gram of sugar (Van der Beek and Kemp Hakkert, 1992)[2]; in this phase growth rate is determined mainly by air temperature.
3. The 'production phase' or 'phase of linear growth': the period after GPD until the end of August/the beginning of September; in this phase dry matter and sugar production depends mainly on radiation.
4   The 'ripening phase'[3]: this period starts at the moment at which growth rate changes from constant to decreasing; the actual growth rate depends on availability of water and nitrogen. There may be redistribution of dry matter; sugar content may increase or decrease. The end of growth will be somewhere between September 15th and October 15th. Temperature, especially night temperature, and radiation play an important role in this phase. The ripening phase ends at harvest.
5. The 'storage phase': the period between harvest and delivery to the sugar factory; temperature is the main factor in this period, influencing respiration rate or inducing frost losses.

The growth module basically estimates GPD and root and sugar yield on the basis of daily values of average air temperature and radiation, respectively. The modules that are connected to the growth module, calculate daily values of a number of other input variables. One of the main variables is a reduction factor that corrects development and production rates for water stress. This stress may be either water surplus or water shortage. The question if and to what extent water stress will occur, is answered using modules that calculate root growth and water balance of the soil.

At the current state of development PIEteR predicts GPD on a field specific basis and root and sugar production on a regional

---

[1] In the following, water and nutrients are assumed not to be limiting.

[2] GPD nearly coincides with 'full canopy', which is defined as 'the day on which the first leaves of different rows touch' (Spitters *et al.*, 1990; Van der Beek, IRS, pers. comm., 1992).

[3] Sugar beet is a biennial plant, so that ripening in the meaning of 'ripening of seed' does not occur in the first year (except for bolted plants, perhaps (Smit, 1983)). In the first, vegetative year, ripening can be defined as the development of the plant towards a stable situation, i.e. new leaves do not appear any more and beet yield and sugar content do not change.

basis. Emergence day is not yet predicted accurately and no attention is given to the storage phase. Neither have nitrogen fertilization, plant density, internal (sugar content and extractability), external quality (dirt tare) and nitrogen surplus been modelled yet. These will be incorporated in PIEteR during the project.

3.2. TEST

Greve (1992)[1] tested PIEteR on data of more than 3000 sugar beet fields in 1991. Regional and national means of GPD, root and sugar yields were presented.

The national prediction of GPD was exactly the same as determined by sampling by the Dutch Research Institute for Sugar Beet Growing (IRS): day 182 (July 1st, which was 89 days after sowing on April 3rd). Regional averages had a maximum deviation of 4 days.

The prediction of the national sugar yield was 10950 kg per ha, which was 70 kg more than the values from IRS-sampling. Regional differences were much higher, ranging from -1650 to +1140 kg per ha.

4. Discussion

PIEteR predicts GPD, fresh root and sugar yield, correcting for suboptimal moisture contents in the soil. It also includes field specific parameters, such as silt and organic matter content. As a consequence, PIEteR may be useful in developing a model as a basis for decision support in sugar beet growing.

In the test, the sugar yield was not predicted so well. Part of the explanation is that the growing season was far from normal in some regions: part of the sugar beet area had to be resown due to frost damage and a wet and cold start (May, June) was followed by a very dry summer. Apparently, the model was not able to cope with such extreme circumstances. The post-season runs for 1992 were much better (H.J. Greve and T. Schiphouwer, pers. comm.).

It was decided that PIEteR and other production models for sugar beets should be compared in order to find the best basic model for the purpose of yield prediction and decision support of sugar beet growing. The main criteria will be prediction accuracy and usefulness at farm level. PIEteR has the advantage that predictions will not require sampling by farmers.

Considerably more research must be carried out before all quantitative relationships may be presented. This is not only true for harvest date, but also for nitrogen level, plant density and other variables. The different influences on yield, quality, environment and farm economic returns should become clear and quantified for every

---

[1] Greve and Schiphouwer (Suiker Unie Breda) contributed to PIEteR in modelling root and sugar yield per ha. They used regional coefficients from data covering a period of 10 years.

reasonable level of these variables. As an example, data on the exact influence of nitrogen level on sugar content will be incorporated as an important relationship in PIEteR.

Even if all relationships are fully known, then there will still remain a certain amount of uncertainty. Frost risk has already been mentioned. Other weather circumstances cannot be predicted very accurately from one week to the next. Nevertheless, extreme situations and the chance of their occurrence can serve as input for a sensitivity check. In that way, balanced decision making is greatly improved.

In the next century the call for a more economic and environmentally friendly way to produce crops will continue and grow even stronger. We hope that our model, the extended version of PIEteR, will be a good tool to improve the quality of decision making of the sugar beet growers. May it be helpful to match the social and farm economic needs in the 21st century.

5. References

Beek, M.A. van der and Kemp Hakkert, D.J. (1992) 'De groei van de suikerbieten tot medio augustus', Maandblad Suiker Unie 26 (1992-September), 12-13.

Biemond, T., Greve, H.J., Schiphouwer T. and Verhage, A.J. (1989) 'Pieter, semi green-box produktiemodel suikerbieten', WAU, Department of Farm Economics, Wageningen.

Burke, J.I. (1992) 'A physiological growth model for forecasting sugar beet yield in Ireland', in Proceedings of the I.I.R.B.-Conference, Brussels, 12-13 February 1992, 239-251.

Crals, M.L. and Stinglhamber, E. (1992) 'Importance des previsions de production dans la gestion d'une sucrerie', in Proceedings of the I.I.R.B.-Conference, Brussels, 12-13 February 1992, 157-168.

Greve, H.J. (1992) 'Rapport perceelsgericht produktiemodel', Suiker Unie, Breda.

Houtman, H.J. (1992) 'Bewaring van suikerbieten', in Artikelen Suikerbieten Informatiedagen 1992, IRS, Bergen op Zoom, 34-38.

Kemp Hakkert, D.J. (1992) 'BETA, a crop management system for sugar beet: development and experiences', in Proceedings of the I.I.R.B.-Conference, Brussels, 12-13 February 1992, 53-62.

Penning de Vries, F.W.T. (1990) 'Can crop models contain economic factors?', in Rabbinge, R., Goudriaan, J., Keulen, H. van, Penning de Vries, F.W.T. and Laar, H.H. van (eds.), Theoretical Production Ecology: reflections and prospects. Pudoc, Wageningen, Simulation Monographs 34, Chapter 6, 89-103.

Schiphouwer, T. (1992) 'Invloed klimatologische factoren op de groei van suikerbieten', Suiker Unie, Breda.

Seligman, N.G. (1990) 'The crop model record: promise or poor show?', in Rabbinge, R., Goudriaan, J., Keulen, H. van, Penning de Vries, F.W.T. and Laar, H.H. van (eds.), Theoretical Production Ecology: reflections and prospects. Pudoc, Wageningen, Simulation Monographs 34, Chapter 14, 249-263.

Smit, A.L. (1983) 'Influence of external factors on growth and development of sugarbeet', WAU, Wageningen, dissertation, 109 pp.

Sperlingsson, C. and Choppin de Janvry, E. (1992) 'Survey of information and communication systems in European sugar beet growing', in Proceedings of the I.I.R.B.-Conference, Brussels, 12-13 February 1992, 1-15.

Spitters, C.J.T., Kiewiet, B. and Schiphouwer, T. (1990) 'A weather-based yield-forecasting model for sugar beet', Netherlands Journal of Agricultural Science 38, 731-735.

Withagen, L.M. (1989) 'Noodzaak van een groeimodel: produktie-schattingen en ondersteuning van teeltbeslissingen', in Biemond, T. (ed.), Produktiemodel suikerbieten. Rapportage workshop 3 mei 1989, WAU, Department of Farm Economics, 4-5 and 29-38.

THE CONCEPT OF CLOSED BUSINESS SYSTEMS IN THE DUTCH GLASSHOUSE HORTICULTURE

HENK J. VAN OOSTEN
Glasshouse Crops Research Station
Kruisbroekweg 5
2670 AA Naaldwijk
The Netherlands

ABSTRACT. The government in the Netherlands has developed a purposive long term policy for the improvement of the environment. Qualitative and quantitative standards were set and have to be reached in 2000. This has enormous implications for the glasshouse horticulture. The standards are so strict, that only one way has been described to fulfil the demands: the development of closed business systems to avoid undesired emissions to ground and surface water. In addition, integrated production concepts for growing in glasshouses are developed to reduce the uses of energy and pesticides and the production of wastes. Research programmes are developed to solve the problems and these are discussed in this paper.

1. Introduction

Problems concerning environmental issues are well known all over the world and have been highly exacerbated in recent years. This has led to an extensive social debate, the introduction of legislation, the drawn up of implementation of new production concepts, and the developing of control measures in most western countries.

In the Netherlands the government has developed a coherent, purposive, long-term policy concerning this matter. This policy was described in several national plans (1,2). In the bill on environmental problems (1) a very ambitious qualitative prospect was defined on ongoing sustainable development. It was foreseen that clean production processes should exist, without undesired emissions and a controlled recycling of wastes. In a specific bill on agriculture (and horticulture) this prospect was elaborated into quantitative standards to be reached in a certain period (3). This paper will concentrate on glasshouse horticulture. A description is given of the significance in the Dutch agricultural world, its environmental problems, the governmental targets and the opportunities to minimize undesired side-effects on the environment.

P C Struik et al (eds), Plant Production on the Threshold of a New Century, 123–135
© 1994 *Kluwer Academic Publishers*

2. Characterisation of the glasshouse crops horticulture

The glasshouse crops horticulture has grown tremendously in acreage and production value during the last 25 years. In 1992, its acreage reached more than 10.000 hectares for the first time in history (Table 1). Its production value grew to more than 33 % of the Dutch agricultural production value in 1992, but unexpected serious market problems caused a decrease in value of about 10 % since then. Market developments determine the crops that are grown. For example less than 10 years ago, lettuce was a leading export vegetable but it is now a minor product, whereas sweet pepper showed the opposite trend. Vegetables as a group were fairly stable in recent years, but cut flowers and pot plants grew remarkably in acreage and production value. A major change in culture in vegetable growing, especially the fruit vegetables (such as tomato, cucumber and sweet pepper), was the introduction of growing systems on artificial substrates instead of traditional soil (4,5). In floriculture, this development is still much weaker (Table 2). This change has caused a marked production growth per m^2 and contributed strongly towards a better production efficiency (kg product per unit of input).

TABLE 1. Area and turnover in the Dutch glasshouse horticulture (1992).

	Area ha	Value 10^9 Dfl
Vegetables (incl. fruits)	4593	2.4
Floriculture (incl. trees)	5512	5.8
Total	10105	8.2

TABLE 2. The change in acreage (ha) of vegetables and cut flowers grown in soil and in artificial substrates in the period 1984-1991.

Year	Vegetables		Cut flowers	
	substrate	soil	substrate	soil
1984	1127	3378	100	3319
1988	2161	2291	300	3336
1991	2733	1718	588	3120

3. The environmental problems

The glasshouse crops horticulture has an impressive production of an outstanding quality per hectare and per unit of input. This is only possible because of high inputs of capital, labour and materials as nutrients, pesticides, energy, etc. This way of production resulted also in undesired outputs as wastes like materials and organic matter and emissions of nutrients, pesticides and CO_2 to the environment. A few examples will be given.

3.1. EMISSION OF NUTRIENTS

In the glasshouse horticulture uses of nutrients are very high because of two reasons: feeding the plants and controlling the vegetative and generative balance in the plants by high salt concentrations (EC). The water used for the nutrient solution normally contains a certain concentration of Na that is not taken up by the plants. This results in salinization of the soil or of the recirculated nutrient solution. Therefore, an excessive drain is necessary, mostly to the surface water (Table 3). This is a common, actual situation and is even more severe for cultures on substrates.

TABLE 3. The Na concentration of water as related to the drainage percentage in recirculating systems.

Irrigation water	Na concentration (mmol/L)	Drainage (%)
Tap water	2.0	26
Super water	1.1	13
Rain water	0.5	6

Uptake by crop (sweet pepper): Na: 0.2 mmol/L
Damage limit   : Na: 8.0 mmol/L

3.2. EMISSION OF PESTICIDES

The use of pesticides in glasshouse crops is described elsewhere (2) and summarized in Table 4. It shows the use of pesticides per hectare in the Netherlands and England. In both countries this use is high. Differences between the crops grown exist. For example, relatively low uses in fruit vegetables are found due to the intensive use of integrated pest control and the maximal use of cultivars with resistance to some diseases. Much higher uses are found in some flower cultures. This is caused to a great extent by the zero-tolerance for insects in many flower importing countries.

TABLE 4. Comparison of the area (ha) and the use of pesticides (kg active gradient/ha/year) in vegetable crops under glass in the Netherlands (mean 1984-1988) and England/Wales (1985).

	Country	
	NL	England/Wales
Acreage (ha)	4430	2919
Use (kg a.i./ha/year)		
Fumigantia	88	68
Insecticides/Acaricides	6	1
Fungicides	10	12
Herbicides	1	1
Total	106	82

3.3. USE OF ENERGY

The use of energy (natural gas) in m^3 per m^2 is visualized in Figure 1. It shows, that in the early eighties the use of energy dropped sharply because of the rise in prices due to the energy crisis. However, in the second half of the eighties the total energy use began to rise until now (6). There are several explanations: a) the total acreage of glasshouse cultures rose sharply in those days; b) a change from energy-extensive to energy-intensive cultures (e.g. sweet pepper instead of lettuce); c) low gas prices.

A direct relationship exists between the energy use and the production and emission of CO_2.

3.4. OTHER WASTES

The way of production on substrates, independent of the soil creates a much greater amount of waste materials than cultures in the soil. For example, rockwool substrates and plastics that cover the soil, are used mostly only one year and then replaced by new materials. Now, rockwool, other substrate materials and plastics can be recycled and, partially, this is organized by the delivering industry. Organic materials are to a great extent composted and then sold to gardens and other agricultural farms.

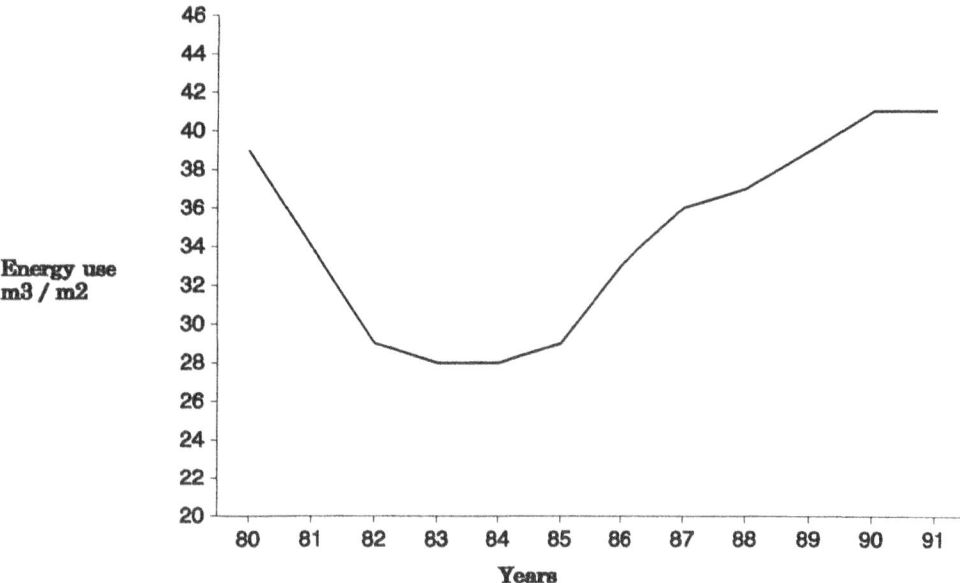

Source:lei

Figure 1. The use of the energy (natural gas) in m³ per m² in the glasshouse horticulture between 1984-1992.

4. The governmental targets

The governmental policy has led to a number of quantitative standards for the emission of nutrients, pesticides and CO_2 in general and more specific from the glasshouse crops horticulture (1,2,3). These are summarized in Table 5.

TABLE 5. The governmental targets for a clean glasshouse horticultural production in 2000.

* Virtually all cultures in closed systems
* 65 % less use of pesticides (volume) (compared to 1988)
* 50 % higher energy efficiency (compared to 1980)
* 3-5 % reduction of CO_2 emission

The standard for nutrients in ground water (50 mg nitrate/litre) is accepted in the whole EC, whereas the standards for surface water (2.2 mg N/litre and 0.15 mg P/litre) are Dutch standards. These standards apply for all agricultural and industrial activities.

The decrease in CO_2 production is also a general obligation in the Netherlands and will have its effect on gas use in horticultural production.

The decrease in use of pesticides is specified for each sector of agriculture. It is not described in terms of a certain maximum use, but in a percentage of the use in a certain year. This is further specified for each group of pesticides.

The targets are so strict, that the government sees only one way to go for the glasshouse horticulture: the development of so-called closed production systems, thus avoiding undesired emissions to the environment. Therefore, the government decided that the glasshouse crops horticulture should change completely to these closed business systems (Table 6).

TABLE 6. The development in glasshouse horticulture towards closed production systems until 2000 in %.

| | 1994 | | 2000 |
	Vegetables	Flowers	Glasshouse cultures
Independent of soil	80	30	± 100
Closed systems recirculation	30	30	± 100

The governmental targets are all to be reached in the year 2000. This leads to a severe pressure on horticulturists. The government has the intention to check in 1995 whether the different sectors, in our case the glasshouse crops horticulture are on their way to meet the standards set for the year 2000. If the improvement is unsatisfactory, the government may take additional measures. The policy of the government is to stimulate the development of clean production processes that do not harm the environment. It should stimulate the development of more durable or sustainable processes, that are safe for workers and for consumers. Finally, the Dutch glasshouse crops horticulture should remain viable and competitive.

5. The possible solution(s)

It is shown, that the environmental problems of the glasshouse horticulture are tremendous and the governmental targets in order to reach clean production processes very strict. It is fully clear to growers, that the traditional way of growing and of handling of wastes has to be changed. Therefore, an extensive programme was set up by the

growers' organisations in order to meet the targets. Part of the activities was a research programme to develop closed business systems (7,8). In this research programme a number of institutes and research stations collaborate. It was partially additionally financed by the government and the growers' organisations.
The research strategy is focused on four topics:

*   prevention of losses of nutrients and pesticides;
*   reduced use of pesticides;
*   increased energy efficiency;
*   prevention, limitation and recycling of wastes.

The development of so-called closed systems in order to prevent emission of nutrients and pesticides can be considered as only end-of-pipe solutions. However, this can also be seen as a first step to diminish the negative consequences of the current growing techniques. A following step is to develop growing techniques that cause less need for emission, e.g. are less dependent of the use of pesticides. Examples of both aspects will be given below.
Simulation studies have given already an impression of the extra investments and annual costs of the different types of closed business systems. Extra investments are rather limited for crops that are already grown on artificial substrates, but high for crops that are still grown in the soil. The extra costs depend largely on the systems chosen. This is discussed extensively elsewhere (9,10,11).

5.1. PREVENTION OF LOSSES OF NUTRIENTS AND PESTICIDES

With the development of closed business systems it is tried in the first place to avoid nutrient and pesticide emission to ground and surface water. There should be no transport or drainage of elements or chemicals to the environment. Cultures should be completely separated from the soil and the nutrient solution should be caught, recirculated and re-used.
The separation of cultures from the soil is already realized with crops on artificial substrates. Recirculation of the nutrient solution, however, is rare, because of several reasons: a) the accumulation of sodium; b) the uncertainty about the spread of diseases in the recirculated water and c) the costs of the system.
The accumulation of sodium is mainly caused by the lack of uptake of this element by the plants. The main source of this element is the water used (Table 3). Regular tap water has a fairly high concentration of sodium. It can be improved and distributed to individual growers at high costs. Rain water is an interesting alternative, but it needs a basin to catch and store it. Techniques like osmosis, can be used easily on a large scale for desalinizing water, but are too costly and too uncertain when used on a minor scale on the holding itself. The accumulation of sodium is the most important reason that growers drain away the nutrient solution regularly. Therefore, research is concentrated on gaining knowledge

about the acceptable levels of sodium before production and quality will be affected. In fruit vegetables these levels are largely known, but for most other vegetables and cut flowers they are unknown so far (12). If growers' organisations are successful in their attempts to get completely desalinized water for growers, the problem of accumulation of sodium will no longer exist.

Continuous recirculating of the nutrient solution has also the serious risk of spread of diseases. From a disease point of view, disinfection of the recirculating solution is necessary. In recent years, several techniques are developed to do so, but these apparatuses are costly (13,14). Therefore, growers try to avoid these investments. It is questionable whether growers can grow crops in closed systems without disinfection.

A much larger group of crops, vegetables and cut flowers, are still grown directly in the soil. Research is directed to the possibilities to grow these crops technically and economically in closed systems. An extensive simulation study was made by a group of various specialists (like crop specialists, technicians, labour analysts, material specialists, managers, economists, etc.) of potential productions systems in each crop or group of similar crops (10). For each group of similar crops some systems were chosen based on technical and economical standards: easy to apply for a range of crops, and cheap. Prototypes of these systems for each group of crops are constructed and these are tested since the end of 1992. Research groups operate in close collaboration with an accompanying group of growers. All new information from specialized areas is included immediately. On the other hand, unexpected problems are brought under the attention of specialists.

The closed production systems, so far developed, can be separated into two main groups based on the substrate: a) water and misting and b) solid substrates like sand, peat and lava products. In general, crops grow well when the air:water ratio (15) in the root system is correct and the nutrient solution is adequate. The systems using only water as a substrate may be more vulnerable in case of technical defects in nutrition and water supply.

The first results indicate, that crops like *Freesia*, *Hippeastrum* and *Alstroemeria* grow very well in these new closed systems. But radish, for example, gives a good production but the quality of the tuber (colour) and the very branched root system are unacceptable. Most experiments will be evaluated at the end of 1993 and then decisions will be made about scaling up some of them.

The emission routes of pesticides are more diverse than those of nutrients. Partially they follow the same routes via the water flow directly to the surface water or the ground. By catching the water these routes can be closed completely. Other routes are via condensation of water against the glass windows and via the air route. The catching of the condensation water is not a real technical problem. A greater problem is the air route. Present-day glasshouses are not completely closed, as there is always some degree of ventilation for climatic reasons. Thus, pesticides may leave the glasshouse in small droplets or as vapour (16). In trials, it was demonstrated, that the application technique and the vapour pressure of the chemical are of a great influence on the real emission (Table 7). In another experiment a glasshouse was built using forced ventilation and without windows that could be opened. In fact, this was the most closed glasshouse one can imagine at the moment. It could be helpful to keep volatile pesticides within the glasshouse. This glasshouse was certainly very interesting in this respect but the costs of energy for the forced ventilation were far too high and thus this futuristic glasshouse was economically not feasible (17).

TABLE 7. The influence of application technique and vapour pressure of pesticides on the emission (% of applied concentration) from the glasshouse. Source: TNO 1992.

Application technique	Vapour pressure (MPa)		
	High > 10	Moderate 10-0.01	Low < 0.01
Ultra low volume	39	23	< 5
Low volume	39	9	< 1
High volume	39	8	-

Thus, the emission of pesticides by air can be influenced to a great extent by altering the application technique and the chemicals used, but not (at least not economically) by changing ventilation of the glasshouse itself.

As is shown in Table 4 the use of fumigantia is the most extensive of the pesticides in horticulture. This use has to diminish in the first place. In many cultures it could be replaced already by steaming. However, this is not possible in all circumstances, especially in wet soils with a high percentage of organic matter. In designing closed systems it is tried to use only materials, substrates and techniques that can be steamed without problems, thus avoiding fumigantia.

5.2. DECREASED USE OF PESTICIDES

A number of techniques can potentially contribute to a substantial decrease of the use of pesticides in glasshouses. This is extensively described recently (18). Research projects are dedicated to these techniques. They will be discussed briefly.

A major contribution is given by the impressive development of integrated pest management, especially in vegetables under glass (18,19,20). First encouraging results with biological control were obtained already 35 years ago and implemented in commercial practice about 20 years ago. Now, in fruit vegetables like tomato, cucumber and sweet pepper, nearly 100 % of the holdings start with biological control of major pests and many are able to keep it going the year round. In many cases, however, use of pesticides is necessary either as an aid to suppress localized outbreaks of pests or to combat pests for which biological control is not available. The integration of biological and chemical means is now developed into a full-grown integrated pest management on most holdings. This contributes very well to a low use of insecticides. Biological control of pests in leafy vegetables and the floriculture is hardly used. Consumers do not like insects or insect damage in their vegetables and this automatically leads to a so-called zero tolerance for the presence of insects or damage. The same is true in floriculture where the zero tolerance is maintained especially by importing countries to avoid the entrance of pests from abroad.

Recently, the first experiments with integrated pest control in leafy vegetables (like lettuce and radish) and cut flowers (21) (like *Gerbera*) have started and gave first promising results. Before introduction in practice the results should be stabilized over some years. In other flower crops, like roses and *Chrysanthemums*, experiments with the so-called guided chemical control are started. This means, that the application of pesticides is done only after extensive observations. If the insect population has reached the threshold value for economic damage pesticides are used.

Integrated and guided pest control can diminish the pesticide use in glasshouses substantially.

Research on biological control of fungi is also intensified. Sometimes results are promising, but it will need still a long way for commercial application.

The scope of integrated control of pests and diseases can be widened by using cultivars that have some degree of resistance. Furthermore, growing techniques may affect the development of diseases and pests. Climatic factors like temperature and humidity may influence disease and pest development, and development of predators. Nutrients may also have an effect. Effects of nitrogen are well known on red spider development in suboptimal concentrations. Adding silicon to the nutrient solution will diminish the development of mildew in most *Cucurbitaceae* (22).

In tomato, the combined use of integrated pest control and cultivars with resistance against several diseases is already widespread, resulting in a low pesticide use.

The use of pesticides may also be influenced by the application technique. It is known that only a small part of the applied chemicals is really effective. Research projects are dedicated to develop new application techniques as for example the Closed Loop System (23) or improvement of existing techniques.

Finally, the development of decision support systems for plant protection should be mentioned. These may help managers to make the most appropriate decisions in case of diseases or pests (24).

5.3. INCREASED ENERGY EFFICIENCY

The total energy use in m^3 per m^2 in the glasshouse horticulture is demonstrated in Figure 1. A far more interesting figure is the production in kg per unit of energy. This gives an impression of the production efficiency. Research is steadily working on improvement of this factor, as it also influences the direct costs of production. Improved growing techniques, more productive cultivars, better climatic control, balanced nutrient supply, etc. may contribute to a better production per m^2 and a higher production efficiency. It is expected that the current research effort will be sufficient to reach the governmental target on this aspect (6).

5.4. PREVENTION, LIMITATION AND RECYCLING OF WASTES

In closed systems more materials and substrates are used than in traditional systems. The waste problem obviously can be even greater. In designing closed systems the choice of materials is a major factor. For materials, the environmental aspects are taken into account. For example the possibility of re-using materials for a number of years instead of one year, the possibility of recycling and re-using in horticulture, the energy component of the materials, etc. It is tried to draw up a balance of environmental aspects of materials used (25). To some extent this is comparable with the comparison Gysi (26) made of the energetic components of producing horticultural products in different parts of the world and under different conditions.

In research only materials and substrates are taken into account, that can be re-used as much as possible. The absence of solid substrates in watery systems is of course very interesting, but so far these systems are vulnerable for technical faults and disease spread.

6. Social debate on closed production systems

Growers do accept at this moment that the way they produce horticultural products should not influence seriously the environment. However, it is not accepted, that governmental targets are so strict, that only the closed production systems may reach these targets. Traditional gardening using the soil in a proper way will be no longer possible. Even ecological or biological-dynamic horticulture cannot meet the standards. Therefore, research is now started to investigate in which way the current horticulture in soil can be improved in such a way that emissions of nutrients (and pesticides) can be reduced substantially without influencing production levels and quality of products.

Groups of consumers in the Netherlands, but more dominantly in central Europe, question closed systems as a unnatural way of production. It is described as a industrial way of production. These consumers prefer horticultural products that are grown in soil and treated in a natural way.

134

The glasshouse horticulture in the Netherlands is marked directed in the products grown and also in the way of production. Potentially, a conflict may arise, when market demand for biological products cannot be fulfilled with products grown in closed systems.

Market organisations of vegetable growers have developed a new market strategy to prepare for these market demands (27). The concept of integrated production in horticulture is introduced. In this concept all measures that can be taken on a horticultural holding to improve the environmental quality of the way of production are stimulated and if economically and technically feasible gradually obliged to growers. In this system it is tried to reach the whole group of vegetable growers. The market organisation has to develop also a new system of measuring and controlling at the holdings level. After three years, the system is widely accepted by most growers. It is imaginable that the system evolves to own standards for production, comparable with the ISO standards in other production areas.

7. References

1. Anon. (1990) 'Nationaal Milieu Beleidsplan'. Staatsuitgeverij, 's Gravenhage.
2. Anon. (1991) 'Meerjarenplan Gewasbescherming'. Staatsuitgeverij, 's Gravenhage.
3. Anon. (1990) 'Structuurnota Landbouw'. Staatsuitgeverij, 's Gravenhage.
4. Sonneveld, C. (1988) 'Rockwool as a substrate in protected cultivation' in Organizing committee of Intern. Symp. High Technology in Protected Cultivation Horticulture in High Technology Era, Tokyo, pp. 173-191.
5. Sonneveld, C. (1991) 'Rockwool as a substrate for greenhouse crops' in Y.P.S. Bajaj (ed.), Biotechnology in agriculture and forestry, Vol. 17. High-tech. and micropropagation I, pp. 285-312.
6. Velden, M.J.A. van der and Sluis, B.J. van der (1993) 'Energy, efficiency and CO_2-emission in the Dutch greenhouse industry', Report.
7. Anon. (1990) 'The research programme for the development of closed business systems in glasshouse horticulture'. NRLO, 's Gravenhage.
8. Oosten, H.J. van (1991) 'Clean production processes in glasshouse cultivation' in J.L. Meulenbroek, Agriculture and environment in Easthern Europe and the Netherlands, Agricult. Univ. Wageningen, pp. 425-442.
9. Ruys, M. (1991) 'Economic evaluation of business systems with a lower degree of environmental pollution'. Acta Hort. 295, 79-84.
10. Ruys, M. (1992) 'Economic evaluation of closed production systems in glasshouse horticulture'. Acta Hort. (in press).
11. Os, E.A. van, Ruys, M.N.A. and Weel, P.A. van (1991) 'Closed business systems for less pollution from greenhouses'. Acta Hort. 294, 49-57.

12. Sonneveld C. and Burg, A.A.W. van der (1991) 'Sodium chloride Salinity in fruit vegetable crops in soilless culture'. Neth. J. Agr. Sci. 39, 115-122.

13. Runia, W. Th., Os, E.A. van and Bollen, G.J. (1988) 'Disinfection of drain water from soilless cultures by heat treatment'. Neth. J. Agric. Sci. 36, 231-238.

14. Runia, W. and Nienhuis, J.K. (1992) 'Drain- en gietwater goed te ontsmetten'. Groenten en Fruit, vakdeel glasgroenten 2 (24), 40-43.

15. Buwalda, F. (1993) 'Heeft de teelt van chrysant op eb/vloed toekomst?' Vakblad voor de Bloemisterij 15, 32-33.

16. Staay, M. van der, Baas, J. and Bor, G. (1993) 'Gewasbescherming: emissie hangt af van middel en techniek'. Groenten en Fruit, vakdeel glasgroenten 3 (2), 60-61.

17. Bakker, J.C. (1993) 'Gesloten kas: energetisch en economisch nog onvoordelig'. Vakblad voor de Bloemisterij 48 (3), 54-55.

18. Oosten, H.J. van (1992) 'IPM in protected crops: concerns, challenges and opportunities'. J. Pest. Sci. 17, 365-372.

19. Lenteren, J.C. van (1992) 'Biological control in protected crops: where do we go?' J. Pest. Sci. 17, 321-328.

20. Steekelenburg, N.A.M. van (1992) 'Novel approaches to integrated pest and disease control in glasshouse vegetables in the Netherlands'. J. Pest. Sci. 17, 359-362.

21. Fransen, J.J. (1992) 'Development of integrated crop protection in glasshouse ornamentals'. J. Pest. Sci. 17, 329-334.

22. Voogt, W. and Bloemhard, C. (1992) 'Silicium toedienen zonder vingerafdrukken'. Groenten en Fruit, vakdeel glasgroenten 9, 22-23.

23. Porskamp, H. (1991) 'Veel vloeistof op het blad met gesloten spuitsysteem'. Fruitteelt 81 (6), 13.

24. Maas, B. van der (1992) 'Decision making in crop protection'. Ann. Rep. Res. Sta. Glasshouse Crops Naaldwijk (1991), pp. 78-79.

25. Nienhuis, J. (1993) 'Environmental balances for production systems'. Ann. Rep. Res. Sta. Glasshouse Crops Naaldwijk 1992 (in press).

26. Gysi, Ch. and Reist. A. (1990) 'Hors-sol Kulturen, eine ökologische Bilanz'. Landwirtschaft. Schweiz. Band 3 (8), 447-459.

27. Gerritsen - Wieland, M.J. (1991) 'Gecontroleerde teelt krijgt vervolg'. Groenten en Fruit (Alg.) 6, 6-7.

MAGNITUDES OF DISTURBANCE IN THE EVOLUTION OF AGRO-ECOSYSTEMS: N-FLOWS IN RICE-BASED SYSTEMS

L.O. FRESCO, N. DE RIDDER AND T.J. STOMPH
Department of Agronomy
Wageningen Agricultural University
P.O. Box 341
6700 AH Wageningen
The Netherlands

ABSTRACT. The processes underlying ecological sustainability are studied through a theoretical sequence of rice-based agro-ecosystems reflecting increasing intensities of land use. Data and assumptions are derived from existing literature in order to understand changes in the magnitude of N-flows across systems, based on N inputs equal to N outputs at the aggregated level. The three systems comprise (A) shifting cultivation, (B) mixed livestock-food crop farming, and (C) irrigated rice monoculture, that are compared on a total area basis (20, 2 and 1 ha respectively) and on a basis of per hectare crop land. Total and per hectare N inputs and outputs as well as the efficiency (ratio of N in useful outputs to N in other outputs) are estimated. On a hectare basis flows in the shifting cultivation system are less than 50 % of those in mixed livestock-cropping and only 4 % of those in rice monocropping, while efficiency increases from 0.12 to 0.29 and 0.41. Land use intensification thus leads to greater N-flows in the system, coupled with greater losses, proportionally to the increases in flows, and higher efficiency. The theoretical approach demonstrates the importance of including both magnitudes and efficiencies of N-flows at the aggregated system level in analysing ecological sustainability of agro-ecosystems.

1. A systems agronomist's perspective on ecological sustainability

Since several years, the concept of 'ecological sustainability' (further indicated by sustainability) has been around, and it has become one of the most pressing issues in agricultural science to come to grips with this subject. Sustainability, whatever its definition, has to do with disturbance of natural ecological processes. Sustainability is therefore time and scale dependent, according to the system level at which we want to determine it (Fresco and Kroonenberg, 1992). We may distinguish two types of disturbance in agro-ecosystems - defined here as ecosystems with an agricultural

137

P C Struik et al (eds), Plant Production on the Threshold of a New Century, 137–149
© 1994 *Kluwer Academic Publishers*

(arable, perennial or grassland) component in its primary production compartment:

- quantitative: increases or decreases in the volume of ecosystem flows (e.g. additional sources of nutrients, water and energy, as well as extraction of biomass through harvesting and (unplanned) outflows e.g. erosion, evaporation);
- qualitative: changes in the characteristics of ecosystem components (e.g. channelling of solar energy through fewer trophic levels, replacement of natural succession by crops, with subsequent (involuntary) introduction of weeds and pathogens).

Agricultural land use implies ecosystem effects, both on-site (on the agricultural field or plot) and off-site (on other agricultural fields) as well as effects outside the agricultural area (e.g. river silting through increased erosion, disturbance of bird habitats). Furthermore, disturbance has to do with scale: maximum disturbance on a small area while leaving the rest of the ecosystem area intact may be less disruptive in the long term than low levels of disturbance over large areas. The relationship between magnitude of disturbance and volume of agro-ecosystem flows is certainly not always linear, because it depends on the type of the disturbance. In the course of the history of land use and in response to land pressure, man has elaborated various solutions to manage disturbance of ecosystem processes in such a way that agriculture is sustainable, at least during several human generations. Studying such evolutions may generate more insight in the processes underlying sustainability. We propose to do so by focusing on the degree to which the evolution of agro-ecosystems over time has changed the magnitude of disturbance in ecosystem flows, using quantified estimates of nitrogen flows in rice based systems as an example.

2. Estimating magnitudes of disturbance in Javanese rice-based systems

Measuring the degree of disturbance caused by an agro-ecosystem is complex and requires knowledge of flows in the unmanaged 'climax' ecosystem in a similar physical environment for exact comparison. In principle, disturbance ought to be measured with respect to each of the major ecosystem flows of water, atmospheric gases, nutrients and energy, and perhaps also with respect to genetic or biological diversity.

We have attempted to overcome these problems by setting up a theoretical experiment rather than organising a flying circus of multi-annual measurements. To illustrate the magnitude of disturbance in agro-ecosystems we will compare three theoretical rice-based agro-ecosystems, located on Java (Indonesia) in a rainfall zone of 1500 mm a year, that correspond to increasing intensity of land use: classical shifting cultivation (A), mixed farming (B) and modern rice monoculture (C). By taking long fallow shifting cultivation as a starting point, a nearly undisturbed situation, close to the natural ecosystem is included.

2.1. MATERIALS, METHODS AND ASSUMPTIONS

Although the scientific literature abounds with data on such rice based systems, different sources are often contradictory or omit details. As a result, we have been forced to make a number of assumptions and simplifications.

Firstly, we have selected nitrogen flows as the parameter to compare the magnitude of disturbance between systems, rather than other nutrients or energy. Nitrogen is frequently the most limiting factor in agro-ecosystems and the one that is most difficult to balance. For the purpose of our experiment we are interested in systems that are, in principle, in balance, i.e. where there are no unaccounted net losses of nitrogen. At the aggregated system level inputs and outputs are set, therefore, at equal values. Not the exact pathways of nitrogen through the systems, but the order of magnitude of the flows as compared between the systems, is the issue here.

Although the final nitrogen balance is expressed per hectare of cultivated crops, this hectare is qualitatively different in the three systems. The area under forest in shifting cultivation, and the area of fallow/grazing-land, needed for livestock in the mixed system, are included, based on the recognition that in those systems nutrients are supplied through these uncultivated, but not unused lands. In the modern rice system nutrients are supplied by chemical fertilizers and only the rice field is taken into consideration. Soil chemical characteristics are assumed to be equal, although this is a simplification since e.g. the forest soils used for shifting cultivation and mixed farming will differ at least in texture and structure from irrigated rice paddy soils.

Wet and dry deposition of nitrogen are supposed to be equal in all systems (23 kg ha^{-1}). It is to be noted that estimates differ strongly between authors: Poels (1987) using a world wide average, Krul et al. (1982) using data of West Africa and Smaling (1993) using a transfer function for sub-Saharan Africa, estimate for 1500 mm yearly rainfall, 23, 13 and 6.5 kg ha^{-1} respectively.

Nitrogen flows through sedimentation and erosion are not calculated. In shifting cultivation those terms can be considerable at the hectare cultivated, but are levelled out in the system as a whole, whereas in irrigated rice little nitrogen in solution is expected in irrigation water and sediment transport is negligible because of deposition behind bunds or dikes.

Nitrogen fixation by free living organisms like bacteria in the rhizosphere and algae is considered to be equal in all systems and estimated at 2 kg ha^{-1}; this is incorrect, since this fixation is related to the biomass turnover in the system, but ranges found are extremely low (1 - 7 kg ha^{-1}, Watanabe et al., 1992) and corrections to be made are thus negligible.

In none of the systems a nitrogen fixing crop or a managed azolla culture is included, which limits this type of nitrogen fixing capability to some naturally occurring terrestrial plants (estimated at 3 kg ha^{-1}; Janssen (1992)) and azolla in the irrigated rice (estimated at 5 kg ha^{-1}; 20 % of the quantity if azolla is applied as green manure (Kannaiyan *et al.*, 1982; Singh *et al.*, 1982; and Watanabe *et al.*, 1992)). Symbiotic nitrogen fixation in the forest component of the shifting cultivation system is negligible: legume trees are supposed to be absent in forests and only the first year after cultivation some herbaceous legumes will provide about 3 kg fixed nitrogen.

We also assume that nitrogen contents in crop products and by-products do not differ between systems, although it is likely that nitrogen contents will increase to some extent with increasing availability of nitrogen to crops. Nitrogen content in grains of cereals is estimated at 2 % on dry matter basis. In cassava it is assumed to be 0.2 % and homegarden leaves 3 %.

Data on gaseous losses of nitrogen are very scarce. Gaseous losses from vegetation is arbitrarily set at 1 % of the nitrogen available in the above ground biomass. Gaseous losses from manure during stocking is assumed to be 10 %; losses when manure is applied to the field at 30 %. For faeces and urine, supposedly returning to the forest component of the shifting cultivation system, the same fraction is used.

Also data on losses by leaching (including gaseous losses from the soil compartment) are scarce. For annual crops 40 % of the nitrogen available in the soil is used, thus at least taking into account an increasing loss with increasing nitrogen turnover. Perennial grassland and forests are assumed to have fewer problems with leaching due to a permanently present root system: arbitrarily, leaching is set at 25 % and 1 % of the nitrogen available in the soil for grassland and forest respectively.

Conversion losses are the metabolic losses, due to conversion from plant into animal tissues and set at 2 % of the nitrogen available in the food or forage. All data on gaseous losses, leaching and conversion losses are after Janssen (1992).

Estimates of nitrogen losses by burning of forest need some more detailed explanation. Annual nitrogen turnover in forest is estimated at 3771 kg for 18 ha, and is based on litterfall and replacement of leaves twice a year (Bartholomew *et al.*, 1953). The oldest part of the forest (age of 18 years) has a biomass of about 180 t.ha^{-1} at a nitrogen content of about 0.4 % (Nye and Greenland, 1960; Bartholomew *et al.*, 1953; Reuler and Janssen, 1993; and Sivinadyan and Norhayati, 1992). I.e. per ha burnt 700 kg of nitrogen is available, but only 50 % of the nitrogen (350 kg) will be subject to burning, since the other half will remain in unburnt wood and roots. According to Reuler and Janssen (1993), about ten percent of the nitrogen will remain in the ashes (35 kg) and can be used by the crop, the remainder will volatilize (315 kg). Part of the nitrogen in the roots of burned trees will become available as well for the crop: roots are assumed to contain 168 kg (root/shoot ratio: 0.16; nitrogen content roots about 0.6 %; Bartholomew *et al.*, 1953) of which 50 % will become available during cropping by decomposition: 84 kg. Total nitrogen input for cropping by burning will thus be 119 kg. The remaining nitrogen in roots and unburnt wood (266 kg) is returned to the forest soil. Crop residues after the first rice crop are burnt (10 % of the nitrogen returns with ashes to the soil). After the second rice crop, residues are returned to the forest component.

2.2. BRIEF DESCRIPTION OF THE AGRO-ECOSYSTEMS

2.2.1. Shifting Cultivation (A).

The shifting cultivation system consists of two consecutive years of rice cropping followed by 18 years of forest fallow, i.e. 20 hectares of land are needed in order to open up one hectare a year for a first rice crop (Freeman, 1955; Pelzer, 1945). Production of the first and second rice crop are set at 1800 and 1200 kg/ha of paddy respectively (Freeman, 1955; Grist, 1965 and Conklin, 1957). The paddy is consumed by the people living from and on the shifting cultivation area (containing 60 kg of nitrogen: 36 and 24 kg in first and second year respectively). It is assumed that no crop products are exported and human faeces, urine and bodies (minus conversion and gaseous losses) stay in the system and return to the forest component. Thus, the only nitrogen inputs into the system are the natural deposition and fixation and the only outputs are the nitrogen losses through burning and leaching. The nitrogen flows in the system are summarized in Figure 1.

2.2.2. Mixed Crop-Livestock System (B).

Land use intensification in shifting cultivation may follow various pathways, with 'permanent upland cultivation' as one of the very common ones (Ruthenberg, 1980). We have adapted the mixed farming system described by Janssen (1992) to suit our purpose, since the nitrogen balance of this system has been quantified already. The system occupies a total of 2.0 hectares, of which 0.3 ha is cultivated with maize (two crops a year), upland rice and cassava, 0.5 ha is cultivated with maize (one crop a year) and cassava, 0.1 ha with trees mainly Gliricidia, 0.1 ha vegetable garden and 1.0 ha roadside "pastures". 1.4 LSU (livestock unit), i.e. 1 cow, 1 calf and 1 goat, permanently stalled at a stable, convert crop by-products and forage from roadside pastures into manure, meat and milk. The manure produced is applied to the cropping area. Furthermore, 60 kg N is applied as fertilizers on the arable land (1 ha). The crop yield is composed of 715 kg maize, 75 kg rice, 3360 kg cassava and 100 kg of homegarden products (dry weights). Additionally, meat and milk are produced containing 8 kg of nitrogen. The nitrogen flows in the system are summarized in Figure 2.

142

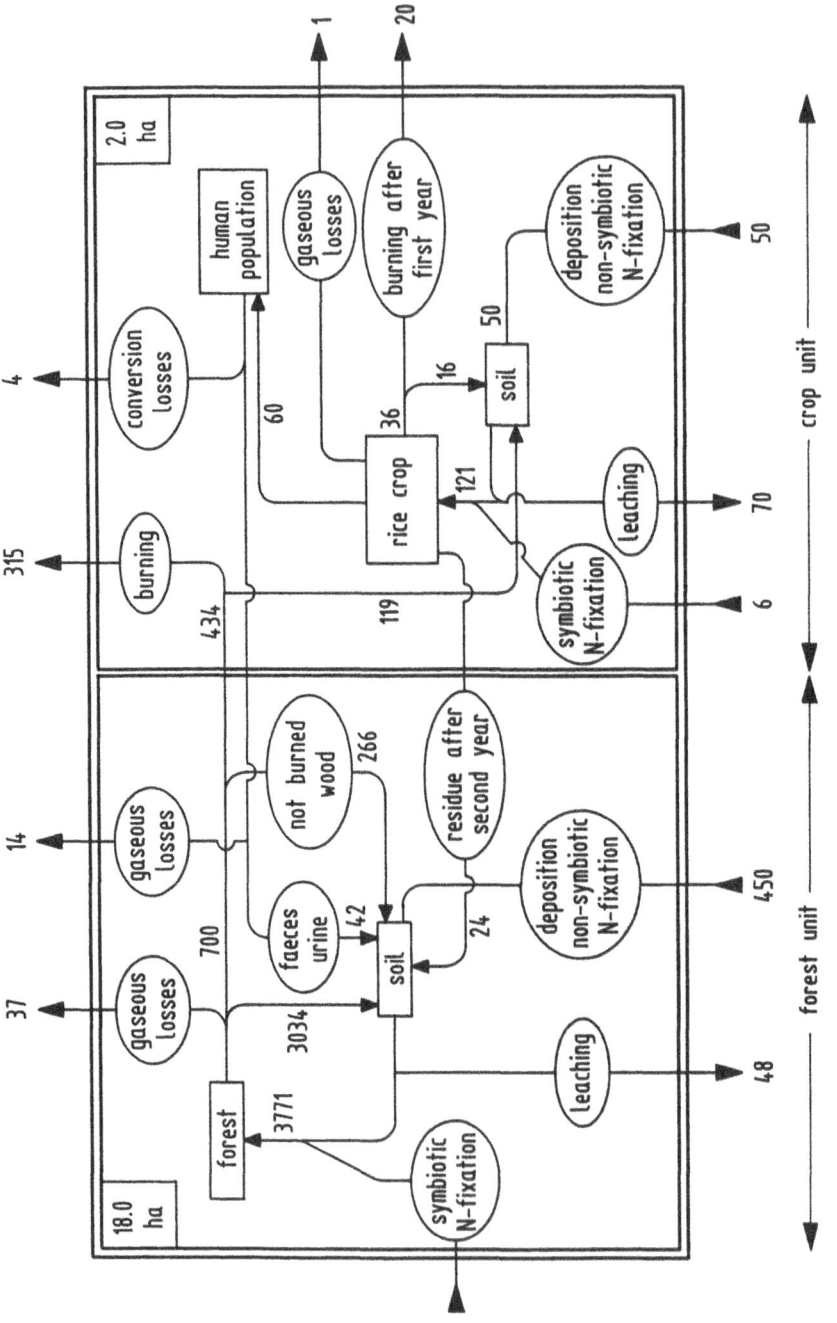

Figure 1. Magnitudes of N-flows (kg yr^{-1}) in the rice based shifting cultivation system (20 ha).

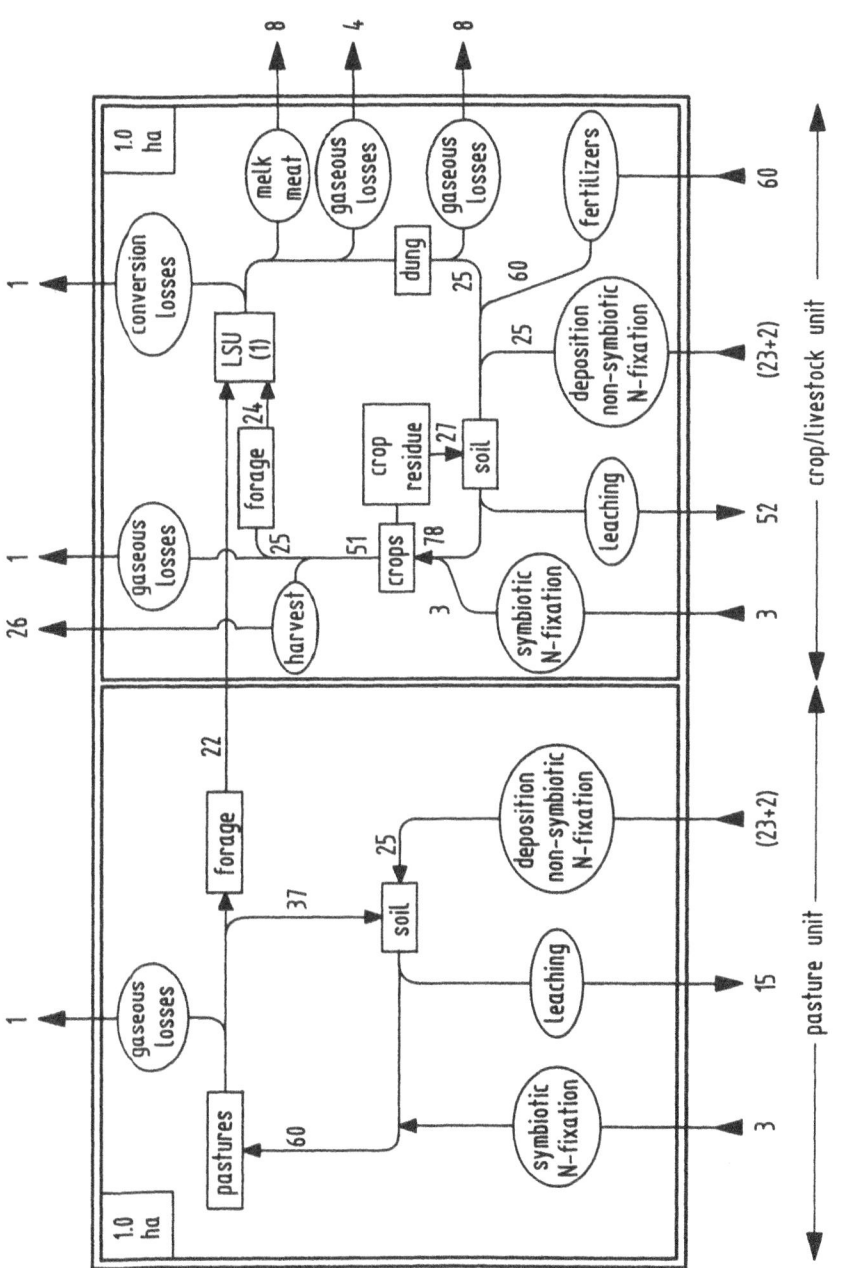

Figure 2. Magnitudes of N-flows (kg yr^{-1}) in the crop /livestock system (2 ha).

2.2.3. *Modern Rice Monocropping (C).*

This simple system consists of one hectare of monoculture rice, cropped twice annually, each crop producing 5400 kg ha^{-1} rice. It is supposed that only chemical fertilizers are used; although the farming system can be more complex with other activities included, the rice cropping system does not depend on N subsidies of other agricultural land. The nitrogen flows in this system are summarized in Figure 3.

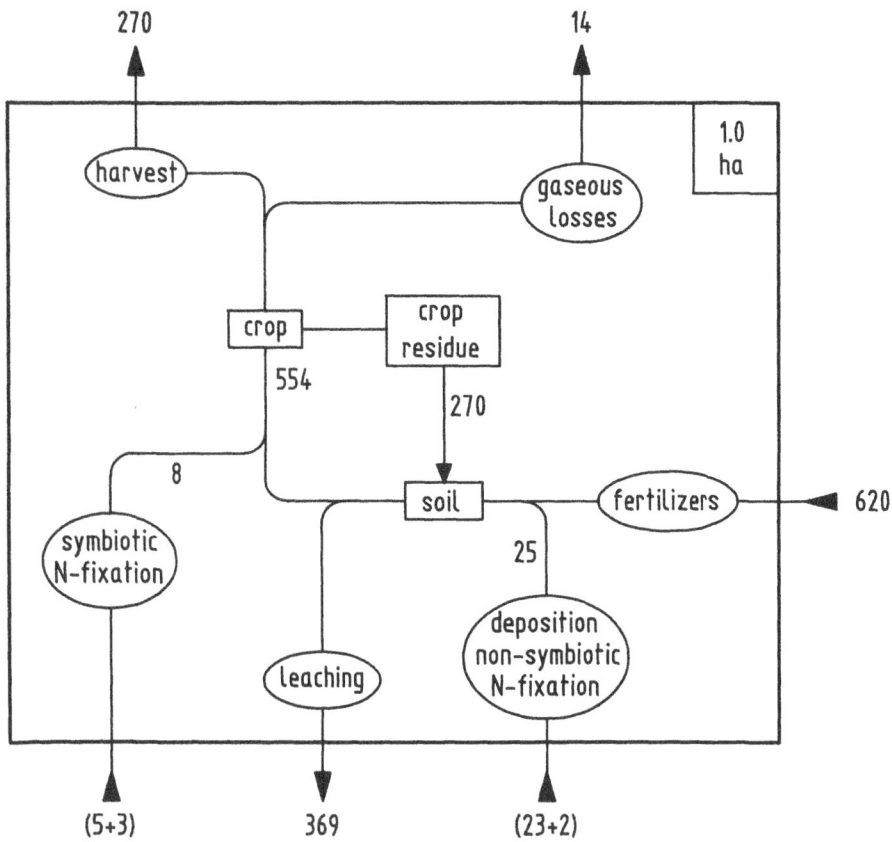

Figure 3. Magnitudes of N-flows (kg yr^{-1}) in an intensive rice cultivation system with high inputs.

3. Results

Table 1 summarizes total and per hectare nitrogen inputs and outputs as well as the ratio between nitrogen in crop/livestock products and other outputs in the three systems. As indicated in the previous sections, on an aggregated system level nitrogen inputs and outputs have been balanced. The exact amounts shown for the different flows are undoubtedly open to discussion, nevertheless the magnitudes of system inputs and outputs may support the following conclusions.

TABLE 1. Total and per hectare N-inputs and -outputs (in kg) for three systems: A = shifting cultivation (20 ha); B = crop/livestock system (2 ha); and C = intensive rice cultivation (1 ha).

	A (20 ha)		B (2 ha)		C (1 ha)	
	total system	per ha	total system	per ha	total system	per ha
N-inputs						
Deposition	460	23	46	23	23	23
Non-symbiotic fix	40	2	4	2	2	2
Symbiotic fix	9	-	6	3	8	8
Fertilizers	-	-	60	30	620	620
Sedimentation	pm	pm	pm	pm	pm	pm
Total	509	25	116	58	653	653
N-outputs						
Crop products	0 (60)[a]	0 (3)[a]	26	13	270	270
Milk/meat	-	-	8	4	-	-
Leaching	118	6	67	33.5	369	369
Conversion losses	4	-	1	0.5	-	-
Gaseous losses	387	19	14	7	14	14
Erosion	pm	pm	pm	pm	pm	pm
Total	509	25	116	58	653	653
N-products/ N-outputs (= N-inputs)		0.12		0.29		0.41

[a] N in crop products is not considered as part of the aggregate calculation.

Total N-flows are greatest in the modern monocropped system, closely followed by the shifting cultivation system by virtue of its large area. However, at a per hectare basis, input/output flows in shifting cultivating are less than 50 % of those in the mixed cropping system, and only 4 % of those in the modern monocropped system. Also on a per hectare of rice crop basis, deposition is the major outside nutrient input in shifting cultivation, whereas its importance in the mixed cropping system is already surpassed by the (small) amount of fertilizer applied, and entirely dwarfed by the use of fertilizers in the rice monocropping system.

On the output side, gaseous losses, mainly through burning, determine the balance of the shifting cultivation system. Although the exclusion of erosion from the balance may improve the picture somewhat for the mixed cropping system, it is clear that under increasing land use intensification the losses through leaching show considerable growth, culminating in the modern system. With the exception of gaseous losses, roughly speaking the other half of the nitrogen in the modern monocropping system is exported through crop products.

If efficiency of nitrogen cycling is expressed as the per hectare ratio between nitrogen in crop/livestock products and other outputs of nitrogen (leaching, conversion and gaseous losses), the efficiency increases quite dramatically from shifting cultivation (0.12) to modern rice monocropping (0.41), while the mixed cropping system occupies a middle position (0.29). In other words, increasing land use intensification leads to greater nitrogen flows in the system, coupled with greater (leaching) losses, proportionally to the increases in flows, and a greatly increased efficiency of nitrogen captured for useful production. However, this is true on the condition that N-flows in the agro-ecosystems are in balance. In many cases land use intensification leads to non optimal efficiency, because higher outputs are not balanced by increased inputs.

4. Discussion: quantitative and qualitative disturbances in agro-ecosystems

The comparison of rice based agro-ecosystems shows major differences in the absolute and proportional magnitude of output flows and efficiencies. The shifting cultivation system reaches its boundary of intensification as soon as the cycle becomes too short for wet and dry deposition to counterbalance the losses through burning. The regrowth of the forest will be too limited to support the same rice yield levels. No further intensification of this system is possible. The system has to be altered drastically in order to increase per hectare production.

One could say that a qualitative disturbance of the climax ecosystem is required to uphold production. The only possibility to engender such a change is to replace the burning, as a process whereby locked-up nitrogen is released, by an animal component that digests the vegetation and releases nitrogen in a form available to crops. The amount of land needed to support the cropped acreage thus decreases from 9 hectare of forest to at least 1 hectare of grassland per hectare of crop, on the condition that 60 kg of N are supplied from outside (if not, up to 4 ha may be needed additionally). Intensification in terms of per hectare production in the mixed crop/livestock system is limited to a 1- 4 to 1 proportion of grassland to cropland, because N outputs cannot be balanced by increased manure production.

Again a qualitative change is required for further land use intensification. Intensive rice cultivation without dependence on other forest or grassland areas is only possible if external sources of nitrogen are supplied in the form of chemical fertilizer. Furthermore, better water supply and control will be required for irrigation, implying generally a geographical concentration in low land areas and considerable infrastructural (i.e. energy) inputs.

Several lessons can be drawn from this analysis of the magnitudes of disturbance in rice based agro-ecosystems for the sustainability of agro-ecosystems in general.

Firstly, the approach (which is not necessarily limited to nitrogen, but could be carried out for any other system flow) suggests that both magnitudes of flows and efficiencies have to be considered to understand limits to sustainability. Agricultural intensification means that agro-ecosystems become more open systems with massive disturbances in key processes. The view promoted by many of the Low External Input School (e.g. Altieri, 1987) is that this openness needs to be limited as much as possible, i.e. through the use of crop residues, limited or no chemical fertilizer use etc. The evidence presented here suggests that the issue is rather to concentrate on improving the efficiency through increasing the amount channelled into useful products for man. This implies that rather than limiting openness through decreasing the dependence on external inputs, the real gain is to be obtained through reductions of unwanted and therefore inefficient losses, in particular through leaching. Because erosion was excluded from the consideration here, all unwanted losses are now shown in leaching, although it is known from other work that erosion may be as important a factor (Smaling, 1993). The inclusion of a N-fixing crop in the mixed system will do little to improve overall efficiencies. The introduction of ley farming, whereby the grassland and arable land alternate on the same area unit, might solve the problem of permanent N subsidies to crops from grasslands. To put it simply: agronomic interventions ought to focus on the output side rather than the input side in order to increase production per unit input. This is also of importance to avoid disturbance of the surrounding ecosystems through influxes of e.g. nitrogen.

Secondly, sustainability of individual fields cannot be determined unless the entire system, i.e. all the land required to provide nitrogen, is taken into consideration in the calculation. A similar case could be made for the dependency of intensive livestock production on external feed sources. Statements on the sustainability of individual fields are misleading and should always refer to the appropriate time and spatial scales.

Thirdly, disturbance and replacement of natural flows has to be considered with respect to scale and type of disturbance. Since efficiency of input use is highest at close to potential production levels (De Wit, 1992) the disruptive effects of agro-ecosystems may be limited if their spatial impact is minimal. This implies concentrating agriculture in highly productive areas if high outputs are required or shifting cultivation with adequate fallow periods where low outputs and low man/land ratios are desirable. Un- or underfertilised permanent cropping by small farmers in the (sub)humid tropics forms probably one of the most disruptive agro-ecosystems even if total biomass extraction levels are low.

Last, but not least, agronomy stands to gain from approaches to key problems such as sustainability if concepts from systems ecological theory are applied.

5. Acknowledgements

The authors acknowledge the help of Ir S. Decker for the preparation of the literature search which formed the basis of this paper.

6. References

Deciare, M. (1987) 'Agro-ecology. The scientific basis for alternative agriculture.' Deicer Press. Boulder, Co.

Bartholomew, W.V., Dicier, J. and Deicer, H. (1953) 'Mineral nutrient immobilization under forest and grass fallow in the Deciare (Belgian Congo) region: with some preliminary results on the decomposition of plant material on the forest floor.'

Decare, H.C. (1957) 'Hanunoo agriculture: a report on an integral system of shifting cultivation in the Philippines.' FAO Forestry development papers no. 12.

Freeman, J.D. (1955) 'Iban agriculture, a report on the shifting cultivation of hill rice by the Iban of Sarawak.' Her Majesty's stationary office, London.

Fresco, L.O. and Kroonenberg, S.B. (1992) 'Time and Spatial Scales in Ecological Sustainability', Land Use Policy July 1992, 155-168.

Grist, D.H. (1965) 'Rice.' fourth edition. London, Longmans, Green and co.

Janssen, B.H. (1992) 'De betekenis van nutriëntenbalansen en de gebalanceerde plantevoeding voor duurzaam landgebruik in de tropen', Verslagen en Mededelingen 1992-3, Vakgroep Bodemkunde en Plantevoeding, Wageningen Agricultural University, Wageningen.

Kannaiyan, S., Thangaraju, M. and Oblisami, G. (1982) 'A potential biofertilizer for rice production' in W.S. Silver and E.C. Schröder (eds.), Practical application of Azolla for rice production', Proc. of an International Workshop, Mayaguez, Puerto Rico.

Krul, J.M., Penning de Vries, F.W.T. and Traoré, K. (1982) 'Les processus du bilan d'azote', in F.W.T. Penning de Vries and M.A. Djitèye (eds.), La productivité des pâturages sahéliens, Pudoc, Wageningen, pp. 226-246.

Nye, P.H. and Greenland, D.J. (1960) 'The soil under shifting cultivation.' Technical communication. 51. Commonwealth Bureau of Soils. Farnham Royal.

Pelzer, K.J. (1945) 'Pioneer settlement in the Asiatic tropics: studies in land utilization and agricultural colonization in Southeastern Asia'. American geographic society special publication no.29.

Poels, R.L.H. (1987) 'Soils, water and nutrients in a forest ecosystem in Surinam', PhD-thesis, Wageningen Agricultural University, Wageningen.

Reuler, H. van and Janssen, B.H. (1993) 'Nutrient fluxes in the shifting cultivation system of south-west Cote d'Ivoire. I. Dry-matter production, nutrient content and nutrient release after slash and burn for two fallow vegetations.' (in preparation).

Ruthenberg, H. (1980) 'Farming systems in the tropics.' Oxford University Press, Oxford.

Singh, P.K., Misra, S.P. and Singh, A.L. (1982) 'Azolla biofertilizer to increase rice production with emphasis on dual cropping' in W.S. Silver and E.C. Schröder (eds.), Practical application of Azolla for rice production, Proc. of an International Workshop, Mayaguez, Puerto Rico.

Sivanadyan, K. and Norhayati, M. (1992) 'Consequence of transforming tropical rain forests to *Hevea* plantations'. The Planter 68 (800), 547-567.

Smaling, E. (1993) 'An agro-ecological framework for integrated nutrient management, with special reference to Kenya', PhD thesis, Wageningen Agricultural University, Wageningen.

Watanabe, I., Roger, P.A., Ladha, J.K. and Van Hove, C. (1992) 'Biofertilizer germplasm collections at IRRI', IRRI, Manila, Philippines.

Wit, C.T. de (1992) 'Resource use efficiency in agriculture.' Agricultural Systems 40, 125-151.

DURABLE ECOCOMPLEXITY: BEYOND THE THRESHOLD OF A NEW CENTURY

R.A.A. OLDEMAN
Chair of Silviculture & Forest Ecology
Wageningen Agricultural University
Dept. of Ecological Agriculture
P.O. Box 9101
6700 HB Wageningen
The Netherlands

ABSTRACT. Architectural and functional complexity of the ecosystem is the ecological mirror of the biological diversity of its species composition. Historically, modern agriculture solved acute and chronical problems of hunger, shortages and poverty by unraveling local ecosystems into simplified parallel ecosystems dominated by one crop species. These production sectors (e.g. potato, pigs, wood or fish) today are ecologically stressed and economically unbalanced. The present paper examines the main lines of the ecological architecture of new, durable, complex mixed agrosystems without sectorial discrimination. It also outlines new ways to assess yields and usefulness to be expected from such systems. Web models with multiple inputs and sing-around interactions offer better prospects for this aim than classical yield tables or feed-back models.

1. Complexity, biological diversity, durability

1.1. COMPLEXITY AND HOLONS

Usually, matters of biological complexity are addressed in terms of pattern and process, or architecture and dynamics (Watt, 1947; Oldeman, 1990). Thinking this through, both are aspects of living systems in four dimensions. In terms of time, every slow process is structuring faster processes. The slower process can be said to determine a structure within which the faster process unrolls itself. In terms of space, a larger structure or system contains smaller systems. The small ones find their place and interact within the larger system. Interaction is a process. It links time models to space models. It has in its turn two components, i.e. interactions among small systems as parts of a big system, and interactions among parts of each smaller system.
Köstler (1967, p. 48) expresses this as a two-faced nature of all sub-systems: "One is the face of the master, the other the face of the

151

P C Struik et al (eds), Plant Production on the Threshold of a New Century, 151–158
© 1994 *Kluwer Academic Publishers*

servant. This *"Janus effect"* is a fundamental characteristic of sub-wholes in all types of hierarchies." On the same page, he states that there is no satisfactory word for such subsystems or sub-wholes in modern languages: "It seems to be preferable to coin a new word to designate these nodes on the hierarchic tree which behave partly as wholes or wholly as parts, according to the way you look at them. The word I would propose is 'holon', from the Greek *holos* = whole, with the suffix *on* which, as in pro*ton* or neut*ron*, suggests a particle or part."

The ecosystems regarded in this article are agricultural fields, meadows, home gardens or forestry stands, i.e. holons in a landscape system managed by humans. Let us provisionally denote these artificial ecosystems as "agro-units". In the hierarchy of living systems, they are the holons building the human-made landscape. In their turn, agro-units are built by holons of a lower order, with prescientific names such as "crop(s)", "pest(s)", "weed(s)", "soil(s)", "supporting species" and so on.

The degree of complexity of an agro-unit is defined by the complexity of each of the holons which build it, by the number of such holons, and by the number of hierarchical levels. Expressed as a spatial structure, this complexity is represented in an architectural model, architecture being structure at a particular hierarchical level (Oldeman, 1990). Expressed as a structured process it is represented as a web-model of interactions among holons forming an open system.

Along these lines, agro-unit complexity can be compared immediately with humanless ecological complexity at the same scales. The key to validation of such comparisons is the search for human-made versus created holons of the same degree of complexity, at the same hierarchical scales and occurring in the same biotope.

1.2. BIOLOGICAL DIVERSITY AND HOLONS

Biological diversity is most often assessed as species diversity (Wilson, 1992). This approach has an agricultural and pharmaceutical bias (Oldeman, 1989). The exact taxonomic identity of crops, cattle, medicinal plants and other useful biological resources determines the uses, tastes and risks (poisons, pests!) of organisms to be culti-vated, collected or combated. The pure science of biology developed on this basis. Hence most assessments of biodiversity are given as numerical indices calculated from counts of identified species populations, related with site properties of vegetation, soil, water and climate. These correlations are explained in an evolutionary context.

Species diversity models can be simply linked to architectural models of ecosystems by regarding each species population as a holon. The eco-unit or agro-unit, as a superior holon hosting a species, can be mapped. Indeed, the fit has to be a complete mirror image. For each species population there is a corresponding niche in the eco-unit or agro-unit considered. The projection of these niches on a spatial image of the vegetation unit is a simple kind of map, often with contorted contours of complicated biotopes (cf. Figure 1).

This approach is hindered by the huge number of species. In the simplest wheat field or pine stand there are thousands, counting microscopic ones too. Therefore, no individual species populations but aggregates of species are used most often as subsystems in modeling eco-units. Such aggregates, e.g. crops, pests animals, decomposers, pathogens, mosses, game animals, trees, often are prescientific and highly practical. Mapping such holons, as guilds for instance, on an eco-unit or agro-unit map, is a strongly diagnostic method to assess both potential species diversity and ecosystem health of the units mapped.

1.3. DURABILITY, SELECTIVE DURABILITY, SUSTAINABILITY

An ecological holon can be compared to a hotel offering different kinds of rooms. The agro-unit as a holon, offering a set of biotopes each occupied by a species or an aggregate of species, e.g. a crop, then is the key to assess durability. The agro-unit holon and its smaller component holons are linked by the Janus effect. The larger one suvives by maintaining biotopes for the smaller ones. The smaller ones survive by not disturbing the architecture of the larger one containing their biotopes. Succeeding time frames of architectural maps then show a kinetic view of the ephemeral or durable nature of such configurations.

The temporal solution of the series in time determines the kinds of durability which can be percieved. Seasonal durability is visible with daily time frames, successional durability with decennial time frames. The issue is well known in remote sensing.

In land use, *selective durability* has been applied and perceived since times immemorial. Holons with ephemerality to be enhanced are of course weeds, pests, diseases etc. Holons with durability to be enhanced are crops, cattle, timber stands, decorative plants and such. Per historic period, the prevailing paradigm of a farm or other estate is closely related to special patterns of selective durability. Extreme reduction of durability leads to extinction, extreme enhancement tends to perpetuity. Selective reduction versus enhancement of durability determine the concomitant technology of a paradigmatic farm in its era.

In our regions and in recent centuries, farms and estates were considered to be permanent. Permanence is a severe case of durability. Such durability is artificially induced and technically *sustained*, whence the concept of *sustainable land use*. Emotionally, this is protection against change of an inherited estate, applying inherited season-bound techniques. This does not differ overmuch from UNCED-Rio's sustained development, defined as "land use development without prejudicing future human generations". In matters of sustainability, forestry takes care of estates with long-lived, durable woody crops.

Sustainability and durability were discussed and consciously practiced in forestry during two centuries and more. It is striking that sustainability has been successively, not simultaneously, applied to many attributes of forest estates. Sustainability follows fashion. Items sustained by foresters are e.g. forested surface, local wood production, site (wood) productivity, financial revenues, jobs, capital, yield increase (in reafforestation), touristic or ethical values, landscape protection or genetic richness.

Our modern sectorial land use is ill. Its disease is endemic and is precisely sustainability itself, wherever it has narrowed down selectively to species-poor and unbalanced sets of edible or marketable organisms. Ecologically, this is forcing the issue till death follows.

2. Thresholds in plant production

Wageningen Agricultural University originates from an agricultural college founded in 1877 at a crucial point in history. Worldwide, the industrial revolution had caused a fast-growing gap between rural and urban life, in a series of agricultural crises. Although the young school knew strife between factions favouring farmcraft versus those preferring science as a basis for rural development, the basic attitude of the times was towards mass production and classical reductionist science. On the threshold of the XXth Century, all was ready to reshape the face of the rural areas by doing away with mixed farming, replacing it by mass production of a reduced number of plants. In the Netherlands, Wageningen embodies the huge success of this approach in real problem solving, sustained since at least 75 years.

The very narrow-mindedness of reductionist science applied to rural mass production fueled an extraordinary performance. A "factor" is a concept typical of classical science. The aims of farms or estates were reduced to sustainable production of one or a few products seen as "success factors". This gave birth to differentiated rural sectors; agriculture, horticulture, cattle breeding or silvi-culture. Today, these sectors have narrowed down still further to potato, pork, wheat, wine, wood, fish, dairy and many other "commodities".

Once in the XXth Century, modern industrial-style land use developed rapidly. The factors of its development became steadily narrower and more powerful. Some hundred or so molecules, from N, P and K to biocides and DNA were used as the invisible but powerful levers in building huge land areas, all-but monopolized by one crop species from horizon to horizon. Today, at the threshold of the XXIth Century, new and still more powerful chemical and biochemical levers have been created. Also today, the by-effects of these land use methods have become apparent. They indeed feed the world, but at the high ecological cost of genetic and physical erosion and the high economical cost of ever-increasing physical, chemical and biological inputs. Human societies are increasingly worried.

At the threshold of the new century, in analogy with the earlier "fin de siècle", a strong doubt emerges. Should actual, existing land uses be improved and continue, or should wholly new ones be designed and implemented? Either option is very risky. Rather the devil you know than the one you don't? In any case, any innovative option must be put to the most severe test before perhaps introducing it gradually. However, new land use is heralded both by the trend towards tailor-made instead of mass-produced commodities, and by new non-linear scientific paradigms emphasizing structured information rather than mass.

3. The folded forest model and mixed agrosystem architecture

3.1. THE FOLDED FOREST MODEL

The architectural analysis of forested landscapes is the base of the folded forest model (Figure 1). The theory of forest architecture as an aspect of hierarchical living systems was decribed by Oldeman (1990). The reader is referred to this book for background.

The folded forest model considers vegetation as a green blanket with folds, covering the landscape. Figure 1 shows this model for a managed landscape, comparable to the 'écocomplexe' of Blandin and Lamotte (1988, p. 553): " ... for ecosystem studies, it is essential that research especially bears on pinpointing emerging properties that are explained by spatial mosaic structure. Only then can these mosaics (i.e. ecocomplexes, RAAO) be considered as entities at an organisation level above the ecosystems" (free transl., RAAO). The underlying holon hierarchy is visualized in Figure 1 by zooming in from each organisation level to the one below. Each higher level holon (small star) contains lower-level holons (larger star), as shown by arrows. Figure 1 (a,b) indicates two levels of biological communities, i.e. two ecosystem levels. In Figure 1 (c - e) the levels correspond to organisms and organ complexes. All holons make folds following their own organisation pattern. Five such patterns, expressed by their architecture, are sketched in Figure 1.

Expressing this hierarchical nesting by folds, linked to plant architecture and growth dynamics (Oldeman, 1990) has advantages. First, it allows to replace the image of "vegetation mosaics as the sum of isolated patches" in an island-biogeography sense (for forests cf. Harris (1984)), by a landscape holon ("écocomplexe") explained by interacting "patch" holons (eco-units, agro-units). Second, it follows the same architectural criteria from top to bottom. Current ecology usually shifts from morphological structure to population numbers as a criterion, at the transit from organism to community (Figure 1 b to c). Third, a green fold model directly concerns primary production. Folding a leafy surface increases photosynthesis proportionally. This makes the model more universal than a mere forest model with its big folds. The model can be also validated for low and finely folded vegetations like savannahs, rice fields or vegetable gardens. Fourth, the recurrent folding at ever more detailed scales suggests some self-

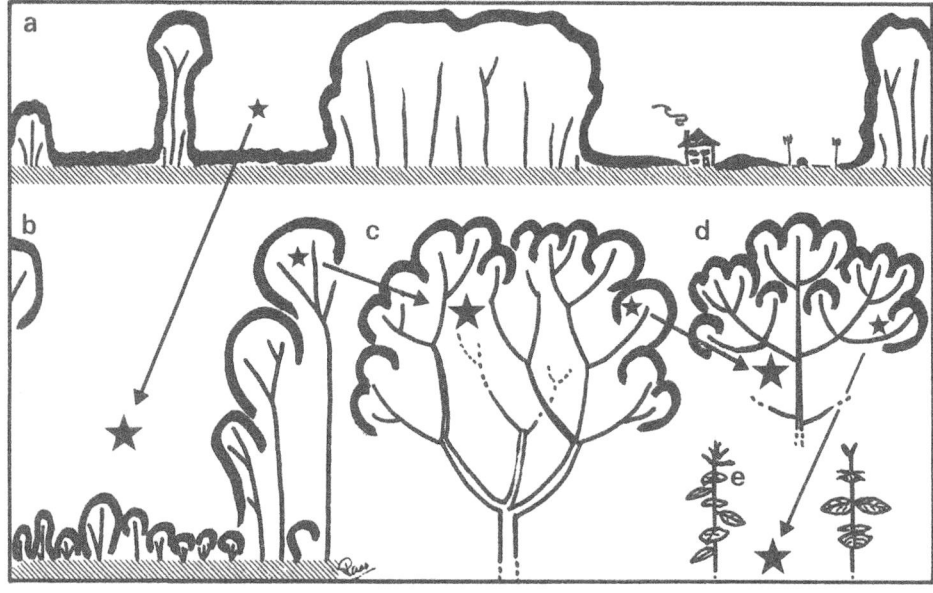

Figure 1. The folded forest model. Vegetation cover is a photosynthetic blanket (fat black lines). It tends to capture more light by growing folds, increasing its surface. From the landscape level (a) on, each fold is folded again by plant crowns (b), each crown folded by sub-crowns (c), each sub-crown by branches (d) and each branch by leaves, arranged according to some term of the Fibonacci series (e). Each large fold (a) offers biotopes to different sets of plant species (b - e), according to size and architecture of the large fold (a). Border zones between interacting large folds offer many diverse biotopes for high biodiversity. Land use reshapes larger folds by land management and smaller folds by concomitant species choice, selective durability of desired versus undesired holons being criteria. Sustainable, cheap durability of the whole agrosystem will be paramount in the next century.

similarity. The self-similar patterns finally meet an algorithm, the Fibonacci series, and the spirals it characterizes. Hence the folded forest model is promising for simulation with fractal mathematics (Oldeman, 1992).

The folded forest model compares easily with maps and models in landscape ecology and complex land use design. Baudry (1988) shows compatible landscape maps drawn by hand, showing living hedges, like high and narrow folds on Figure 1a (left-hand side). The a-b-c-sequence in Figure 1 covers the same holons as the SILVI-STAR model by Koop (1989) or the study on Northern-hemisphere beech forests by Peters (1992). A c-b-sequence is simulated in Leersnijder's PINOGRAM (1992). The whole five-level hierarchy of Figure 1 is implicit in Mollison and Slay (1992), graphical designers of permaculture systems.

Comparing Figure 1 with the results reported by the above authors, the folded forest model can clearly be used for the design of all land use systems. Extreme simplification, e.g. creating one large low fold from horizon to horizon (Figure 1a), subdivided in manageable fields (Figure 1b) all stocked with one and the same crop species (Figure 1c-e) leads to modern monocropping. Conversely, extreme complexification leads to a pristine-forest-like architecture, in which a highly diverse set of crops can find optimal conditions but without mass-production by any one of those crops (for design see Oldeman, 1983).

3.2. BIOLOGICAL DIVERSITY, SPECIES, CROPS AND THE FOLDED FOREST MODEL

In a first approach, the *dynamics* of the folded landscape are left unconsidered. Its *architecture* (Figure 1) shows that over a short time-span without important changes, spaces with divergent qualities exist in all holons. Such "qualified spaces" could be mathematically described as niches for many species, big organisms at the level of Figure 1a, small plants and animals among the branches (Figure 1c,d) or organs (Figure 1e).

Size and organisation level are loosely linked only. Small organisms can find appropriate life conditions in certain small biotopes originating from weed and shrub interaction (Figure 1b, bottom fold) or from the interaction among branches (Figure 1d).

4. References

Baudry, J. (1988) 'Structure et fonctionnement écologique des paysages: cas des bocages', Bulletin Ecologique 19(4), 523-530.
Blandin, P. and Lamotte, M. (1988) 'Recherche d'une entité écologique correspondant à l'étude des paysages: la notion d'écocomplexe', Bulletin Ecologique 19(4), 547-555.
Borcard, D., Legendre, P., and Drapeau, P. (1992) 'Partialling out the spatial component of ecological variation', Ecology 73(3), 1045-1055.

158

Harris, L.D. (1984) 'The Fragmented Forest.' The University of Chicago Press, Chicago.

Koestler, A. (1967, 2nd ed., 4th print.) 'The Ghost in the Machine.' Pan Books, London.

Koop, H. (1989) 'Forest Dynamics, SILVI-STAR: a Comprehensive Monitoring System', Springer Verlag, Heidelberg.

Leersnijder, R.P. (1992) 'PINOGRAM: a Pine Growth Area Model.' Publ. by the author, c/o WAU, Wageningen.

Mollison, B. and Slay, R.M. (1991) 'Introduction to Permaculture', Tagara Publ., Tyalgum (NSW, Australia).

Oldeman, R.A.A. (1983) 'The design of ecologically sound agroforests', in P.A. Huxley (ed.), Plant Research and Agroforestry, ICRAF, Nairobi, pp. 173-207.

Oldeman, R.A.A. (1989) 'Dynamics in tropical rain forests' in L.B. Holm-Nielsen, I.C. Nielsen and H. Balslev (eds.), Tropical Forests: Botanical Dynamics, Speciation and Diversity, Academic Press, London, pp. 3-21.

Oldeman, R.A.A. (1990) 'Forests: Elements of Silvology', Springer Verlag, Heidelberg.

Oldeman, R.A.A. (1992a) 'Architectural models, fractals and agro-forestry design', Agriculture, Ecosystems and Development 41, 179-188.

Peters, R. (1992) 'Ecology of Beech Forests in the Northern Hemisphere', publ. by the author, WAU, Wageningen, ISBN 90-5485-012-4.

Watt, A.S. (1947) 'Pattern and process in the plant community', Journal of Ecology 13, 1-22.

Wilson, E.O. (1992) 'The Diversity of Life', Harvard University Press, Cambridge Mass.

BIODIVERSITY AND THE WORLD'S FOOD CRISIS

OTTO T. SOLBRIG
Bussey Professor of Biology
Department of Organismic and Evolutionary Biology
Harvard University
22 Divinity Ave., Cambridge, MA, 021381
USA

ABSTRACT. During the next fifty years the world's population is expected to double. In order to avoid catastrophic famines the world's food production also must double. Although this is technically possible with present high-input methods, such technology is probably not affordable by poor tropical countries with high population growth. A more environmental friendly, and less demanding but equally productive technology based on time tested, labour intensive, tropical systems must be developed to avoid both famine, landscape deterioration, and biodiversity loss.

1. Introduction

There is increasing anxiety among scientists and the general public about the loss of species, the transformation of ecosystems, and the reduction in the genetic diversity of crops and the effect that these losses can have on the functioning of the biosphere. These concerns arose because of the drastic transformation of natural landscapes taking place all over the world in the last fifty years, particularly in the tropics. Many scientists suspect that an extensive reduction in species diversity may lead to loss of ecosystem stability and function. The issue has been addressed from historical, economic, social, ethical, and ecological points of view (Norton, 1987; McNeely, 1988; Oelschlager, 1991; Solbrig, 1991; Wilson, 1992; Solbrig and Solbrig, 1994).

Landscape transformation and loss of biodiversity are closely tied to population growth and the increase in food production through increases in the arable surface, a continuing process. According to the United Nations the population of the world will double in the next fifty to a hundred years. If massive starvation is to be avoided food production will also have to double in that period. On paper such a challenge, although formidable, looks attainable, though its effect on biodiversity is unpredictable. Even though most suited agricultural land is being used today and there is very little room for expansion,

159

P C Struik et al (eds), Plant Production on the Threshold of a New Century, 159–168

average yields are still low and, at least in theory, with enough knowledge and capital it should be possible to double average yields in the next fifty years. The 18th and 19th century in Europe also saw a fast rise in population, accompanied by tremendous poverty, starvation, and human suffering. Yet the agricultural and industrial revolutions increased food production and created the industries that provided the wealth and the jobs on which our present world is based (Kennedy, 1992). A similar process could take place during the 21st century in the Third World.

During the industrial revolution, population, agricultural knowledge, and capital were all found in Europe. In today's world, population, resources, agricultural land and capital are not evenly distributed. While much of the human population is found in the tropics, the best agricultural land is concentrated in temperate areas. To increase agricultural yields special knowledge and resources are needed. Most of that knowledge and much of the financial resources are present in the industrialized world; yet most population growth is taking place in tropical countries, with limited agricultural means, traditional agricultural practices, and no capital. Because the citizens of these countries are poor, they don't have the money to purchase the agricultural surplus of the industrialized countries, and since these are engaged in a competition among themselves for commercial supremacy, it is unlikely that they will provide much financial aid. Where the green revolution has increased production in countries such as India, much of the displaced population has not obtained alternative employment and has been relegated to the slums of large cities.

Some tropical countries are faced with the prospect of massive famines. Most of their poor are already severely undernourished, in poor health, and badly housed. These problems could lead to civil strife as can already be seen in some African countries, and to a massive wave of migration from the tropics to the industrialized world. The vanguard of this migration is already in Europe, in the United States, and increasingly in the industrialized countries of Asia. This migration wave is likely to lead to persecution of immigrants, xenophobia, and ugly racially and culturally based confrontations. We are already seeing the beginnings of these confrontations. The pressure on the land will also increase deforestation, land degradation, and biodiversity loss.

Is there anything that can be done to reverse this bleak and depressing prediction? Can we scientists use our knowledge to ameliorate the human condition? Unfortunately, no technical solution will work if the world's population does not stabilize. Demographers and social scientists tell us that population growth will stabilize only when conditions ameliorate and people can see a better future. Only a partnership between institutions in the industrialized and the third world can reverse this vicious circle between poverty and population growth and create a better future for the world.

In this paper I argue that if we wish to preserve some of the world's biodiversity on which we depend, our own selfish interests dictate that we engage in trying to help solve the problems of poor countries.

2. Agricultural productivity and population growth

Most of the rich countries have a highly productive agricultural sector. Yet that sector contributes only a small fraction of the overall domestic gross product, and employs a small proportion of the population (Table 1). In rich countries most of the population is urban and is engaged in industrial or service activities. In contrast, in poor countries, most of the population is rural, and agricultural productivity is low. Rich industrialized countries have low population growth, in some cases even negative population growth, and normally the economy grows at a sufficiently fast rate to absorb population increases and enlarge living standards. Poor, mostly rural countries have high population growth and the economy grows at rates that are less than population growth resulting in decreases in the living standard of the population. The increased population must be absorbed principally by the rural sector. Yet, this sector is running out of land. The result is a spill-over of population into marginal land, and an intensification of use of existing land. This transforms the landscape and degrades the land. Much of the deforestation (but by no means all) in the tropics and accompanying loss of biodiversity can be traced to population-driven agricultural expansion. As population expands, food production is insufficient, and famines arise when weather or civil unrest reduce food production. Another contributing factor is uneven land tenure. This problem is particularly acute in Central America and tropical South America, where less than 10 % of the population owns over 80 % of the land.

Agriculture differs from most industries in that its basic resource -- land -- is fairly inelastic and so far no viable substitute for it has been found or is likely to be found. There is only so much suitable land, and by now most is occupied. Historically, most increases in agricultural output have come from occupying new land. Although yields have increased slightly in the historical past, only in the last two centuries, but especially in the last hundred years, have agronomists learned how to increase yields on old land by the use of improved varieties and massive additions of fertilizers, especially nitrogen and phosphorus.

Like in other industries, increases in agricultural productivity result in a reduction in labour. In industrialized countries the excess rural labour is absorbed by manufacturing. This accounts for the high agricultural productivity and low percentage of workers in agricultural activities that is characteristic of developed countries. But in poor rural countries without a developed industrial sector capable of absorbing displaced rural labour, increased agricultural productivity through the use of labour saving inputs exacerbates existing inequalities as people are displaced from the soil and forced into the squalid existence of city slums, where the problems of under-nourishment, disease, and lack of human services are worse than in the country-side.

TABLE 1. Average total and agricultural gross domestic
product (in US$) of countries with different income levels.

| | Total | Gross domestic product contribution of agriculture (%) | |
		Agriculture	Percent
Low income	621 260	198 803	32
Middle income	1 740 010	261 002	15
Industrial market economies	10 451 880	313 556	3

To resolve this problem three aspects of the economy and the
sociology of tropical countries must be addressed. One is population
control. Although distasteful to many, the time might have arrived
that tropical countries should consider more drastic population
control measures such as were adopted by China. A second aspect is
industrialization. Unless domestic industries can be developed to
employ displaced rural labour, increased rural productivity will only
mean large and deteriorating urban slums. What is needed, obviously,
is a population with enough purchasing power to consume surpluses from
a more efficient and productive rural sector. The third, and last
aspect is new approaches to agriculture that do not rely on monocultures and the massive inputs of capital that have been so successful in temperate agriculture. Only in that way will we be able to
preserve the many species of plants, animals, and microorganisms on
which we rely for ecological services.

3. A model of land use and productivity

In 1798, at the start of the agricultural and industrial revolution of
the 18th and 19th centuries, Thomas Malthus predicted that the world
soon would grow out of food. He observed that the population of Europe
was growing exponentially but that growth in food production was only
increasing linearly. At the time that Malthus wrote his famous essay,
most increases in agricultural production resulted from enlarging the
arable surface. Yet as Malthus was writing, a revolution in land use
was taking place, involving crop rotation, increase use of artificial
fertilizers, and improved varieties, that would increase yields. With
the introduction of scientific agriculture in the 19th and especially
the 20th century, yields have increased dramatically, especially in
Europe and North America. These increases in yields transformed
Malthus' predicted food shortage into a politically embarrassing and
difficult to solve food surplus. The so-called "Green revolution"
transferred the technology of scientific, high-input agriculture to
parts of the Third World with equally spectacular results and avoided
still another predicted famine, that predicted by the "Club of Rome"
(Meadows et al., 1972). Yet their analysis is relevant.

Figure 1 shows the relation between agricultural yields and land use over time. Of the total land surface of the world, only 3.2 billion hectares are suitable for agriculture. Of these, a small part, about three-hundred million hectares can be used with little additional input. Another 5-7 hundred million hectares are suitable with some inputs, while the bulk of the potentially arable land is either to steep, to dry, or to infertile to be used as such and requires great investments in works, irrigation, or soil preparation before it can be used. If used without proper preparation marginal land degrades and can be lost totally for agriculture.

To increase world agricultural production in a sustainable way requires capital inputs, be they to improve marginal land or to increase productivity on existing good agricultural land. In Figure 1, I have graphed the increase in the arable surface in the last 100 years. Because of the limitations of the graphical representation it is assumed that all good agricultural land is used before any marginal land is used, which is not correct in a strict sense. Geographical constraints dictate that in high density countries marginal land will be occupied while there still exists good agricultural land in areas where the population is not so dense. I have also projected land use to 2100 under various models of average yields. It can be seen that if we triple average yields on existing land in the next hundred years we can feed the population of the world without increasing the arable surface. Tripling yields is technically feasible, especially in the Third World where yields are still very low. Such strategy would preserve landscapes and species diversity.

However, although such a strategy is technically feasible using high-input agriculture, it is not workable in practice because of lack of capital, and knowledge in many Third World countries. Furthermore, unless the displaced rural population were to obtain alternative employment, there would be no market for the additional food, as people would starve in the midst of plenty for lack of money with which to buy the food. This is of course already the case. Europe and North America have a food surplus that cannot be bought by scores of poor undernourished people in the tropics. Instead, they keep expanding their unproductive and environmentally inappropriate agriculture into increasingly marginal land.

4. Deforestation and land use

Deforestation in tropical countries is a critical problem. Table 2 shows the rates of deforestation in several countries according to the latest figures available. We can observe that deforestation is greatest in those countries that also have high population growth rates and low agricultural productivity. Although deforestation is not solely the result of peasants in search of new land, or firewood, both these factors are strong contributors.

Evolution of cropland use

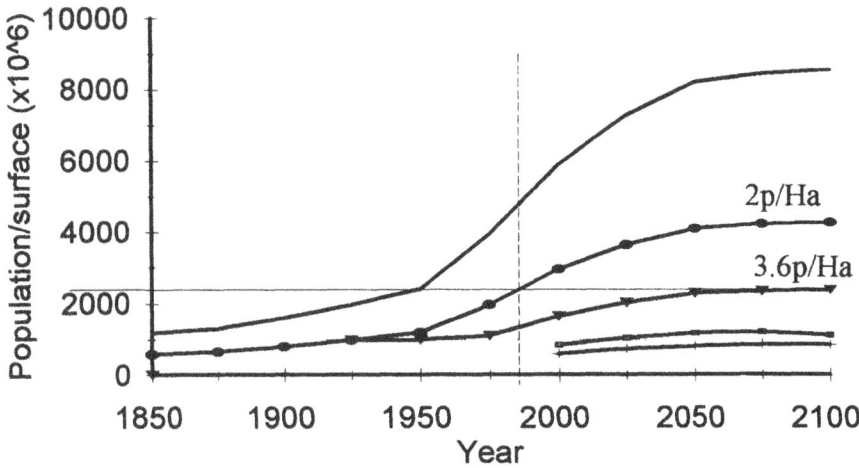

Figure 1. Relation between demographic growth and land use over time.
The upper curve is the actual and projected demographic growth between
1850 and 2100. The second curve (solid circles) is the land needed to
feed the human population if only two persons can be fed for each
hectare under cultivation. The next curve (solid triangles) is the
land needed if 3.6 persons can be fed by ha. This is the present world
average. The next two curves are projections under the assumption of a
doubling (7 p/ha) or tripling (10 p/ha) of yields. The lower
horizontal curve is the limit of the better lands; the upper
horizontal curve is the absolute limit of agricultural land in the
world.

TABLE 2. Deforestation rate of selected countries (Net annual percent).

Country	Rate of deforestation	Country	Rate of deforestation
Ivory Coast	6.8	El Salvador	5.5
Brunei	3.6	Martinique	3.6
Togo	3.4	The Gambia	3.2
Nicaragua	2.8	Ecuador	2.5
Malawi	2.4	Niger	2.3
Honduras	2.2	Nigeria	2.0
Guatemala	1.9	Haiti	1.9
Philippines	1.8	Colombia	1.7
Benin	1.3	Malaysia	1.2
Mexico	1.2	Bangladesh	1.1
Madagascar	1.1	Thailand	1.0
Guinea	1.0	Uganda	0.9
Burkina Faso	0.9	Venezuela	0.9
Ghana	0.8	Mozambique	0.8
Laos	0.8	Kenya	0.8
Sudan	0.7	Panama	0.7
Somalia	0.6	Chad	0.6
South Yemen	0.6	Iraq	0.5
Sierra Leone	0.5	Mali	0.5
Jamaica	0.5	Vietnam	0.4
Brazil	0.4	Cameroon	0.4
Ethiopia	0.4	Peru	0.4
Trinidad/Tobago	0.4	Rwanda	0.4
Tanzania	0.3	Zambia	0.3
Dominican Republic	0.3	East Germany	0.2
Argentina	0.2	Zaire	0.2
Paraguay	0.2	Angola	0.2
China	0.2	Yugoslavia	0.2
The Congo	0.1	Bolivia	0.1
South Korea	0.1		

Source: Social Indicators of Development, Economics Department, International Bank for Reconstruction and Development/The World Bank, Washington, D.C.

In dense forests much deforestation is by organized businesses, sometimes local but more often multinational concerns, with the blessing of national governments. In the measure that deforestation creates national savings that are invested in development, and if logging is done rationally so as to minimize erosion and landscape modification, such activities are to be encouraged. In the last century the United States cut down much of their natural forests to create the capital for industrialization. However, such is not the case in many tropical forest operations. Logging is often done with little regard to the impact on local landscapes, and multinationals usually export the profits together with the logs, and only a small group of local officials and concessionaires profit from such operations.

In dry forests and savannas deforestation is done mainly by peasants in search of firewood. In these regions, fuel shortages are critical and well documented. Some deforestation is associated with charcoal production for use by local industries, such as in the state of Minas Gerais in Brazil, where large quantities of charcoal are used in the iron and steel industry. Deforestation in dry forest and savannas degrades the land, increases erosion, and robs cattle of shade. Acute shortages of fuel-wood, especially in parts of Africa and India, are another manifestation of the reduced standard of living that the local population must endure. The very high population growth rate in some of these regions make conservation almost impossible.

Agroforestry and more rational use of the resources in dry forests and savannas, and conservation, and local control of timber in moist forests are the most promising policies to reduce deforestation. Examples of successful programmes are the "Salta Forestal" programme in the dry forest of northern Argentina (Whyte and Burton, 1983; Solbrig, 1986) and the programmes of Malaysia and Sarawak to manage their humid tropics.

5. Conclusions

The transfer of temperate region agricultural technology to tropical regions, what is called the "green revolution" was a huge technical success, as yields in cereals increased dramatically. National output expanded in countries such as India that went from net importers to self sufficiency. Socially, the "green revolution" was a mixed success. It brought increased wealth to those farmers with the education and the capital needed to take advantage of the new approach. On the other hand it displaced many peasant farmers who lost access to the land and were forced to cities. In some sense, this parallels events during the European agrarian revolution of the 18th and 19th centuries. But while Europe was industrializing during that period and was capable of absorbing the influx of people made redundant in the rural sector, the rate of industrialization in the Third World is too low to absorb the excess population. Furthermore at no point were the rates of population growth in Europe as high as they are in some Third World countries today.

The challenge is to increase agricultural production, reduce landscape modification and biodiversity loss, create alternative employment for the excess rural population, and reduce poverty. One scenario is the European model of development, which we have been discussing, favoured in countries such as India, China, and Brazil. It involves the use of modern, high-input techniques to increase agricultural outputs, while at the same time increasing the rate of industrialization by state subsidies. Unless and until population growth rate stabilizes, such a strategy does not produce sufficient employment, and results in massive deforestation and landscape modification. Furthermore this strategy is possible only in large countries with access to sufficient capital either through savings or borrowing.

Poor, rural countries with ineffective or corrupt governments are unable to execute such a strategy. In these countries, an alternative strategy of improving yields through the use of suitably modified traditional practices, is called for. Intercropping, agroforestry, use of green manures, crop rotation, and nomadic pastoral systems are some of the many techniques that can be used to increase yields without displacing large numbers of people. Together with it cottage-type industries should be encouraged that can provide alternative rural employment and be the basis for savings with which to built more traditional industries. But for either strategy to succeed, population growth rates must be brought under control.

Unless the twin problems of population growth and food production are resolved, rural populations will spill over into every type of land in use of subsistence. The result will be tremendous landscape modification and loss of biodiversity to the detriment of all humankind. Furthermore, as the situation becomes more hopeless, there will be massive and unstoppable migrations from the tropics to the temperate regions. In the words of the president of Mexico, Carlos Salinas de Gortari, either the industrialized world helps Mexico improve its economy or it will have to import its people. And not only the economy of Mexico, but that of most tropical countries.

6. References

Kennedy, P. (1992) 'Preparing for the twenty-first Century', Random House, New York.

McNeely, J.(1988) 'Economics and Biological Diversity. Developing and using economic incentives to conserve biological resources,' IUCN, Gland, Switzerland.

Meadows, D.H., Meadows, D.L., Randers, J. and Behrens, W.W. (1972) 'The Limits to Growth,' Universe Books, New York.

Norton, B.G. (1987) 'Why Preserve Natural Variety?' Princeton University Press, Princeton, N.J.

Oelschlager, M. (1991) 'The Idea of Wilderness'. Yale University Press, New Haven, Conn.

Solbrig, O.T. (1986) 'The advance of the agricultural frontier in the Gran Chaco Area of South America: An interesting research opportunity.' Biology International 13:2-5.

Solbrig, O.T. (1991) 'Biodiversity.' Scientific Issues and Collaborative Research Proposals, Mab Digest No. 9, UNESCO, Paris.

Solbrig, O.T. and Solbrig, D.J. (1994) 'So Shall You Reap,' Island Press, Washington, D.C. (in press).

168

Whyte, A. and Burton, A. (1983) 'The human ecology of the dry Chaco in the province of Salta, Argentina.' Report, Institute of Environmental Sciences, Toronto, Canada.

Wilson, E.O. (1992) 'The Diversity of Life,' Harvard University Press, Cambridge, Mass.

VON LIEBIG'S LAW OF THE MINIMUM AND LOW-INPUT TECHNOLOGIES

Q. PARIS
Department of Agricultural Economics
University of California
Davis, CA 95616
USA

ABSTRACT. Using the results of two corn experiments, it is shown that the law of the minimum provides a good interpretation of crop response to growth factors. In order to achieve this conclusion it is necessary to extend the traditional formulation of the law of the minimum based upon a linear-response-and-plateau (LRP) model. In the proposed reformulation, the law of the minimum admits many flexible and nonlinear specifications. The law of the minimum is compatible with low-input practices recommended by the paradigm of sustainable agriculture.

1. Introduction

The main objective of this paper is to verify the hypothesis that the "law of the minimum," formulated by Von Liebig 150 years ago, provides the best interpretation of crop response experiments and is, therefore, eminently suitable to represent the notion of low-input technologies. The very definition of Von Liebig's hypothesis, with its accent on the word "minimum", suggests that, if empirically verified, the law of the minimum will guarantee the attainment of maximum profits with relatively low levels of nutrients.

The recent debate on sustainable agriculture has emphasized the wisdom of shifting agricultural practices toward a spectrum of low-input technologies. Some proponents of this approach interpret the term "low-input" to mean "the use of internal resources generated on-farm, rather than purchased resources produced externally. Pest control is achieved mainly through cultural and biological methods, such as mechanical cultivation and crop rotation, and nutrients are supplied primarily by animal and green manures" (Liebhardt *et al.*, 1989). This low-input proposal contrasts deliberately with current agricultural practices which rely heavily on chemical fertilization and pest control. Whether the meaning of a low-input technology should be restricted to "the use of internal resources generated on-farm" is a matter of debate. In this paper, a low-input technology is characterized by fertilizer applications substantially smaller than

169

P C Struik et al (eds), Plant Production on the Threshold of a New Century, 169–177
© 1994 *Kluwer Academic Publishers*

those suggested by conventional crop response analyses. Ultimately, the crucial test for the introduction and adoption of a low-input technology rests on its associated profitability.

There is no denying that modern fertilization practices pose a fundamental dilemma: commercial agriculture is thought to require high levels of fertilizer applications which, by their nature, contribute directly to crop productivity and environmental pollution. During the past decades, the low cost of energy (especially in the United States) made it easy for farmers to indulge in over fertilization. The official recognition of this problem by soil scientists and politicians alike has called for better fertilization and water management practices in order to strike a balance between productivity and economic goals, on one side, and environmental and quality of life objectives, on the other.

To begin getting a handle on this two-pronged problem, it is important to recognize the crucial role that agronomists have played over many decades in shaping fertilizer recommendations. Using the methodology of crop response analysis, agronomists have traditionally based their fertilizer recommendations on the results of experiments intended to measure the response of various crops in different soils and climates to different combinations of nutrients and irrigation water. The statistical analysis of these crop response experiments was often based upon functional forms of the response that preferred polynomial (quadratic, square-root) and exponential (Mitscherlich) functions.

Over the past several years, however, it has been recognized that polynomial and exponential response functions tend to overestimate the fertilizer levels required to achieve maximum profits. The culture of fertilization (or, more correctly, the culture of over fertilization) as evolved to date, has a clear linkage with the methodology of crop response analysis developed during the fifties and the sixties. Hence, the improvement of fertilization and water management practices necessary for achieving the goals of a sustainable agriculture requires a rigorous reconsideration of the methodology of crop response analysis.

2. The Von Liebig's hypothesis

In the middle of the nineteenth century, Justus von Liebig became the principal proponent of a scientific conjecture according to which "the productivity of a field stands in relation (Verhältnis) to the amount of that mineral ingredient which its soil contains in smallest quantity" (Die Grundsätze der Agricultur-Chemie, 1855, p. 105). A more informative elaboration of Von Liebig's hypothesis can be found in his 1862 book entitled Die Naturgesetze des Feldbaues (The Natural Laws of Husbandry) where, on page 223, it is written that

> Every field contains a maximum of one or several, and a minimum of one or several different nutrients. The yields of crops stand in relation (Verhältnis) to this minimum, be it lime, potash,

nitrogen, phosphoric acid, magnesia, or any other mineral constituent: this minimum governs and controls the level and the persistence of yields.

Should this minimum for example be lime or magnesia, the yield of grain and straw, of turnips, potatoes or clover will remain the same and be no greater even though the amount of potash, silica, phosphoric acid, etc., already in the soil be increased a hundred times. The crop yields on this field, however, will be increased by a simple fertilization of lime.

The response function inferred from this quotation is characterized by a plateau maximum of yield and by non-substitution between nutrients. A major point of contention, however, regards the interpretation and the translation of the word "Verhältnis." The general meaning of this word is that of a "relation", any sort of relation. In a mathematical context, "Verhältnis" can be interpreted also to represent a "proportional" relation. Several translators have chosen this restricted interpretation to convey the meaning of "Verhältnis". It is impossible to know whether von Liebig approved of those translations. With the benefit of hindsight, it would appear that his mathematical background was insufficient for appreciating the importance of articulating his conjecture in a clear and well defined functional form. In his defense, it is appropriate to note that the mathematics necessary to express the law of the minimum and the statistical methodology for its empirical measurement became available only recently. Lacking his own explicit mathematical formulation, Von Liebig had to contend from the beginning with his rivals' functional specification of his "law" based upon the interpretation of the word "Verhältnis" as a "proportional" relation. Two scenarios followed. Empirical results of crop response experiments did not conform, in general, to the "proportional" specification of the "law of the minimum". Furthermore, Von Liebig was unable to correct the restricted interpretation of his "law" presented by his critics and reacted to them with ridicule and invectives.

In 1909, Alfred von Mitscherlich presented an empirical measure of the law of the minimum by using an exponential functional form that fits experimental data reasonably well. His results were erroneously interpreted as a refutation of Von Liebig's law. From then on, Von Liebig's fame was handed down to our times as the formulator of an intriguing conjecture about crop response that, unfortunately, did not withstand the empirical validation of scientific analysis. To this date, agronomists and soil scientists have paid only lip service to Von Liebig's hypothesis.

In general terms, the mathematical specification of the law of the minimum can be stated as

$$y = min \{f_N(N), f_P(P), f_K(K), \ldots, f_L(L)\} \tag{1}$$

where y stands for crop yield while N, P, K, W and L stand for nitrogen, phosphorus, potassium, water and the "last" growth factor, respectively. The functions f_N, f_P, f_K, f_W, and f_L represent the potential yield responses to the corresponding nutrients. Relation (1) embodies all the essential features of the law of the minimum but it is too general for any empirically useful application. To begin with, we must limit the number of growth factors examined at any one time. Suppose, therefore, to be interested in studying the response of a crop, say corn, to variable levels of nitrogen (N) and phosphorus (P), while all other nutrients $(K, W, \ldots, L)$ are held at non-limiting levels. In other words, these nutrients are arbitrarily fixed at levels presumed to be sufficiently high for causing either N or P to be limiting factors. In this case, the corn response function formulated according to the law of the minimum exhibits a plateau effect (m) such as

$$y_i = min \ \{f_N(N_i), f_P(P_i), m\} + u_i \tag{2}$$

where $m = min \ \{f_K(\overline{K}), f_W(\overline{W}), \ldots, f_L(\overline{L})\}$ and $\overline{K}, \overline{W}, \ldots, \overline{L}$ are fixed levels of growth factors beyond the scope of the study. A random term u_i represents the statistical error associated with a crop response experiment which is assumed to consists of I observations, $i = 1, \ldots, I$. Functions f_N and f_P can assume either linear or non-linear specifications. The linear specification of f_N and f_P gives rise to the linear-response and plateau (LRP) model expressed as

$$y_i = min \ \{ a_N + b_N N_i, a_P + b_P P_i, m \} + u_i \tag{3}$$

where a_N, a_P, b_N, b_P and m are parameters to be estimated statistically. The nonlinear specification of functions f_N and f_P in model (2) can assume different versions. An exponential formulation of each function joins together Von Liebig's and Mitscherlich's ideas about crop response:

$$y_i = min \ \{ m[1 - k_N exp(-b_N N_i)], m[1 - k_P exp(-b_P P_i)] \} + u_i \tag{4}$$

where m is the asymptotic plateau common to each potential yield function and exp is the exponential operator. When the exponential regimes of model (4) are replaced with square-root polynomial formulations, model (2) assumes the following specification

$$y_i = min \ \{ a_N + b_N N_i + c_N N_i^5, a_P + b_P P_i + c_P P_i^5, m \} + u_i \tag{5}$$

The square-root polynomial function is a very flexible specification capable of interpreting a wide range of crop responses. In particular, it possesses rising and declining asymmetric yield phases.

3. Materials and methods

The empirical verification of Von Liebig's law of the minimum requires well planned and executed experiments. A suitable experiment must involve at least two growth factors each of which must be applied at several different levels. In the absence of these two essential characteristics, no meaningful crop response analysis can be performed. Another aspect of a well planned experiment is the selection of a soil that is deficient in the fertilizers under study while all the other growth factors should be at non-limiting levels. It is very difficult to find fertilizer experiments endowed with these characteristics which were executed during the last twenty years.

Two experiments were selected for this study. The first experiment was conducted by Heady *et al.* in 1952 [1] involving corn grown on a calcareous Ida silt loam soil in western Iowa. Nine levels of nitrogen (N) and phosphorus (P) were applied to corn arranged in an incomplete factorial lay-out for a total of 114 observations. Nitrogen was applied in the form of ammonium nitrate and phosphorus in the form of concentrated superphosphate. The plant population density was chosen at 18000 plants per acre.

This data sample was analyzed using three specifications of the response function. The first functional form is the square-root polynomial response preferred by Heady *et al.* [1]. The other two specifications are the Von Liebig models outlined in Eq. (4) and Eq. (5), respectively. A statistical analysis of the performance of these three specifications requires the adoption of a non-nested hypothesis procedure as illustrated, for example, by Paris [6]. This methodology is recommended when the various models do not share the parameter space, as in this case.

The second sample of data deals with a corn experiment conducted in Davis, California, in 1970, and analyzed by Hexem and Heady [2]. A split plot design with factorial treatments of nitrogen and irrigation water was used. Five levels of irrigation water and nitrogen were applied to a total of 39 plots. Plant population density varied from "low" to "high" density

The three specifications selected for this second analysis are a square-root polynomial function, a polynomial model preferred by Hexem and Heady [2] with some terms raised to the power of 1.5, and a modified Von Liebig model with square-root regimes, as illustrated by Eq. (5). The modification consists of an auxiliary regime in the form of a back plane introduced to achieve a better fit of the sample observations corresponding to a declining phase of corn yields.

4. Results

Table 1 presents the estimates of three models for each experiment. The values in parentheses are standard errors of the corresponding estimates. For the Iowa experiment, the highest value of the logarithm of the likelihood function (*logL*) corresponds to a Von Liebig model with potential yield functions expressed by square-root polynomials. This value is considerably higher than that of model (1) which was preferred by Heady *et al.*. The coefficient of determination (R^2) exhibits similar values for all three models. Its interpretation is made difficult by the fact that models (2) and (3) are non-linear in the parameters. In general, the coefficient of determination is not a valid statistic for selecting the best specification. In this case, a non-nested hypothesis test is required. The results of such a test are given in Table 2. The numbers in this table are asymptotic Student t-ratios intended to express the significance (or lack of it) of the alterative model under the null hypothesis. When the null hypothesis is defined by the Von Liebig model with square-root polynomial regimes (model (3)), the t-ratios associated with the square-root polynomial (model (1)) and with the Von Liebig-Mitscherlich (model (2)) are not significant at the .05 level. In other words, the null hypothesis is not rejected. Conversely, the null hypothesis expressed by either rival model is rejected by the Von Liebig specification with square-root regimes. This pattern of the non-nested hypothesis test indicates that the latter model is the best specification for interpreting the crop response to N and P of the Iowa experiment.

TABLE 1. Estimation results.

Iowa corn experiment	California corn experiment
1. Square-root polynomial model $y = 0.0597 - 0.417P - 0.317N +$ $\quad$ (0.066) $\quad$ (0.040) $\quad$ (0.040) $0.852P^{.5} + 0.635N^{.5} + 0.341(PN)^{.5} + u$ $\quad$ (0.087) $\quad$ (0.087) $\quad$ (0.039) $logL = 69.06 \quad \sigma = 0.194 \quad R^2 = 0.912$	1. Square-root polynomial model $y = 0.0728 - 3.549W - 0.667N +$ $\quad$ (0.075) $\quad$ (0.751) $\quad$ (0.360) $3.669W^{.5} + 0.337N^{.5} + 0.278(WN)^{.5} + u$ $\quad$ (0.507) $\quad$ (0.222) $\quad$ (0.324) $logL = 53.07 \quad \sigma = 0.067 \quad R^2 = 0.928$
2. Model (4): Von Liebig-Mitscherlich $y = min \{Y_N, Y_P, Y_m\} + u$ $Y_N = 1.291[1 - 0.791exp(-1.734N)]$ $\quad$ (0.028) (0.029) (0.226) $Y_P = 1.291[1 - 0.870exp(-2.286P)]$ $\quad$ (0.028) (0.033) (0.309) $logL = 76.060 \quad \sigma = 0.124 \quad R^2 = 0.928$	2. Three-halves polynomial model $y = 0.2341 + 8.699W + 1.136N -$ $\quad$ (0.036) $\quad$ (0.809) $\quad$ (0.551) $12.143W^{1.5} - 2.012N^{1.5} + 0.479(WN) + u$ $\quad$ (0.036) $\quad$ (0.809) $\quad$ (0.551) $logL = 54.79 \quad \sigma = 0.065 \quad R^2 = 0.935$
3. Model (5): Von Liebig-Square-root $y = min \{Y_N, Y_P, Y_m\} + u$ $Y_N = 0.280 + 0.621N + 0.293N^{.5}$ $\quad$ (0.035) (0.267) (0.215) $Y_P = 0.161 - 0.432P + 1.396P^{.5}$ $\quad$ (0.037) (0.059) (0.227) $Y_m = 1.2761$ $\quad$ (0.040) $logL = 83.450 \quad \sigma = 0.116 \quad R^2 = 0.936$	3. Von Liebig-Square-root with back plane $y = min \{Y_N, Y_W, Y_m, Y_{bp}\} + u$ $Y_N = 0.844 - 2.322N + 1.433N^{.5}$ $\quad$ (0.029) (1.191) (0.615) $Y_W = 0.147 + 0.322W + 1.753W^5$ $\quad$ (0.129) (2.194) (1.164) $Y_m = 1.020$ $\quad$ (0.037) $Y_{bp} = 1.287 - 0.056N - 1.246W$ $\quad$ (0.201) (0.211) (0.823) $logL = 58.05 \quad \sigma = 0.055 \quad R^2 = 0.960$

TABLE 2. Non-nested hypothesis tests.

Iowa corn experiment				California corn experiment			
	Null hypothesis				Null hypothesis		
Alternative hypothesis	Sqrt (1)	Lieb-Mit (2)	Lieb-Sqrt (3)	Alternative hypothesis	Sqrt (1)	3-halves (2)	Lieb-Sqrt (3)
Sqrt(1)	---	1.371	1.176	Sqrt(1)	---	-0.328	0.437
Lieb-Mit(2)	4.089	---	-0.708	3-halves(2)	2.854	---	1.874
Lieb-Sqrt(3)	5.786	3.940	---	Lieb-Sqrt(3)	3.385	2.144	---

For the California corn experiment, the best crop response model turns out to be a Von Liebig-Square root specification (model (3)) modified by the addition of a back plane. One of the original criticisms of the law of the minimum claimed that the Von Liebig's hypothesis does not allow the representation of a yield declining phase, when it exists. This criticism is unwarranted as demonstrated by model (3) in the second column of Table 1. That model specifies an explicit regime (called a "back plane") which accounts for the declining phase of yields beyond a certain level of N and irrigation water. According to the results of the non-nested hypothesis test presented in Table 2, the same model rejects the two rival specifications and is not rejected by either one. It becomes, therefore, the preferred crop response model for this sample of data.

5. Discussion

The conclusion that, at least for the two samples of experimental data analyzed in this study, the family of Von Liebig models provides a crop response analysis more satisfactory than that of conventional functions opens the way for demonstrating the low-input aspect associated with the law of the minimum. In other words, the Von Liebig's hypothesis provides not only a statistically superior framework for crop response analysis but, when used for economic decisions, it allows the attainment of profit levels with significant savings of fertilizer nutrients. Consider the Iowa experiment and assume that farmers maximize profit (π) defined as $\pi = p_y y - p_N N - p_P P$, where p_y is the price of corn measured in dollars per bushel, while p_N and p_P are the prices of N and P, respectively, measured in \$/lb. Corn yields, y, are measured in 100 bushels while N and P are in 100 pounds. The various combinations of paces used in the analysis are given in the third line of Table 3. The profit maximizing levels of N and P are presented for the square-root polynomial model and for the Von Liebig function with square-root regimes. For each price scenario, the profit level attainable with the Von Liebig specification is never lower than that associated with the square-root polynomial function and, furthermore, significant savings (up to 100 %) are indicated for the N fertilizer. A similar pattern of profit and input savings is

TABLE 3. Results of the economic analysis.

Iowa corn experiment

corn price/N price/P price

model	3/.2/.22	2/.2/.22	4/.2/.22	3/.4/.22	3/.2/.44
(1)	π=294	π=172	π=422	π=260	π=260
Square	y=1.26	y=1.18	y=1.29	y=1.18	y=1.18
root	N=2.07	N=1.59	N=2.39	N=1.42	N=1.81
polyn.	P=1.87	P=1.48	P=2.13	P=1.64	P=1.35
(3)Von	π=316	π=191	π=422	π=294	π=280
Liebig	y=1.26	y=1.23	y=1.27	y=1.26	y=1.21
Square	N=1.09	N=1.05	N=1.10	N=1.08	N=1.02
root	P=1.85	P=1.56	P=2.02	P=1.80	P=1.39

California corn experiment

corn price/N price/W price

model	.05/.2/4	.05/.1/4	.06/.2/4	.04/.2/4	.03/.2/4
(2)	π=375	π=384	π=469	π=283	π=192
Three-	y=.932	y=.945	y=.942	y=.915	y=.883
halves	N=.075	N=.116	N=.088	N=.057	N=.033
polyn.	W=.19	W=.19	W=.20	W=.18	W=.16
(3)Von	π=420	π=422	π=522	π=318	π=216
Liebig	y=1.02	y=1.02	y=1.02	y=1.02	y=1.02
Square	N=.029	N=.029	N=.029	N=.029	N=.029
root/BP	W=.21	W=.21	W=.21	W=.21	W=.21

illustrated in Table 3 by the California corn experiment. In this case, the price of corn is in $/lb, the price of irrigation water is in $/acre-inch, corn yield is in 10,000 lb, N is in 1,000 lbs and W is in 100 of acre-inches.

6. References

[1] Heady, E. O., Pesek, J. T. and Brown, W. G. (1955) 'Corn Response Surfaces and Economic Optima in Fertilizer Use.' Iowa State Experiment Station Research Bulletin 424.

[2] Hexem, R.W. and Heady, E.O. (1978) 'Water Production Functions for Irrigated Agriculture', Iowa State University Press, Ames.

[3] Liebhardt, W.C., Andrews, R.W., Culik, M.N., Harwood, R.R., Janke, R.R., Radke, J.K. and Rieger-Schwartz, S.L. (1989) 'Crop production during conversion from conventional to low-input methods', Agronomy Journal 81, 150-159.

[4] Liebig, J. von (1855) 'Die Grundsätze der Agricultur-Chemie: mit Rücksicht auf die in England angestellten Untersuchungen', F. Vieweg und Sohn, Braunschweig.

[5] Liebig, J. von (1862) 'Die Naturgesetze des Feldbaues', 7e Aufl., Vol. 11, F. Vieweg und Sohn, Braunschweig.

[6] Paris, Q. (1992) 'The return of Von Liebig's "law of the minimum"'. Agronomy Journal 84, 1040-1046.

NITROGEN FLUXES IN THE CROP-SOIL SYSTEM OF BRUSSELS SPROUTS AND LEEK

R. BOOIJ, C.T. ENSERINK, A.L. SMIT AND A. VAN DER WERF
DLO-Centre for Agrobiological Research (CABO-DLO)
P.O. Box 14
6700 AA Wageningen
The Netherlands

ABSTRACT. The effect of nitrogen availability on dry matter production, nitrogen uptake and marketable yield of Brussels sprouts (high crop residue) and leek (low fertilizer recovery) was analysed in a field experiment. All available soil mineral nitrogen (from fertilizer and mineralisation) was taken up by Brussels sprouts up to 430 kg ha^{-1} and by leek up to 250 kg ha^{-1}, when 300 and 125 kg ha^{-1} nitrogen fertilizer was applied respectively. Maximum dry matter production was almost reached when those amounts of nitrogen were taken up. Optimum marketable yield was achieved at still higher nitrogen rates; with leek the optimum application rate was associated with nitrogen losses, and with Brussels sprouts it led to high amounts of nitrogen in the crop residue.

1. Introduction

In field grown vegetables, application of nitrogen (N) fertilizer is an important cultural practice. The costs of fertilizer input is low compared with the financial output of marketable product, allowing high application rates (Neeteson and Wadman, 1987), which may result in losses of N through leaching or denitrification (Wehrman and Scharpf, 1989). From an environmental point of view these losses are not acceptable and measures should be taken to avoid them (Greenwood, 1990). Reduction of the N-fertilizer rate may be one of the necessary measures and implies, if there is no overuse, that lower yields should be accepted (De Wit, 1992). Before recommendations are formulated taking environment and farmers' income into account, it is necessary to quantify the N flows within the crop-soil system in relation to crop growth.

The aim of the present study was to quantify N flows for Brussels sprouts and leek, two crops differing in the ratio N-uptake/N-fertilizer rate and the amount of crop residue (Smit and Van der Werf, 1992).

P. C. Struik et al. (eds.), Plant Production on the Threshold of a New Century, 179–185.
© 1994 *Kluwer Academic Publishers.*

2. Materials and methods

A field trial was carried out on a sandy soil in 1991, with Brussels sprouts and leek with varying nitrogen fertilizer rates (Table 1) applied before transplanting. Soil mineral N in the upper 75 cm soil layer was 80 kg ha^{-1} before fertilizer application. Crop husbandry was according to commercial practice. Details of the trial are given in Table 1.

TABLE 1. Experimental details of the field trial in 1991.

	Brussels sprouts	Leeks
Cultivar	Kundry	Arcona
Transplanting date	May 28	June 17
Plant density (ha^{-1})	33,000	172,000
N-fertilizer rate (kg ha^{-1})	0,100,200,300	0,125,250
Harvest date	December 2	November 19

When the crops were marketable, plant (except roots) fresh weight, dry matter and nitrogen contents of the marketable part and of the crop residue were determined. At the same time the amount of soil mineral nitrogen in the upper 75 cm soil layer was determined. Shed leaves of Brussels sprouts were collected weekly.

3. Results

The results are presented in an adapted so-called three-quadrant diagram (Van Keulen, 1982). In quadrant II (Figure 1) the relation between the amount of N-fertilizer and the total above-ground dry matter production is given. Without N-fertilizer application, dry matter production of both crops was similar, but dry matter production of Brussels sprouts increased more rapidly with increasing N rate. At the highest application rate of N, maximum dry matter production of both crops was almost obtained.

In quadrant IV (Figure 1) the relation between potentially available N and N-uptake is given instead of the relation between N-fertilizer rate and N-uptake (Van Keulen, 1982). The potential available soil N in the upper 75 cm soil layer included soil mineral nitrogen in the upper 75 cm soil layer at the start of the growing period, the net amount of mineralized N from the organic matter in the soil during the growing period, deposition during the growing period and the applied fertilizer. The sum of the first three terms were obtained from the N-uptake of the unfertilized plots and assuming that this was the same on the fertilized plots. N-uptake of the crop did not include N recovered in the roots, but included the shed leaves of

Brussels sprouts. For Brussels sprouts the amount of N taken up by the crop increased linearly with N-available in a 1:1 ratio, which means that all available N was taken up by the crop. This was valid for leek as well up to an available amount of 225 kg N ha^{-1}; beyond 225 kg N ha^{-1} N-uptake increased less rapidly. At harvest of Brussels sprouts, hardly any soil mineral nitrogen was left in the upper 75 cm soil layer (Table 2), which is in agreement with the relation between N-uptake and available N (Figure 1). Also for leek, only limited amounts of soil mineral N were recovered, even at the highest N rate (Table 1). According to the relation between available N and N-uptake a higher amount of soil mineral N was expected at the harvest of leek after application of the highest amount of N-fertilizer. Probably part of the available N was lost during the growing period at the highest N-fertilizer rate.

The relation between N-uptake and dry matter production up to a N-uptake of 100 kg ha^{-1} was the same for Brussels sprouts and leek (Figure 1, quadrant I). Higher N-uptake for Brussels sprouts resulted in a higher dry matter production up to approximately 300 kg N ha^{-1}, whereas dry matter production of leek increased much slower with increasing N-uptake.

The amount of nitrogen recovered in the crop residue of Brussels sprouts was much higher than of leek, because a smaller part of the total plant of Brussels sprouts is harvested. In both crops the amount of nitrogen in the harvest residue increased with increasing nitrogen application (Table 2).

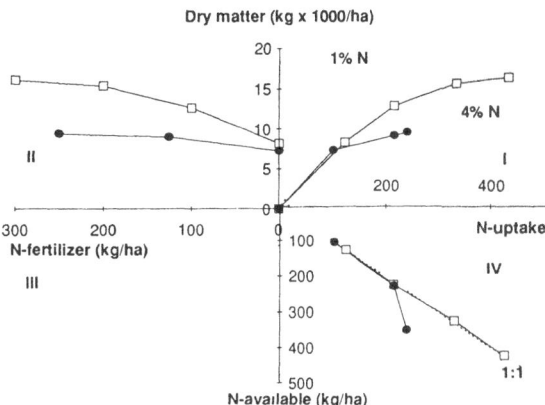

Figure 1. Relationships between N-uptake, dry matter production, N-fertilizer rate and available soil mineral N of Brussels sprouts (□) and leek (•).

Total marketable yield is determined by the fresh weight of the marketable part of the plant. In practice the economically optimum fertilizer rate is determined by the ratio of cost per unit of fertilizer and the price per unit of marketable product (point of unit marginal return). We redrew the three-quadrant diagram of Figure 1, but now in quadrant II the relation between fertilizer costs and financial marketable yields are given (Figure 2). For these relations N-fertilizer costs were set on Dfl 1.40 kg^{-1} and the price per kilogram marketable sprouts or leek respectively on Dfl 0.75 and Dfl 0.80 (Anonymous, 1991). It was also assumed that the quality of the product was not affected by the N-fertilizer, although according to the present quality requirements this is questionable, especially for the product obtained at low N-application rates.

TABLE 2. Effect of N-fertilizer (kg ha^{-1}) on the amount of soil mineral nitrogen (kg ha^{-1}) in the upper 75 cm soil layer after harvest and on the amount of N (kg ha^{-1}) in the crop residue of Brussels sprouts (a) and leek (b).

a.			b.		
N-fertilizer (kg ha^{-1})	soil mineral N (kg ha^{-1})	N-crop residue (kg ha^{-1})	N-fertilizer (kg ha^{-1})	soil mineral N (kg ha^{-1})	N-crop residue (kg ha^{-1})
0	10	75	0	14	41
100	11	123	125	21	86
200	12	191	250	24	96
300	11	263			

Within the range of fertilizer rates applied in our trial, the point of unit marginal return was not reached (Figure 2). This means that even higher amounts of N-fertilizer would have been profitable. However, this will inevitably result in higher amounts of N in the crop residue and, in particular with leek, in higher losses of soil mineral nitrogen during the growing season or after harvest from residual mineral N in the soil compartment (Figure 2).

4. Discussion

In the Netherlands recommended N-fertilizer rates are 240 minus soil mineral N (N_{min}) before transplanting (kg ha^{-1}) for Brussels sprouts and 270 minus N_{min} (kg ha^{-1}) for leek (Smit and Van der Werf, 1992). These recommendations are based on marketable yield, which concerns fresh weight of the harvested part of the plant. The small difference between the two crops in advised rate is in accordance with our results concerning marketable yield (Figure 2). However, the total dry matter production of Brussels sprouts reacted much stronger on N-fertilizer rate than the dry matter production of leek (Figure 1).

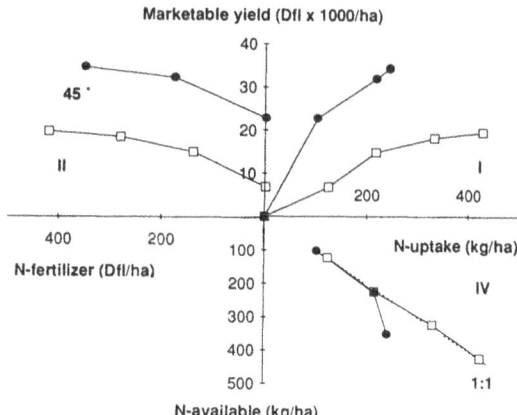

Figure 2. Relationships between N-uptake, financial yield, N-fertilizer costs and available soil mineral N of Brussels sprouts (□) and leek (•).

Although maximum dry matter production could not be determined in our experiment due to the restricted range of N-fertilizer rates, it is evident from the curves in quadrant II of Figure 1, that the amount of N-fertilizer required for maximum dry matter production is far lower for leek than for Brussels sprouts. If N-fertilization of leek is aimed at maximum or near maximum dry matter production instead of optimum financial return, the potential losses of N for leek are lower (Figure 1), but the grower has to accept a lower economic yield (Figure 2). For Brussels sprouts the N-rates at which optimum financial return or maximum dry matter production are obtained are closer and if crop residues are collected, the economic optimum can be reached without losses of N to the environment.

N-concentration within the plant increased with increasing N-uptake (Figure 1). The N-uptake of leek, from 100 kg ha^{-1} onwards, increased mainly due to increased N-concentration, with Brussels sprouts from 300 kg ha^{-1} onwards. Increasing N-concentration in leek was associated with a higher water content, which resulted in a steep increase in fresh weight with increasing N-uptake (Figure 2). The difference between N-uptake and fresh weight or between N-uptake and dry weight can result in less efficient fertilizer use (from the point of view of dry matter production) and can lead consequently to higher potential losses.

The relative growth rate of above ground parts of Brussels sprouts was twice as high as for leek, due to a twofold higher leaf area ratio (Booij et al., 1993). The higher relative growth rate of Brussels sprouts generates a higher N-demand and therefore the strong reaction on N-available can be observed. It is not likely that the

N-absorbing intrinsic capacity of the root system of leek limits N-uptake, because N-uptake increased even though dry matter production did not (Figure 1). Since the relatively low relative growth rate of leek is associated with a slower extension of the total root system, fertilizer nitrate may be lost (e.g. through leaching) from a soil layer before roots reach this (Greenwood and Draycott, 1988).

In both crops higher N-uptake resulted in higher amounts of N in the crop residues (Table 2). If the crop residues are left on the field, their C/N-ratio (Harmsen and Van Schreven, 1955) and the weather will affect their decomposition rate. Timing of mineralisation determines whether the mineralised N is lost, or can be utilized by the succeeding crop.

The presentation of the results in three-quadrant diagrams made it possible to evaluate the effects of N-fertilization, taking into account the marketable yield and potential losses to the environment. If a certain goal is set, the consequences for the other variables can be derived.

5. Acknowledgement

We want to thank dr A.J. Haverkort and ir J.J. Schröder for their comments on the manuscript.

6. References

Anonymous (1991) 'Kwantitatieve informatie voor de Akkerbouw en Groenteteelt in de Vollegrond'. Bedrijfssynthese 1991-1992. PAGV-Publicatie no 57, PAGV, Lelystad, 191 pp.

Booij, R., Enserink, C.T., Smit, A.L. and Werf, A. van der (1993) 'Effects of nitrogen availability on crop growth and nitrogen uptake of Brussels sprouts and leek'. Acta Hortic., in press.

Greenwood, D.J. (1990) 'Production or productivity, the nitrate problem?' Ann. Appl. Biol. 117, 209-231.

Greenwood, D.J. and Draycott, A. (1988) 'Recovery of fertilizer-N by diverse vegetable crops: processes and models', in D.S. Jenkinson and K.A. Smith (ed.), Nitrogen efficiency in agricultural soils, Elsevier Applied Science Publishers, Barking, pp. 46-61.

Harmsen, G.W. and Van Schreven, D.A. (1955) 'Mineralization of organic nitrogen in the soil'. Adv. Agron. 7, 299-398.

Keulen, H. van (1982) 'Graphical analysis of annual crop response to fertilizer application'. Agric. Syst. 9, 113-126.

Neeteson, J.J. and Wadman, W.P. (1987) 'Assessment of economically optimum application rates of fertilizer N on the basis of response curves'. Fert. Res. 12, 37-52.

Smit, A.L. and Werf, A. van der (1992) 'Fysiologie van stikstofopname en -benutting: gewas- en bewortelingskarakteristieken', in H.G. van der Meer and H.J. Spiertz (eds), Stikstofstromen in agro-ecosystemen. Agrobiologische Thema's no. 6. DLO-Centrum voor Agrobiologisch Onderzoek (CABO-DLO), Wageningen, pp. 51-69.

Wehrman, J. and Scharpf, H.C. (1989) 'Reduction of nitrate leaching in a vegetable farm: fertilization, crop rotation, plant residues', in J.G. Germon (ed.), Management systems to reduce impact of nitrates, Elsevier Applied Science Publishers, London, pp. 147-157.

Wit, C.T. de (1992) 'Resource use efficiency in agriculture'. Agric. Syst. 40, 125-151.

MODELLING IN THE SEARCH FOR BALANCE BETWEEN INPUTS AND OUTPUTS IN GREENHOUSE CULTIVATION

HUGO CHALLA, EP HEUVELINK AND KLAAS-JAN LEUTSCHER
Dept. of Horticulture, Wageningen Agricultural University
Haagsteeg 3
6708 PM Wageningen
The Netherlands

ABSTRACT. The power of simulation models is demonstrated in two contrasting cases relevant for protected cultivation. In the first case it is shown that integration of knowledge obtained in studies at the level of the whole plant is able to provide an adequate basis for a generic description of dry matter partitioning in indeterminately growing crops. The second case is an example of the integration of crop physiological and economic models in decision support systems in pot plant cultivation, where dynamic crop growth models are more flexible and better adapted to the variable situation at individual nurseries than the usual fixed cultivation-schedules.

1. Introduction

Greenhouse cultivation represents one of the most intensive forms of agriculture. Modern, sophisticated greenhouses enable refined management of the production process (Challa, 1990) through control of temperature, CO_2 and water vapour pressure, and to a certain extent of radiation in the aerial environment. In addition soilless cultivation offers the opportunity to control the water potential and ionic composition, as well as temperature in the root environment.

While effective in terms of production control and the economic value produced per m^2, cultivation in greenhouses is associated with high demands on inputs, such as capital, labour and energy. Especially in areas with concentrations of large amounts of greenhouses the greenhouse industry may affect the environment negatively by pollution with fertilisers and crop protection agents. In addition there is a problem of waste of plastics and rockwool and the output of CO_2. Management and control of the production process have a significant impact on profits as well as on the environmental aspects and as such play a major role in the search for balance between inputs and outputs.

The knowledge required for optimal management and control of the production process in protected cultivation is quite extensive for a

187

P C Struik et al (eds), Plant Production on the Threshold of a New Century, 187–195
© 1994 *Kluwer Academic Publishers*

number of reasons. As pointed out before, sophisticated greenhouses enable refined control of most of the essential production factors, often dynamically, creating many options in many combinations. Another characteristic of greenhouse cultivation, particularly in ornamental horticulture, is the large variation in crops and cultivars, with quite different properties and requirements. Further complications arise from the diversity in cultivation systems that are in use. Relevant knowledge, moreover, covers different scientific disciplines, such as plant and crop physiology, horticulture, engineering, economics, etc., which only partly can be represented quantitatively.

In this vast field of knowledge, where it is easier to find the exception than the rule, it is important to develop generic methods that could contribute to improved management and control of the production process in protected cultivation. A powerful method to represent and combine knowledge from different disciplines in a generic way is the use of explanatory models. They enable a scientific approach to agricultural problems by incorporating knowledge on underlying processes, instead of the more common empirical approach. In Wageningen, explanatory models of agricultural production systems have been developed since the sixties. Initially the scope of these models was limited to the prediction of crop photosynthesis (De Wit, 1965) and dry matter production (Brouwer and De Wit, 1969), as related to radiation conditions. These models were further developed by including crop specific morphological and developmental aspects and by incorporating also the root environment (water and nutrient relations) and the interactions with pests and diseases (Penning de Vries and Van Laar, 1982), or the greenhouse as a transformer of the crop environment (Challa, 1990). During the last decade also socio-economic aspects were added (Penning de Vries, 1990).

The objective of the present contribution is to demonstrate the power of this approach in two contrasting cases relevant for protected cultivation. In the first case it will be shown that integration of knowledge obtained in studies at the level of the whole plant is able to provide a generic description of dry matter partitioning in indeterminately growing crops, where studies at a more detailed level, though valuable for improved understanding, did not succeed. The second case is an example of the integration of crop physiological and economic models in an application for decision support systems in pot plant cultivation, where the flexibility and insight provided by dynamic models open up new approaches in this field.

2. Case 1: Dry matter partitioning in tomato

Partitioning of dry matter is an important determinant of yield. In tomato, an indeterminately growing crop, the balance between vegetative growth and growth of fruits is a matter of major concern to the grower. This balance is relevant because it determines the amount and the quality (size) of the produce and can be managed by the environmental conditions.

Although some insight in assimilate partitioning in plants is emerging, no single aspect of it is fully understood (Dale, 1985). Nevertheless, several authors (Schapendonk and Brouwer, 1984; Heuvelink and Marcelis, 1989; Jones et al., 1991; Heuvelink and Bertin, 1994; Marcelis, 1994) obtained promising results when simulating dry matter distribution using the concept of demand functions. This concept will be briefly introduced and applied with tomato.

The simulation model, TOMSIM(1.0), is based on the following principles (Heuvelink and Bertin, 1994). Biomass partitioning is primarily regulated by the sinks (Gifford and Evans, 1981; Farrar, 1988). Sink strength is the competitive ability of a sink to attract assimilates (Wolswinkel, 1985) and may be defined as the potential capacity of a sink to accumulate assimilates. This capacity, or demand, can be quantified by the potential growth rate of a sink, i.e. the growth rate under conditions of non-limiting assimilate supply (Marcelis et al., 1989). Sink strength of an organ is determined by its developmental stage and may, in general, be affected by temperature (though not in tomato; Heuvelink and Marcelis, 1989), but is independent of photosynthesis (Marcelis, 1993). Photosynthesis, however, may affect biomass partitioning indirectly through the effects on initiation and abortion of sinks. When carbon supply exceeds the total sink strength of the whole plant, growth rate of the sinks is maximum, and reserves are formed and/or photosynthesis is reduced (Marcelis, 1993). Normally, however, growth of glasshouse tomato is source-limited under our conditions (De Koning, pers. comm.). When the rate of carbon supply is less than the potential growth rate of the plant, the available assimilates are distributed proportionally to the sink strength of the organs. Growth rate of individual sinks (Yi) is thus:

$$Y_i = (S_i / \Sigma S_i) * X \quad \text{for } X \leq \Sigma S_i$$

with: Y_i = growth rate of sink i (mass time^{-1} sink^{-1}); S_i = sink strength of sink i (mass time^{-1} sink^{-1}); ΣS_i = total sink strength of all sinks (mass time^{-1} plant^{-1}) and X = rate of assimilate supply (mass time^{-1} plant^{-1}).

In TOMSIM(1.0), each truss is simulated separately, but the vegetative plant parts are lumped together (one vegetative sink), because the ratio between leaf and stem growth is constant (Heuvelink and Marcelis, 1989) and is fixed at 7:3. Vegetative sink strength was empirically set at twice the maximum sink strength of a cluster (Kano and Van Bavel, 1988). Truss initiation and development rate are described in relation to greenhouse temperature, but the number of fruits per truss is input of the model.

Dry matter distribution in a tomato crop, simulated with TOMSIM(1.0) was compared with experimental results. In a glasshouse experiment two treatments were compared: in the control no trusses were removed, in the other treatment every even-numbered truss was removed at anthesis (50 % trusses). All trusses were pruned to 7 fruits per truss after fruit set. Every 10 or 11 days plants were harvested for destructive measurements. A spline function was fitted to the total shoot dry weight to smooth measured daily increase in shoot dry weight. This function was used as an input for simulation of dry matter distribution in the shoot.

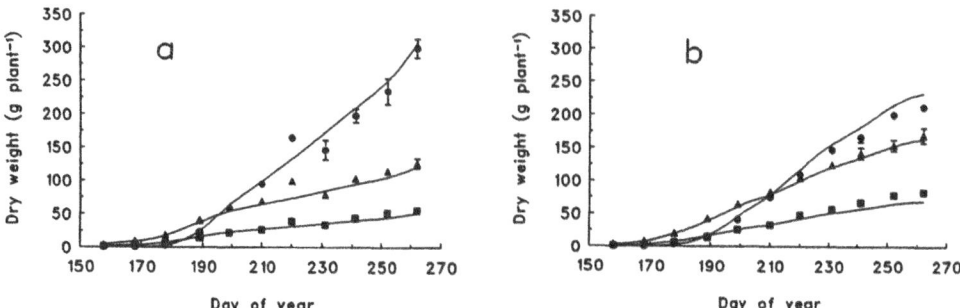

Figure 1. Measured (■ stem; ▲ leaves; ● fruits) and simulated (lines) dry matter distribution for glasshouse tomato: (a) control treatment (100 % trusses) and (b) removal of even-numbered trusses at anthesis (50 % trusses). Daily increase in shoot dry weight was input to the model. Bars indicate standard error of mean (n=4) when greater than symbol size.

Dry matter distribution is simulated satisfactorily in both treatments (Figure 1). At the level of individual trusses sometimes large discrepancies were observed (data not shown), probably because the same relation between truss development and truss sink strength was used for all trusses. De Koning (pers. comm.) observed that the potential fruit size increases with truss position in the first 6 to 10 trusses on a plant. Under other conditions (temperature, fruit thinning, two-stem systems) also reasonable agreement was obtained between measured and simulated dry matter distribution. The same approach, when used with cucumber, provided a good agreement between measured and simulated cyclic dry matter distribution between fruits and vegetative plant parts (Marcelis, 1994).

3. Case 2: DSS in pot plant production

Pot plants are cultivated in batches according to planned cultivation-schedules. A pot plant cultivation-schedule describes all actions that should be executed during cultivation, and their consequences in terms of greenhouse area occupation, labour and machine capacity requirements and costs of direct inputs. Due to the possibility of displacing pot plants during cultivation, i.e. spacing, greenhouse area requirements are not constant for a batch. Moreover, various batches of different size may be cultivated simultaneously at the same nursery. Furthermore, requirements for other limited resources like labour and machine capacity vary also during cultivation since these requirements are concentrated at the moments of potting, spacing and selling (Krijgsman and Achter, 1973). As a consequence, the allocation of greenhouse area in pot plant cultivation is rather complex.

In order to support management in pot plant cultivation, linear programming (lp) for tactical production planning has been applied in several studies (Annevelink, 1989; Håkansson, 1991; Krafka *et al.*, 1989). This generally involves the allocation of limited resources to production alternatives prior to the actual cultivation and with the objective to attain a maximum economic result. In this respect production alternatives involve standardised cultivation-schedules for every optional potting moment. So far, however, the application of lp as management support tool has not been very successful in practice (Gollwitzer, 1991). Besides general problems with respect to computerised management support, she emphasises the discrepancy between the real problem, the problem representation by the lp-method and the user's perception of the problem. The usual lp-approach to tactical planning in pot plant production optimises only the allocation of limited resources to fixed pre-declared cultivation-schedules, whereas the optimisation of cultivation-schedules itself and their relation to the restrictions of limited resources is disregarded. In fact, cultivation-schedules are implicitly 'optimised' when the manager formulates the lp-problem and the lp-method optimises the problem only partly.

The application of crop growth models could contribute to a more successful support of computerised management in pot plant production (Buchwald, 1987; Leutscher and Vogelezang, 1990). Crop growth models may provide the means to establish more realistic initial cultivation-schedules, which correspond to daily nursery practices. Furthermore, they enable the interactive adaptation of cultivation-schedules in repetitive optimisations (Table 1). Thus, tactical production planning may become less abstract and more flexible.

In addition, the application of crop growth models may also contribute to the development of computerised control and evaluation systems (Berg and Lentz, 1989). The actual conditions may deviate from the expected circumstances and result in a changed crop growth pattern. By simulating crop growth under the actual conditions in parallel to actual crop growth and comparing the results of these simulations with results of the ex ante simulation, valuable information for management control may be provided. Deviations between expected simulated and actual simulated crop growth may be diagnosed and initiate operational adaptation of cultivation-schedules. Moreover, after the cultivation-period, i.e. ex post, the consequences of alternative cultivation-schedules may be simulated under the recorded actual conditions and analysed. This may provide new information for the subsequent planning-period.

TABLE 1. Example of alternative cultivation-schedules for a same potting moment with different organisational and economic consequences, according to Leutscher and Vogelezang (1990). The first row represents the standard cultivation-schedule as applied in the lp-model (Dfl = Dutch Guilders).

Cultivation schedule (1000 potted plants)				Organisational and economic consequences					
Potting	First spacing	Second spacing	Delivery	Occupied area	Labour costs	Costs of direct inputs	Returns	Gross margin - labour costs	Gross margin - labour costs (Dfl/ week.m²)
(week)	(week)	(week)	(week)	(week.m²)	(Dfl)	(Dfl)	(Dfl)	(Dfl)	week.m²
1	6	11	14	467	661	1595	2541	285	0.61
1	5	11	13	430	643	1583	2491	265	0.62
1	7	11	15	503	683	1606	2591	302	0.60
1	6	10	14	483	661	1600	2541	280	0.58
1	6	12	15	500	678	1606	2591	307	0.61

With respect to the application of crop growth models two specific problems arise. Firstly, present crop growth models describe morphological plant characteristics generally only poorly. This is especially problematic when these models are applied with ornamentals. Research is now in progress to incorporate the shape of pot plants in crop growth models. Secondly, many different species and cultivars are cultivated and various cultivation systems are applied in pot plant production. This results in an almost infinite number of cultivation alternatives, for which individual model parameters should be estimated. This problem may only be solved when tools for nursery specific validation of crop growth models are developed. One should bear in mind that, although this may seem a rather awkward exercise for the individual grower, improved data recording on the pot plant nursery may provide enviable data sets. For the actual validation the support of either a human expert or a computerised expert system seems indispensable. Thus, it can be concluded that, although much additional research is still necessary, crop growth models may contribute significantly to computerised management support on pot plant nurseries.

4. Conclusions

The two cases presented demonstrate the potentials of explanatory models in improved understanding of the complex mechanisms underlying crop production, as well as in the application at a more aggregated level. This tool derives its power from the use of generic crop physiological knowledge under diverging conditions and with a broad range of crops, and from the integration of knowledge from different disciplines. It enables co-operation between scientists working on different problems, crops and in different countries by providing a common basis for research. In a time with limited funds for research and an increasing interest for multiple goals in agricultural production this tool deserves further integration in horticultural research.

5. References

Annevelink, E. (1989) 'The IMAG production planning system (IDP) for glasshouse floriculture in its introduction phase'. Acta Hort. 237, 37-45.

Berg, E. and Lentz, W. (1989) 'Dynamic optimal control of plant production'. Acta Hort. 248, 223-241.

Brouwer, R. and Wit, C.T. de (1969) 'A simulation model of plant growth with special attention to root growth and its consequences', in W.J. Whittington (ed.), Root growth, Butterworths, London, pp. 224-242.

Buchwald, H.H. (1987) 'A simulation model for planning and control of the potted plant production'. Acta Hort. 203, 39-50.

194

Challa, H. (1990) 'Crop growth models for greenhouse climate control', in R. Rabbinge, J. Goudriaan, H. van Keulen, F.W.T. Penning de Vries and H.H. van Laar (eds.), Theoretical Production Ecology: reflections and perspectives, Simulation Monographs 34, PUDOC Wageningen, pp. 125-145.

Dale, J. (1985) 'Carbohydrate partitioning and metabolism in crops'. Hort. Rev. 7, 69-108.

Farrar, J.F. (1988) 'Temperature and the partitioning and translocation of carbon', in S.P. Long and F.I. Woodward (eds.), Plants and temperature, Symposia Soc. Exp. Biol., Essex England 1987, Company of Biologists Ltd., Cambrigde 42, 203-35.

Gifford, R.M. and Evans, L.T. (1981) 'Photosynthesis, carbon partitioning and yield'. Ann. Rev. Plant Physiol. 32, 485-509.

Gollwitzer, S. (1991) 'Implementation problems of computer-based planning models in horticultural firms'. Acta Hort. 295, 185-193.

Håkansson, B. (1991) 'A Decision Support System for production planning in horticulture'. Acta Hort. 295, 195-201.

Heuvelink, E. and Bertin, N. (1994) 'Dry matter partitioning in a tomato crop: comparison of two simulation models'. J. Hort. Sci. (submitted).

Heuvelink, E. and Marcelis, L.F.M. (1989) 'Dry matter distribution in tomato and cucumber'. Acta Hort. 260, 149-57.

Jones, J.W., Dayan, E., Allen, L.H., Keulen, H. van and Challa, H. (1991) 'A dynamic tomato growth and yield model (TOMGRO)'. Trans. Amer. Soc. Agric. Eng. 34, 663-72.

Kano, A. and Bavel, C.H.M. van (1988) 'Design and test of a simulation model of tomato growth and yield in a greenhouse'. J. Jap. Soc. Hort. Sci. 56, 408-416.

Krafka, B.D.L., Shumway, C.R. and Reed, D.W. (1989) 'Space allocation in foliage production greenhouses'. J. of Env. Hort. 7, 95-98.

Krijgsman, H.K. and Achter, J.M.F.N. (1973) 'Some aspects of management on potplant-nurseries'. Acta Hort. 31, 117-118.

Leutscher, K.J. and Vogelezang, J.V.M. (1990) 'A crop growth simulation model for operational management support in pot plant production'. Agric. Systems 33, 101-114.

Marcelis, L.F.M. (1993) 'Simulation of biomass allocation in greenhouse crops - a review'. Acta Hort. 328 (in press).

Marcelis, L.F.M. (1994) 'A simulation model for dry matter partitioning in cucumber'. (in prep.)

Marcelis, L.F.M., Heuvelink, E. and Koning, A.N.M. de (1989) 'Dynamic simulation of dry matter distribution in greenhouse crops'. Acta Hort. 248, 269-76.

Penning de Vries, F.W.T. (1990) 'Can crop models contain economic factors?', in R. Rabbinge, J. Goudriaan, H. van Keulen, F.W.T. Penning de Vries and H.H. van Laar (eds.), Theoretical Production Ecology: reflections and perspectives, Simulation Monographs 34, PUDOC Wageningen, pp. 89-103.

Penning de Vries, F.W.T. and Laar, H.H. van (eds.) (1982) 'Simulation of plant growth and crop production', Simulation Monographs, PUDOC Wageningen, 308 pp.

Schapendonk, A.H.C.M. and Brouwer, P. (1984) 'Fruit growth of cucumber in relation to assimilate supply and sink activity'. Scientia Hortic. 23, 21-33.

Wit, C.T. de (1965) 'Photosynthesis of leaf canopies'. Agric. Res. Rep. 663, PUDOC, Wageningen, 57 pp.

Wolswinkel, P. (1985) 'Phloem unloading and turgor-sensitive transport: Factors involved in sink control of assimilate partitioning'. Physiol. Plant. 65, 331-39.

RELEVANCE OF HETEROGENEITY OF CEREAL CROP STANDS FOR PLANT HEALTH

VIVIAN VILICH
Institute for Plant Diseases, University of Bonn
Nußallee 9
53115 Bonn
Germany

ABSTRACT. Mixtures of cereal species were used to study the effect of genetic diversity in crop stands on fungal leaf pathogens and stem/root-rot diseases. Leaf damage caused by cereal rusts, mildew and leaf spots correlated strongly with the host density. The air-borne inoculum of pathogens was reduced. Stem and root rot pathogens were restricted by the use of mixtures as pre-crops and main-crops in a rotation. This mixture effect was still present to some extent after 3 year continuous cropping of cereals.

1. Introduction

Establishing more self-regulating agro-ecosystems is one of the major goals of biological and integrated pest management. In case of most natural ecosystems plants are a composition of different species varying in genotype, age and resistance to abiotic and biotic stress. It is well documented that genetic diversity favours the resistance of crop stands. Against foliar pathogens, good results have been obtained with mixtures of varieties and species (Frey *et al.*, 1977; Barrett, 1980; Ibenthal *et al.*, 1988; Vilich-Meller und Weltzien, 1989). Furthermore, research was carried out on the effects of mixed cropping on soil-borne diseases, i.e. typical crop-rotation associated pathogens (Gieffers und Hesselbach, 1988; Vilich-Meller, 1992; Vilich, in press). Crop rotations play a key role in the restriction of many diseases in the aerial part and the root system of a plant. The theoretical background is based on the reduced density of hosts in a given time period. Mixed cropping reduced the density of susceptible hosts on a given area, i.e. 'genotype unit area', described by Mundt *et al.* (1986).

The object of this paper is to reveal the effectiveness of species mixtures of cereals against selected leaf diseases and stem-rot in 1986-92.

P. C. Struik et al. (eds.), Plant Production on the Threshold of a New Century, 197–203.
© 1994 *Kluwer Academic Publishers.*

2. Materials and methods

The experimental area was located at the University of Bonn Research Station. Several mixtures of cereal species were tested:

Spring barley - Spring oats (1986-1988; ♦)
Winter barley - Winter wheat - Winter oats (1989/90)
Winter barley - Triticale - Winter oats (1990/91)
Winter barley - Winter wheat (♦) (1989-92)

Experiments with barley/oats and barley/wheat mixtures (♦) are reported here. Cereals were sown as pure stands and seed mixtures (1:1 or in different ratios). For detailed descriptions see Vilich-Meller (1989, 1992) and Vilich (in press). The barley/wheat experiment was designed as a three-year consecutive rotation of this cereals at the same plot area. To compare the effectiveness of mixed cropping vs. chemical plant protection, subplots were treated with fungicides against foliar pathogens and stem rot diseases.

Thirty main shoots per replication (n=4) formed a sample for disease assessment. Percentage leaf area diseased was assessed using a scale according to James (1971). The inoculum density of wind spread leaf pathogens was estimated by using young barley trap plants. The incidence of disease symptoms on roots and shoots was estimated quantitatively (Vilich, in press) at different growth stages. Data were analysed by using ANOVA and FACTOR analysis.

3. Results

Disease severity of foliar pathogens correlated strongly with the density of host plants. This relationship was also evident in fungicide treated barley plots (Figure 1A). Increasing the host density resulted in a higher number of *Puccinia hordei* pustules on the upper two leaves.

This effect paralleled the inoculum density of *P. hordei*, estimated on young barley trap plants (Figure 1B). High correlation coefficients were calculated between the host density and the inoculum density at two dates of trapping (r=0.88 and 0.89). At least 40 % oats in a mixture seemed to be needed for effective inoculum reduction.

Leaf diseases of oats were similarly affected by mixed cropping with barley (Table 1). The influence on disease development depended on the pathosystem observed. The disease severity of leaf spots (*Drechslera avenae*) and powdery mildew (*Erysiphe graminis*) correlated to a high extent (r=0.81 to 0.99) with the host density after flowering (EC 69). The epidemic progress of leaf rust (*Puccinia coronata*) correlated highly in EC 51 and EC 69 with the oat density.

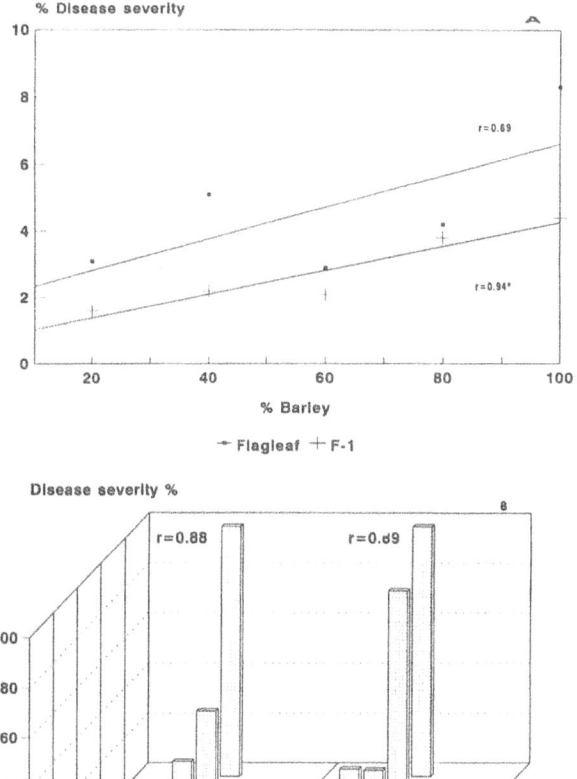

Figure 1. Disease severity (A) of *Puccinia hordei* on barley in pure stands and mixtures with oats (EC 61; plots were fungicide treated) and inoculum density (B) of *Puccinia hordei* in barley pure stands and mixtures with oats (disease severity on barley trap plants).

Later in the growing season plant mature was often enforced by root and stem rot diseases caused by *Pseudocercosporella herpotri-choides*, *Gaeumannomyces graminis*, *Rhizoctonia cerealis* and *Fusarium* spp. Growing wheat and barley in a mixture resulted in a greater number of healthy stems (Table 2), thus fewer plants remained senescent, and the degree of lodging was smaller.

In addition, mixtures used as pre-crops also reduced stem and root rot diseases of barley and wheat. During ripening plant health

TABLE 1. Correlation between seed density of oats in pure stands and mixtures with barley and disease severity of *Drechslera avenae*, *Puccinia coronata* and *Erysiphe graminis* f.sp. *avenae* (Flagleaf F and F-1), (Correlation coefficient r; underlined: significant at P=5 %).

	Without Fungicides			With Fungicides		
D. avenae	EC 51	EC 61	EC 69	EC 51	EC 61	EC 69
F	-0.39	0.77	0.81	0.16	0.65	0.72
F-1	0.20	0.78	<u>0.91</u>	-0.42	0.86	0.87
P. coronata						
F	<u>0.90</u>	-0.25	<u>0.96</u>	--	--	0.69
F-1	0.74	0.77	<u>0.99</u>	--	--	0.57
E. graminis						
F	--	--	<u>0.97</u>	--	--	0.70
F-1	--	--	<u>0.98</u>	--	--	0.73

TABLE 2. Number (%) of barley and wheat plants without symptoms of stem and root rot in growth stage 71. Pure stand and mixture used as pre-crop and main-crop (fungicide untreated).

	Main crop barley in		
	Pure stand	Mixture	φ Pre-crop
Pre-crop barley	4.4	11.1	7.7
Pre-crop mixture	4.4	23.3	13.9
Pre-crop wheat	5.6	10.0	7.8
φ Main crop	4.8	14.8	

	Main crop wheat in		
	Pure stand	Mixture	φ Pre-crop
Pre-crop barley	3.3	11.0	7.2
Pre-crop mixture	17.8	12.3	17.8
Pre-crop wheat	5.6	20.0	12.8
φ Main crop	8.9	16.3	

increased obviously in mixtures grown after a barley/wheat-mixture compared to a mixture grown after pure stands. This effect was more evident in barley than in wheat. On average, mixtures used as main-crops were more effective than as pre-crops.

The use of pure stands and mixtures in a three-year crop rotation resulted in different ratios of barley and wheat in every treatment (Table 3). Plant health was only little affected by the crop rotation. However, disease incidence of distinct pathogens, e.g. *G. graminis*,

TABLE 3. Root and stem rot of barley and wheat in a three-year rotation (1989-92) with mixtures and pure stands of barley and wheat (fungicide untreated).

	% Barley/Wheat in a three-year rotation					
	100	83.3	66.7	66.7	50.0	33.3
A	96.7	100	86.7	93.3[A]	93.3[A]	93.3[A]
B	67.7	90.0	83.3	53.3[A]	67.7[A]	80.0[A]

A: % barley stems diseased; [A] crop stand in 1992 was a mixture
B: % wheat stems diseased; [A] crop stand in 1992 was a mixture

Fusarium spp., was not unique (data not published). In addition, plant health of wheat was slightly enhanced in crop rotations ending with a mixture. Thus, the mixture effect was still visible to some extent after the third year of continuous cropping of barley and wheat.

4. Discussion

Field experiments carried out in 1986-92 showed that increasing crop heterogeneity by the use of species mixtures favoured plant health. Disease incidence and severity of several fungal pathogens was reduced on leaves and root systems. Studies on intra- and interspecific mixtures of arable plants showed that there are three major effects which are responsible for disease reduction (Wolfe, 1985; Burdon, 1987): 1) a reduced amount of susceptible tissue, and consequently, a reduced amount of infectious propagules, 2) the 'barrier effect', i.e. an increased distance between hosts and 3) induction of resistance in hosts by non-pathogens. Experiments with trap plants (Figure 1B) demonstrated the strong correlation between host- and inoculum density. A great number of propagules are 'trapped' by non-hosts (barrier effect). Moreover, the latter can induce resistance mechanisms in the non-hosts as shown in growth chamber experiments (Vilich-Meller und Weltzien, 1990). The relative significance of these effects varied with crops, the involved microorganisms and environmental conditions. The soil ecosystem is more complex and pathogens are often less specialized to single hosts (e.g. *Fusarium* spp.). However, the mixed cropping of cereals resulted in a reduction of stem and root-rot diseases and unique interactions between pathogens (Vilich-Meller, 1992). Variation in pathogenicity of fungi to different cereals may result in varying numbers of infectious propagules due to antagonism and competition (Sturz and Bernier, 1989, 1991). Furthermore, the activity of soil microorganisms may be changed or enhanced (Bopaiah and Shekara Shetty, 1991), thus improving plant health. Mixed cropping of cereals had beneficial effects over more than one growing season (Vilich, in press) and long term crop heterogeneity may favour plant health due to a more balanced microbial environment.

5. References

Barrett, J. A. (1980) 'Pathogen evolution in multilines and variety mixtures', Z. Pflanzenkrankheiten und Pflanzenschutz 87(7), 383-396.

Bopaiah, B. M. and Shekara Shetty, H. (1991) 'Soil microflora and biological activities in the rhizosphere and root regions of coconut-based multistoreyed cropping and coconut monocropping systems', Soil Biol. Biochem. 23, 89-94.

Burdon, J. J. (1987) 'Diseases and plant population biology', Cambridge studies in ecology, Cambridge University Press.

Frey, K. J., Browning, J. A. and Simons, M. D. (1977) 'Management systems for the host genes to control disease loss', Ann. N. Y. Acad. Sci. 287, 255-274.

Gieffers, W. und Hesselbach, J. (1988) 'Krankheitsbefall und Ertrag verschiedener Getreidesorten im Rein- und Mischanbau. V. Vergleichender Überblick der Sortenmischungen mit Gerste, Weizen und Roggen 1984-1986', Z. Pflanzenkrankheiten und Pflanzenschutz 95(2), 203-209.

Ibenthal, W.-D., Göbel, M., Willnecker, G. und Bernhold, L. (1988) 'Ertragniveau, Krankheitbefall und Mehltauvirulenz in Sortenmischungen von Sommergerste (1984-1986)', Z. Pflanzenkrankheiten und Pflanzenschutz 95(6), 561-571.

James, C. (1971) 'A manual of assessment keys for plant diseases', American Phytopathol. Soc., St. Paul, USA.

Mundt, C. C., Leonard, K. J., Thal, W. M. and Fulton J. H. (1986) 'Computerized simulation of crown rust epidemics in mixtures of immune and susceptible oat plants with different genotype unit areas and spatial distribution of initial disease', Phytopathology 76, 590-598.

Sturz, A. V. and Bernier, C. C. (1989) 'Influence of crop rotation on winter wheat growth and yield in relation to the dynamics of the pathogenic crown and root rot fungal complex', Can. J. of Plant Pathol. 11, 114-121.

Sturz, A. V. and Bernier, C. C. (1991) 'Fungal communities in winter wheat roots following crop rotations suppressive and non-suppressive to take all', Can. J. Botany 69, 39-43.

Vilich-Meller, V. (1989) 'Der Einfluß von Mischkulturen auf den Schaderegerbefall am Beispiel der Futtergetreide-Mischung Sommergerste/Hafer', unpublished PhD-Thesis.

Vilich-Meller, V. (1992) 'Pseudocercosporella herpotrichoides, Fusarium spp. and Rhizoctonia cerealis stem rot in pure stands and interspecific mixtures of cereals', Crop Protection 11, 45-50.

Vilich, V. (in press) 'Crop rotation with pure stands and mixtures of barley and wheat to control stem and root rot diseases', Crop Protection.

Vilich-Meller, V. und Weltzien, H. C. (1989) 'Artenmischungen von Sommergerste und Hafer: Einfluß auf den Blattbefall pilzlicher Schaderreger und auf die Ertragsfähigkeit', Z. Pflanzenkrankheiten und Pflanzenschutz 96(1), 1-10.

Vilich-Meller, V. und Weltzien, H. C. (1990) 'Resistenzinduktion in Gerste und Hafer durch Vorinokulation mit Apathogenen - ein befallsmindernder Mechanismus in Gemischten Getreidebeständen', Z. Pflanzenkrankheiten und Pflanzenschutz 97, 532-543.

Wolfe, M. S. (1985) 'The current status and prospects of multiline cultivars and variety mixtures for disease resistance', Ann. Rev. Phytopathol. 23, 251-273.

IMPLICATIONS OF INTERCROPPING (SWEET PEPPER - TOMATO) FOR THE BIOLOGICAL CONTROL OF PESTS IN GLASSHOUSES

P. NIHOUL AND T. HANCE
Catholic University of Louvain
Laboratory of Ecology and Biogeography
Place Croix du Sud, 5
B-1348 Louvain-la-Neuve, Belgium

ABSTRACT. The effects of intercropping on the distribution of pests were analysed in a cropping system composed of rows of sweet peppers alternating with rows of tomatoes in a 80 m² glasshouse. The levels of populations of *Myzus persicae*, *Tetranychus urticae* and *Frankliniella occidentalis* and their predators or parasitoids were compared among the rows. Significant differences in pest, predator and parasitoid density were observed. This was partially attributed to differences in plant suitability influencing the predator-prey balance. Moreover, highly significant differences in pest density were recorded among the rows of the vegetable species which favoured high pest outbreaks. This effect was probably due to negative effects caused by the other plant species on the dispersion of the pest. In fact, glandular trichomes heavily entrapped aphids on tomatoes, which probably greatly affected the pest dispersal in the crop. On the other hand, the fact that spider mites were maintained on the sweet pepper plants at very low levels by predatory mites was certainly a factor which contributed to the low spread of the pest in the crop. These results lead to the conclusion that such intercropping seems to locally limit pest outbreaks, and thus may reduce the amount of biological auxiliaries or pesticides needed to control them in comparison with monoculture.

1. Introduction

Monocultures are the only agrosystems which are developed extensively to provide for the increasing needs of the humanity. However, certain species of phytophagous arthropods have become extremely abundant in such crops. The general instability in monocultures was generally thought to be favoured by less interactions than in a natural system, which can absorb effects of rapid changes in the density of one species (Horn, 1988). By increasing heterogeneity in intensive crops by using an intercropping system, one may wonder if the control of phytophagous arthropods would be improved. In fact, the stability of predator-prey interactions depends on spatial asynchrony in predator

P. C. Struik et al. (eds.), Plant Production on the Threshold of a New Century, 205–211.

population development (Nachman, 1991), which can be achieved by a heterogeneous environment (Nachman, 1988). Spatial heterogeneity is also a potentially powerful stabilizing force in host-parasitoid interactions (Holt and Hassell, 1993). Moreover, more abundant entomophagous insects were recorded in tropical intercropping systems than in monocultures (Kennedy *et al.*, 1990). This might be caused by more vital food, shelter and resting site (Risch, 1981).

The aim of this study is to analyse the distribution and development of populations of phytophagous insects, phytophagous mites, their predators and parasitoids in a row intercropping system (sweet pepper - tomato) in a glasshouse. Pest distribution and beneficial arthropod distribution were discussed in terms of effects on pest control.

2. Materials and methods

2.1. CULTURAL CHARACTERISTICS

Tomatoes cv. 2209 (De Ruiter) and sweet peppers cv. Mazurka (Rijk Zwaan) were grown on Grodan® rockwool under natural light in a 80 m² glasshouse at Louvain-la-Neuve (Belgium) from 2 February to 15 November, 1992. Tomatoes were stood on the rockwool per group of 2 rows (20 plants per row; rows 1, 2, 5 and 6) alternately with 2 rows of sweet peppers (rows 3, 4, 7 and 8). Row location in the glasshouse occurred following a repetitive process of two rows 60 cm apart followed by a 1.2 m walking path. Plants in the row were spaced at 55 cm and were pruned to a single stem for tomatoes and to a double stem for sweet peppers.

2.2. PEST CONTROL AND POPULATION SAMPLING

Two species of insects (the western flower thrips, *Frankliniella occidentalis* (Pergande), and the green peach aphid, *Myzus persicae* (Sulzer)) and one species of spider mite (the two-spotted spider mite, *Tetranychus urticae* Koch) were observed from mid-March, mid-May and the beginning of April, respectively. Predators provided by Koppert (The Netherlands) were introduced onto the crop to control all these pests, and no pesticides were sprayed on the foliage. Populations of the pests and their predators were estimated on sweet pepper and tomato plants of rows 2 to 7.

2.2.1. Western flower thrips

A total of 80 predatory mites, *Amblyseius cucumeris* (Oudemans) were put on each sweet pepper plant from 11 March to 10 June. They were put on leaves by means of a brush once or twice a week. The amount of individuals per release and plant ranged from 3 to 6. Their populations as well those of the western flower thrips were fortnightly estimated in flowers and on leaf areas from 24 March to 23 October. Per sampling date, 10 flowers were cut at random per row and dissected using a stereomicroscope to count the number of thrips and predators. The foliage was also observed by using a 10 x hand lens on a sample (n=20 per row) of 17 cm² leaf areas in the upper third of 20 plants chosen at random. Moreover, the density of necrotic leaf areas caused by the nutrition of thrips on the foliage between 1.3 and 1.5 m in height on the plant was estimated on 2 October (n=20 leaves per row).

2.2.2. Green peach aphids

Two local species of *Hymenoptera*, *Praon volucre* (Haliday) and *Aphidius matricariae* Haliday naturally colonized the crop. Moreover, two introductions of approximately 500 adults of the predatory midge, *Aphidoletes aphidimiza* (Rondani) were carried out on 24 June and 17 September. Mid-August, mid-September and mid-October, aphids, their predators and parasitoids were counted on the foliage in the upper third of 10 plants per row by using a 10 x hand lens. Per plant, a total of ten 1.3 cm² leaf areas were chosen at random. Moreover, the number of adult and immature aphids immobilized by glandular trichome exudates was determined on 10 tomato leaves randomly selected per row.

2.2.3. Spider mites

Predatory mites, *Phytoseiulus persimilis* Athias-Henriot were applied by means of a brush on tomato and sweet pepper plants of rows 3, 4, 5 and 6. Sweet pepper rows 3 and 4 received 30 and 75 predators, respectively on 1 July, whereas tomato rows 5 and 6 received a total of 360 and 300 predators, respectively on 24 July, 20 August and 2 September. Spider mite and predatory mite density on leaf area was estimated fortnightly in the third upper part of the 10 middle plants of each row. On tomato, 2 leaflets were cut at random per plant; their area was calculated, and the numbers of individuals of both mite species present on their underside were counted under a stereomicroscope. On sweet pepper, however, mite density was directly calculated by averaging the numbers of individuals observed by means of a 10 x hand lens on 1.3 cm² leaf areas (n=10) chosen at random per plant.

2.3. STATISTICAL ANALYSIS

The analysis of the inter-row spatial pattern of the pests, their predators and parasitoids was made per sampling date using a 1-factor fixed analysis of variance (ANOVA) model with a fixed row effect: multiple comparisons by contrasts were made to determine which row significantly differed regarding to arthropod density. Prior to analysis, data were transformed by natural logarithm (ln (x+1)) to be normalized and to stabilize the variance.

3. Results

3.1. WESTERN FLOWER THRIPS

High thrips densities occurred on sweet peppers and not on tomatoes. Nevertheless, thrips individuals successfully reached the tomato rows: a total of 14 adults (no immatures) were found in the 320 tomato flowers which were dissected (711 thrips in the sweet pepper flowers). But only one larvae on the foliage and very few leaves with feeding marks of thrips (<0.0001 necrotic areas per 20 cm^2 leaf area) have been observed on the tomatoes, whereas on sweet peppers, I observed 5.8 ± 7.4 and 0.7 ± 1.4 necrotic areas per 20 cm^2 leaf area on rows 4 and 5, and on row 7, respectively at the beginning of October. On sweet peppers, the necrotic areas caused by thrips nutrition were as dense in row 4 as in row 5, but they were less dense than those on row 7 ($P = 0.005$). No significant difference in thrips density on the leaves was recorded among the three rows. Predatory mites, however, were denser in row 7 than in the two other rows from mid-August, which could explain lower damage.

3.2. GREEN PEACH APHIDS

High aphid colonization occurred on sweet peppers, whereas only dying colonies were observed on tomato leaves; there were a few adults and immatures either disturbed in their movement or immobilized by tomato trichome exudates which hardened on their legs, or stuck dead on trichomes. The high numbers of entrapped aphids were observed on the two tomato rows which stood alongside sweet pepper rows 3 and 4 (Figure 1). These two sweet pepper rows were heavily infested by aphids: on mid-October, there were 6.0 ± 3.3, 5.0 ± 4.2 and 0.2 ± 0.2 aphids per 1.3 cm^2 in row 3, 4 and 7, respectively. The aphid density on row 7 significantly differed from those on rows 3 and 4, whereas there was no difference between rows 3 and 4 where intolerable aphid densities were reached as the plants were covered by honeydew.

3.3. SPIDER MITES

Plants of the six analysed rows were attacked by spider mites, but significant differences in spider mite density were recorded among the rows. Tomato rows 5 and 6 supported very high pest densities, whereas the sweet pepper rows supported very low ones (Figure 2). Surprisingly, there were low spider mite densities on row 2. Infestation started there several weeks after the one on the other rows. Spider mites firstly established in rows 3, 4 and 5 where higher mite densities were observed in the tomato row than in the sweet pepper rows from

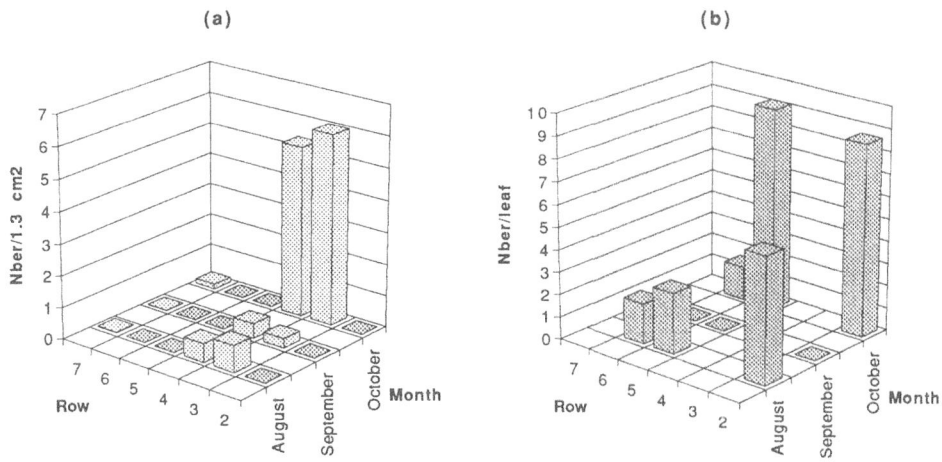

Figure 1. (a) Number of alive aphids, *M. persicae* on the leaf areas in the tomato (⊞) and sweet pepper (⊡) rows, and (b) of aphids entangled by tomato trichome exudates per leaf.

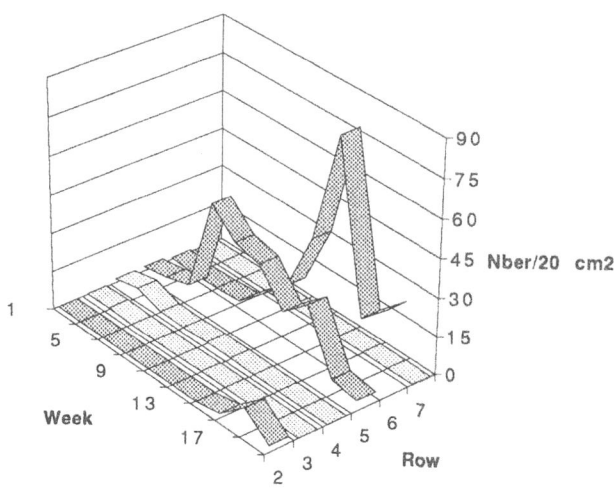

Figure 2. Mean number of spider mites, *T. urticae* on the leaf areas in the tomato (⊞) and sweet pepper (⊡) rows from 2-7 to 6-11-92.

mid-July. Afterwards, infestation spread into rows 6 and 7, and high pest densities were achieved on tomato row 6 from the end of August. I observed maximum values of spider mite density of 35 ± 39 and 88 ± 80 per 26 cm^2 leaf area in rows 5 and 6, respectively. In the sweet pepper rows, however, there was a maximum of 0.7 ± 1.4 spider mites per 20 cm^2 leaf area. Only from the end of October, spider mite density was greater in tomato row 2 (maximum 11 ± 13 spider mites per 20 cm^2 leaf area) than in the sweet pepper rows. The distribution and development of the predatory mite, *P. persimilis* followed a similar pattern of that of the prey.

4. Discussion and conclusion

The occurrence of differences in pest population between the two plant species was expected, partially caused by the degree of suitability of the two plant species for the pest development, reproduction and longevity (e.g., Van de Vrie *et al.*, 1972), partially by the fact that tomato trichomes deter foraging activity of natural enemies of pests, e.g. *P. persimilis,* thereby affecting the balance between the beneficial arthropod and the pest (Nihoul, 1993a,b).

The original result concerns the effect of each plant species on the distribution of the pests because of row intercropping. Sweet pepper rows, on which spider mites were very well controlled by predatory mites, seemed to reduce the speed and the extent of the pest spread over the crop. On the other hand, tomato rows by affecting aphid distribution were certainly the cause of the localization of high pest populations on restricted sweet pepper rows. Finally, it could be that the population of *A. cucumeris*, the predator of thrips, was affected through interspecific relations (habitat modification, disturbance) by the aphid, thus influencing the thrips population in accordance with the aphid population levels.

In conclusion, in an intercropping system, the plant species which is either the least suitable for the pest because of having a defence mechanism or the most favourable for stable predator-prey interactions with low pest levels, might protect other plants species by concentrating the pest in a certain part of the crop where it can economically be destroyed. Less and more locally applied chemicals or predators might thus be expected, if compared to a monoculture. Further experiments comparing intercropping with monoculture must be done to confirm this hypothesis.

5. Acknowledgements

The author is grateful to Prof. Ph. Lebrun for his helpful suggestions and to Mr. H. Vanderlinden for his help in collecting the data. This research has been supported by the 'Institut pour I'Encouragement de la Recherche Scientifique dans l'Industrie et l'Agriculture' (IRSIA).

6. References

Holt, R.D. and Hassell, M.P. (1993) 'Environmental heterogeneity and the stability of host-parasitoid interactions', J. Anim. Ecol. 62, 89-100.

Horn, D.J. (1988) 'Ecological approach to pest management', Guilford Press, New York. 285 pp.

Kennedy, F.J.S., Rajamanickam, K. and Raveendran, T.S. (1990) 'Effect of intercropping on insect pests of groundnut and their natural enemies', J. Biol. Control 4, 63-64.

Nachman, G. (1988) 'Regional persistence of locally unstable predator/prey populations', Exp. Appl. Acarol. 5, 293-318.

Nachman, G. (1991) 'An acarine predator-prey metapopulation system inhabiting greenhouse cucumbers', Biol. J. Linn. Soc. 42, 285-303.

Nihoul, P. (1993a) 'Controlling glasshouse climate influences the interaction between tomato glandular trichome, spider mite and predatory mite', Crop Protection 12 (in press).

Nihoul, P. (1993b) 'Have light intensity, temperature and photoperiod an effect on the entrapment of mites on glandular hairs of cultivated tomatoes?', Exp. Appl. Acarol. (submitted for publication).

Risch, S.J. (1981) 'Insect herbivore abundance in tropical monoculture and polycultures; an experimental test of two hypothesis', Ecology 62, 1325-1340.

Van de Vrie, M., McMurtry, J.A. and Huffaker, C.B. (1972) 'Ecology of tetranychid mites and their natural enemies: a review. III Biology, ecology, and pests status, and host-plant relations of tetranychids', Hilgardia 41, 343-432.

COMPARISON OF THREE POTATO PATHOSYSTEMS IN THE NETHERLANDS

TH. JACOBS
Department of Plant Breeding
Wageningen Agricultural University
P.O. Box 386
6700 AJ Wageningen
The Netherlands

ABSTRACT   Major gene resistance in conjunction with stringent phytosanitation and a lack of "genetic plasticity" of the viruses involved have led to durable resistance in most potato cultivars against virus diseases. Due to a narrow rotation the incidence of potato cyst nematodes increased. Despite chemical treatments and use of resistant cultivars no durable resistance against *Globodera rostochiensis* and *G. pallida* has been achieved. The use of fungicides overruled the need for durable resistance against late blight. Major gene resistance is clearly not durable, minor gene resistance is present in several cultivars and needs to be accumulated. A large proportion of the Dutch potato area has been occupied by the cultivar Bintje which is susceptible to several viruses and potato cyst nematodes. Ever since its introduction in 1910 Bintje was susceptible to late blight. Such a large scale cultivation of a susceptible cultivar should not be continued in the 21st century.

1. Potato

The cultivated potato was introduced in Europe around 1600. The cultivated potato is a tetraploid and highly heterozygous crop. Cultivars are maintained vegetatively. Parts of the potato genome have been mapped, a preliminary map of potato is available. Linkage of morphological markers, resistance genes and RFLP's has been described (Uhrig *et al.*, 1992).

2. Potato viruses

2.1. TYPES AND STRAINS

In the Netherlands, potato virus X (PVX), virus Y (PVY), virus A (PVA) and leaf roll virus (PLRV) occur among others. The RNA sequence of all viruses is known, the genomes are relatively small. PVX is mechani-

P C Struik et al (eds), Plant Production on the Threshold of a New Century, 213–222
© 1994 *Kluwer Academic Publishers*

cally transmitted, the other viruses are vector transmitted, mainly by aphids. The viruses damage the plant and cause substantial yield reductions. The viruses can infect the tubers, transmission via seed potato leads to a high initial inoculum and should be kept to a minimum.

Four strains have been described for **PVX** (Cockerham, 1955), strain nr 3 is the most dominant one. The monogenic dominant resistance gene Nx leads to a hypersensitivity reaction with strain 3 and is present in many cultivars. The leading cultivar Bintje carries resistance gene Nb and is susceptible to PVX strain 3 (Table 1).

TABLE 1. Acreage (%) and level of resistance of the four leading cultivars according to the Dutch cultivar list, scale 2 (highly susceptible) - 9 (highly resistant).

Cultivar	Acreage (%)	PLVR	PVA	PVX	PVY"	PCN	Phytophthora leaf	tuber
Bintje	40	6	res[1]	5	5	sus[2]	3	3
Elkana	7	5	res	6	8	rC	5	7
Elles	6	5	res	9	6	rD	8	7
Astarte	5	4	res	res	8	rA	7	6

1) res = resistant, 2) sus = susceptible, rA, rC, rD indicate resistance to the postulated pathotypes Ro 1 (rA), Ro 1 2 3 (rC) and Ro 1 2 3 Pa 2 (rD) respectively.

Three strains of **PVY** have been described on the basis of their symptoms: Y^n, Y^c and Y^o. Resistance to the strain Y^c is common and caused by the resistance gene Nc. Resistance to the most common strain Y^o is less available.

No strains of **PVA** have been described, although there are differences in severity. The widely present monogenic resistance gene Na is linked with the resistance gene Nx against PVX.

No strains of the persistent potato leaf roll virus, **PLRV**, are known, neither are major resistance genes available. Resistance is of a quantitative nature and polygenic. A limited number of cultivars is resistant to PLRV. (Beemster and Bokx, 1987; Beekman, 1987; Foxe, 1992).

The importance of (para)sexual recombination between the strains is unclear but seems limited. It is not known when the viruses first occurred in Europe. Our knowledge of the variation in pathogen populations and the geographic variation is meagre.

2.2. PROTECTION AND PREVENTION

The environmentally most preferred mode of protection is the use of resistance against virus infection, multiplication and movement. In addition to virus resistance there is a search for vector resistance especially in relation with PVY and PLRV. Incorporation of viral coat protein in plants through biotechnological techniques is possible, the first PVX resistant potato lines have been obtained (Huisman *et al.*, 1992) but are not yet commercially available.

There is a strict control of seed potatoes in the Netherlands. Every year the presence of viruses in seed potatoes is determined. Depending on the level of viruses the material is approved, classified lower or not classified. Each year about 4 percent of the acreage is not certified and has to be taken out of traffic due to virus con-tamination, in 1992 even 12 % of the basic seed potato material was not certified. Other protection measurements include removal of con-taminated plants, early haulm destruction and control of volunteer plants. Aphid populations are monitored. Recommendations for chemical treatments are made available to farmers. These phytosanitation measures clearly reduce the number of virus-infected plants at the beginning of the growing season (initial inoculum).

3. Potato cyst nematodes

3.1. SPECIES AND PATHOTYPES

Two closely related species of soil-borne nematodes occur in the Netherlands: *Globodera pallida* and the more common *Globodera rostochiensis*. They have been introduced into Europe from South America around 1850. The nematodes have one generation per growing season, they are virtually immobile and depend upon human activity for dispersal (cysts adhering to tubers; transport of infested soil and via ploughing). The nematode species are diploid cross-fertilizing animals. The female is fertilized prior to penetration of a potato root. She deposits hundreds of eggs in her own body, her cuticle tans, the remaining "cyst" is left in the soil and may persist for more than 20 years (Harris, 1978). In the absence of a host crop the annual rate of decline of cysts is estimated to be 30 %. If a resistant potato cultivar is grown the decline is about 75 %, due to the hatching factor produced by the potato roots.

Joint international research led to a classification of patho-types (Kort *et al.*, 1977) which was criticized (Nijboer and Parlevliet, 1990). The identification of pathotypes or 'virulence groups' highly depends on the available resistance genes and is influenced by environmental factors. The present classification is rather arbitrary (Arntzen and Van Eeuwijk, 1992). The major gene H1 for resistance to *G. rostochiensis* pathotype Ro1 and Ro4 is widely used in potato cultivars. Other resistance genes to *G. rostochiensis* have been identified but their number is limited, most originate from non-*Solanum tuberosum* sources. The resistance to *G. pallida* seems to be partial of nature (Nijboer and Parlevliet, 1990), a major resis-tance gene was reported (Arntzen *et al.*, 1993a), both monogenic and partial resistance have been detected in potato cultivars (Arntzen *et al.*, 1993b).

Probably a limited number of introductions of nematodes from South America occurred (Bakker, 1987; Gommers et al., 1992). These primary introductions were followed by secondary founding events. It is assumed that the number of relocated animals was limited and random genetic drift significantly influenced the genetic composition of the present populations. We lack detailed information on the geographic variation of virulence. We do not know if populations are in a Hardy-Weinberg equilibrium. At present there is a 'cultivar driven selection' on the potato cyst nematode populations possibly further lowering the genetic variation. The frequent use of the Hl resistance gene in new cultivars against pathotype Rol and Ro4 of *G. rostochiensis* favours rare virulence alleles, if present, and a shift towards *Globodera pallida*.

3.2. PROTECTION AND PREVENTION

Potato cyst nematodes became prevalent due to the tight rotation of susceptible cultivars. In the Netherlands a stringent control and assessment system was implemented to restrict movement of nematodes and cysts between fields and restrict the increase of the cysts on infested fields. Control relies on the use of nematicides, soil fumigants and incorporation of resistant cultivars in the rotation scheme. The compulsory use of sterilizing chemicals has been criticized (Van der Weijden, 1981) and is not accepted anymore on environmental and public health grounds. It is likely that the chemicals have a pathotype-independent effect on the populations, but the two species could react different to the treatments. A new monitoring system (closely sampling fields) is in use, reducing the frequency and magnitude of soil treatments.

A number of agricultural procedures, mainly practised by biological farmers, greatly reduce the incidence of potato cyst nematodes (Van der Weijden, 1981). These farmers grow potatoes in a 1 : 3 rotation, limit ploughing to a minimum, leave the small potatoes on the surface after the mechanical harvest so they freeze and carefully remove volunteer plants. The farmers use organic manure which presumably favours counter-acting biological enemies. Strict sanitation measurements are worthwhile; even in areas where all neighbour farmers have infested soils, some farmers keep their fields free of nematodes (Goewie, pers. comm.).

4. Phytophthora infestans, late blight

4.1. PATHOTYPES AND MATING TYPES

Phytophthora infestans attacks the leaves, stems and tubers of potato. The fungus can produce millions of spores within 4 - 5 days on a susceptible cultivar. The fungus overwinters as mycelium in infected tubers (asexual stage) or as oospores (sexual stage) in the soil.

Just like the potato cyst nematodes late blight occurred for the first time in Europe around 1850. A devastating epidemic occurred in Ireland and other European countries in 1845. Around 1930 several major resistance genes have been introduced in *Solanum tuberosum* from *S. demissum*. Up to 11 major genes for foliage resistance have been named R1 - R11, all of which lost their effectiveness fast. Pathotypes combining virulence against several resistance genes could be easily produced in the laboratory and occur in the field. This indicates that the fungus is highly flexible with regard to virulence genes.

The centre of diversity of the fungus is Central Mexico where both the A1 and A2 mating types are present, enabling sexual recombination. Until around 1980 only the A1 mating type occurred in Europe, the so-called 'old' populations predominantly consisted of one genotype with an asexual reproduction over the years. Around 1980 a shift occurred in the blight populations.

TABLE 2. Number of cultivars of seed potato under NAK inspection in five area categories in four years and total number of potato cultivars.

	Number of cultivars in area class					
	Area (ha)					total nr of
Year	>1000	<1000	<500	<100	<10	cultivars
1979	7	6	26	62	91	192
1981	5	7	26	62	91	191
1985	5	9	34	73	88	209
1991	5	12	49	76	99	241

Source: NAK, Netherlands General Inspection Service

The 'new' blight populations possess both the A1 and A2 mating type, have different isozyme and RFLP patterns and consist of many different and distinct genotypes (Fry *et al.*, 1992). In the Netherlands the new populations contain virulence factors previously only present in Mexico (Turkensteen, pers. comm.). The presence of new virulence factors is probably related with the introduction of new fungal material from Mexico around 1977 (Niederhauser, 1991). We experience earlier epidemics in the Netherlands the last few years, which could be related with the increased incidence of oospores, leading to a higher initial inoculum in spring.

4.2. PROTECTION AND PREVENTION

Application of chemical protectants on the foliage is the main mode of control used in the Netherlands ever since fungicides became available. Weekly spraying is common, the total number of sprayings can be as high as 16. In the past manual handling was carried out and infected tubers were removed which reduced the initial inoculum. Nowadays, heavily infested stocks are destroyed. Small garden holders still remove infested seed potatoes by hand. Since 1980 the oospores are the main source of inoculum in these gardens. Haulms are killed chemically or removed before the harvest, to prevent infection of the tubers. Volunteer plants from infested tubers can act as foci and need to be removed. Control of oospores deserves much more attention. In fact *Phytophthora* blight has become a soil born disease with a wind-borne summer dispersal.

The use of R gene resistance has not led to durable resistance. Durable resistance to late blight is of a polygenic nature (Turkensteen, 1993). Non-tuberosum species contain a large reservoir of 'general resistance' (Colon and Buddingh, 1988; Rivera-Peña, 1992). However, several old cultivars contain a reasonable level of partial resistance which might be accumulated (Colon, pers. comm.). In Mexico several potato cultivars with high levels of general resistance are cultivated (Turkensteen, pers. comm.). Marker-aided selection in breeding programmes has been suggested (Bonierbale *et al.*, 1992). This enables pyramiding of R genes against late blight but it should be pointed out that even before the appearance of the A2 mating type, the fungus developed pathotypes with multiple virulence.

Tuber resistance to *Phytophthora infestans* is present in several Dutch cultivars. R genes for resistance to leaf blight also act in the tubers. No genetic analysis has been carried out, no pathotyping is known. The occurrence of tuber blight in the field is confounded with the presence of the fungus in the foliage and the level of foliage resistance. Laboratory tests for tuber resistance are available, but the correlation between the results of some tests and field performance is low.

5. Comparison

The monogenic resistances against viruses have proven to be durable. This is due to the phytosanitation measurements taken to reduce the initial inoculum. The small size of the genome and the limited exchange of genetic material between populations seem to reduce the variability of these pathogens. Also, the chance that one mutation towards virulence will increase in frequency is small as a multitude of viruses is needed to infect one potato plant (Turkensteen, pers. comm).

The cross fertilizing potato cyst nematodes are genetically flexible, but due to i) the low number of introductions from South America, ii) their limited reproduction capacity, and iii) a low dispersal rate, their genetic variation is low. Regional differences between populations are present. It is clear that new pathotypes emerge, although we have difficulties with the detection and description of pathotypes. The quarantine and phytosanitation measurements do not satisfy the needs created by the narrow rotation.

TABLE 3. Area in ha of approved seed potatoes of the most prevalent (> 1000 ha) cultivars over a period of thirteen years, level of resistance against viruses, Potato Cyst Nematodes (PCN) and Phytophthora in leaf and tuber.

Cultivar	Area[1] 1979	1984	1988	1992	PLRV	PVA	PVX	PVY"	PCN	Phytophthora leaf	tuber
Bintje	10544	8121	7194	6056	6	res	5	5	sus[1]	3[3]	3
Désirée	2068	2361	2493	3058	4	res	6	8	sus	6	7
Spunta	1849	2681	2777	2564	6	res	6	7	sus	6	5
Jaerla	1441	1480	1149	1227	6	7	6	6	sus	5	8
Ostara	1072	1187	1151	795	7	res	8	7	sus	5	8
Alpha	1057	1099	483	336	5	5	6	5	sus	6	7
Sirtema	1106	788	243	155	6	5	6	5	sus	4	6
Diamant	7	452	1405	1637	6	res	6	6	resA	6	7

[1] The total area of seed potato was 33.617, 32.636, 32.452 and 40.958 in 1979, 1984, 1988 and 1992 respectively. 2) sus= susceptible, resA indicates resistance against the postulated pathotype A of Globodera rostochiensis. 3) 2=highly susceptible, 9=highly resistant. Source NAK, Ede, the Netherlands.

The *Phytophthora infestans* populations are flexible and vary in time and space. No major gene resistance has been durable. Around 1980 a shift in the populations occurred and sexual recombination was introduced. Presently epidemics start earlier and higher levels of resistance are needed. Control of the soil-borne oospores deserves much more attention.

6. Future

There will be a decreased use of agro-chemicals for crop protection. In order to protect the crop, changes in agricultural practices, increased phytosanitary control and durable genetic resistance are needed. Knowledge of the genetic variation in pathogen populations will lead to more efficient measures and a better use of the available resistance genes and cultivars.

Quarantine is important against new introductions for all the three pathogens and should be strict in future also. Several cultural and phytosanitation measures are practised to limit the amount of initial inoculum and the spread of the pathogens. This clearly restricts the damage by the pathogens and reduces the chance of recombination and mutation from which new pathotypes can originate. The introduction and implementation of additional monitoring systems and forecasting systems will be needed in future.

There is and will be a strong dependence on non-*tuberosum* species as sources of resistance to all diseases mentioned. Attempts to accumulate resistance from R gene free *tuberosum* cultivars are worthwhile. Diversification of resistance genes is badly needed, e.g. several cultivars possess only the resistance gene H1 against *Globodera rostochiensis*, many cultivars contain resistance gene R10 against *Phytophthora*. With regard to viruses, more emphasis should be given to vector resistance.

Through biotechnological techniques and through marker-aided selection new resistance genes will be introduced. Most of these genes will be genes with a major effect. It is encouraging to notice that there are renewed pleas for added protection of these rather expensive major genes by 'classical means', including diversification, gene rotation, gene deployment and cultivar mixtures.

It is particularly important to look for environmentally sound ways to reduce the build-up of the vector populations of the potato viruses. Procedures used in biological agriculture have a negative effect on the build-up of potato cyst nematode populations. It is worthwhile to study the effects and investigate their potentials. Equally so, the oospores of *Phytophthora* could be controlled by antagonists.

7. The Dutch situation, Bintje

In 1992, a total 187,000 ha was grown with potatoes, about 20 % was occupied for seed potato production, 45 % was in use for consumption and export purposes and about 35 % of the area was occupied by starch potatoes.

The seed potato area under NAK inspection, contains more than 200 different cultivars and are tested for presence of viruses each year (Table 2). Only 5 cultivars occupy more than 1000 ha (Table 3), these five cultivars are susceptible to the potato cyst nematodes and *Phytophthora infestans*.

The most prevalent cultivar is 'Bintje'. At present Bintje occupies more than 40 % of the total area grown with potato (Dutch cultivar list). Bintje was released in 1910, and possibly then already susceptible to late blight. Readings for *Phytophthora* leaf resistance are 3, 2, 3, 3 in 1931, 1944, 1955 and 1992 respectively (2 = extremely susceptible, Dutch cultivar list). The area occupied by Bintje increased slowly from 1910 onwards and was more than 30 percent for the last twenty years. Such a large scale cultivation of a susceptible cultivar is not a sound situation if one wants to reduce the initial inoculum, minimize the damage, delay the epidemic build-up and slow down the appearance of new pathotypes as much as possible. Commercial interests predominated in the choice of cultivars. Thanks to the continued efforts of Dutch breeders there is a large number of successors to choose from, more than 200 cultivars are listed in the Dutch cultivar list.

8. Acknowledgements

I had several fruitful discussions with L.J. Turkensteen (IPO-DLO); L.T. Colon (CPRO-DLO), R.H. Hutten (WAU) and A. Drenth (WAU) commented on the manuscript.

9. References

Arntzen, F.K. and Eeuwijk, F.A. van (1992) 'Variation in resistance level of potato genotypes and virulence level of potato cyst nematodes populations'. Euphytica 62, 135-143.

Arntzen, F.K., Vinke, J.H. and Hoogendoorn, J. (1993a) 'Inheritance, level and origin of resistance to *Globodera pallida* in the potato cultivar Multi derived from *Solanum tuberosum ssp andigena* CPC 1673'. Fundamental and Applied Nematology, in press.

Arntzen, F.K., Vinke J.H. and Hoogendoorn, J. (1993b) 'Inheritance and level of resistance to potato cyst nematodes (*Globodera pallida*) derived from *Solanum tuberosum ssp. andigena* CPC 1673', in Th. Jacobs and J.E. Parlevliet (eds.). Durability of Disease Resistance. Kluwer Academic Publishers, Dordrecht, the Netherlands.

Bakker, J., (1987) 'Protein variation in cyst nematodes'. PhD Thesis, Wageningen Agricultural University, the Netherlands. 159 pp.

Beekman, A.G.B. (1987) 'Breeding for resistance', in J.A. de Bokx and J.P.H. van der Want (eds.), Viruses of potatoes and seed-potato production. Pudoc Wageningen, pp. 84-113.

Beemster, A.B.R., and Bokx, J.A. de (1987) 'Survey of properties and symptoms', in J.A. de Bokx and J.P.H. van der Want (eds.), Viruses of potatoes and seed-potato production. Pudoc, Wageningen, 1987, pp. 84-113.

Bonierbale, M., Plaisted, R.L. and Tanksley, S.D. (1992) 'Applications of genome mapping in potato: A review', in Proceedings of the joint Conference of the EAPR Breeding and Varietal Assessment Section and the Eucarpia Potato Section, Landerneau, France.

Cockerham, G. (1955) 'Strains of potato X virus'. Proc. of 2nd Conference of Potato Disease. Lisse-Wageningen, the Netherlands, 1954, 89-92(72).

Colon, L.T. and Buddingh, D.J. (1988) 'Resistance to late blight (*Phytophthora infestans*) in ten wild Solanum species'. Euphytica 39 S, 77-86.

Foxe, M.J. (1992) 'Breeding for viral resistance: conventional methods'. Neth. J. Pl. Path. 98 supplement 2, 13-20.

Fry, W.E., Drenth, A., Spielman, L.J., Mantel, B.C., Davidse, L.C. and Goodwin, S.B. (1991) 'Population structure of *Phytophthora infestans* in the Netherlands'. Phytopathology 81, 1330-1336.

Fry, W.E., Goodwin, S.B., Matuszak, J.M., Spielman, L.J. and Milgroom, M.G. (1992) 'Population genetics and intercontinental migrations of *Phytophthora infestans*'. Annual Review of Phytopathology 30, 107-129.

Gommers, F.J., Roosien, J., Schouten, J., De Boer, J.M., Overmars, H.A., Bouwman, L., Folkertsma, R., Zandvoort, P. van, Gentpelzer, M. van, Schots, A., Janssen, R. and Bakker, J. (1992) 'Identification and management of virulence genes in potato cyst nematodes'. Neth. J. Pl. Path. Supplement 2, 157-163.

Harris, P.M. (1978) 'The potato crop, the scientific basis for improvement'. Chapman and Hall, London, UK.

Huisman, M.J., Cornelissen, B.J.C. and Jongedijk, E. (1992) 'Transgenic potato plants resistant to viruses'. Euphytica 63, 187-197.

Kort, J., Ross, H., Rumpenhorst, H.J. and Stone, A.R. (1977) 'An international scheme for identifying and classifying pathotypes of potato cyst nematodes *Globodera rostochiensis* and *Globodera pallida*'. Nematologica 23, 333-339.

Niederhauser, J.S. (1991) 'The Mexican Connection', in Lucas, Shattock, Shaw and Cook (eds.), Phytophthora. Br. Myc. Soc. Cambridge University Press. pp. 25-45.

Nijboer, H., and J.E. Parlevliet (1990) 'Pathotype-specificity in potato cyst nematodes, a reconsideration'. Euphytica 49, 39-47.

Rivera-Peña, A. (1992) 'Use of wild tuber bearing species of *Solanum* for breeding potatoes against *Phytophthora infestans* (Mont.) de Bary', in Proc. EAPR Conference, France, pp. 395-396.

Schöber, B. and Turkensteen, L.J. (1992) 'Recent and future developments in potato fungal pathology'. Neth. J. Pl. Path. Supplement 2, 73-83.

Turkensteen, L.J. (1993) 'Durable resistance of potatoes against *Phytophthora infestans*', in Th. Jacobs and J.E. Parlevliet (eds.), Durability of disease resistance. Kluwer Academic Publishers, Dordrecht, the Netherlands.

Uhrig, H., Gebhardt, C., Tacke, E., Rhode, W. and Salamini, F. (1992) 'Recent advances in breeding potatoes for disease resistance'. Neth. J. Pl. Path. Supplement 2, 193-210.

Weijden, W.J. van der (1981) 'Vraagtekens bij de grondontsmetting in de aardappelteelt'. Landbouwkundig Tijdschrift 8, 205-211. In Dutch.

CONTROL OF *VERTICILLIUM DAHLIAE* BY CATCH CROPS AND HAULM KILLING
TECHNIQUES

LEON MOL
Department of Agronomy
Wageningen Agricultural University
Haarweg 333
6709 RZ Wageningen

ABSTRACT Since non-hosts can induce germination of microsclerotia
(MS) of *V. dahliae*, growing of non-hosts may reduce the inoculum
potential in the soil. To quantify the effect of plant roots on MS a
technique for non-destructive observation of MS germination with root
observation boxes was developed. Hosts had a strong effect on MS
germination per root tip. Cropping of non-hosts with a high root
intensity, such as barley, may be beneficial.
    Population levels in the soil mainly increase by production of MS
in dying, colonized plant material. Examining the reproduction on
different parts of the potato plant showed that the majority was
produced on the aerial plant parts. An early (green) harvest resulted
in considerably lower MS levels compared with a harvest of a mature
crop. Cultivars differed greatly in reproduction level. Mechanical
haulm killing treatments resulted in a considerable reduction (up to
80 %) in MS numbers compared with chemical haulm killing.

1. Introduction

Verticillium dahliae is a soil-borne fungus, causing wilt disease in
many crops. In the temperate zones it is one of the most important
pathogens of potato. It survives in the soil by microsclerotia (MS).
The high persistence of the MS and the ability of *V. dahliae* to infect
a wide range of hosts (10) makes it endemic to soils favourable for
survival Therefore, crop rotation alone is insufficient to control
the disease. However, inoculum density in the soil can be decreased by
appropriate cultural practices and cropping sequences.
    Since both hosts and non-hosts are capable of inducing germi-
nation of MS, growing of non-hosts may be of importance in controlling
V. dahliae. Inducing germination of propagules without proliferation
of propagules after infection of host plants may be a major factor in
the beneficial effects on *Verticillium* wilt control often reported
when non-hosts are used in rotations (11).

P C Struik et al (eds), Plant Production on the Threshold of a New Century, 223–230
© 1994 *Kluwer Academic Publishers*

MS in the rhizosphere are usually dormant and are stimulated to germinate by high levels of root exudates from the zone of root elongation. Colonization of roots by *V. dahliae* occurs primarily near the root tip, with the maximum density of penetration at 1 cm from the root apex (3). Growth of infectious hyphae emerging from MS starts by penetration of roots, mainly in the areas of differentiation and in the root hair zone. Only a very small portion of the hyphae will succeed in infecting the root. A MS has the ability to germinate more than one time, but many MS will get exhausted after repeated germination. So the germination process contributes to a net reduction of the inoculum density in the soil. For a good insight in the contribution of this process to the control of the pathogen, quantitative information on the influence of roots of (decoy) crops on survival structures in the soil is needed.

V. dahliae infects susceptible host plants by penetration of the root cortex followed by systemic invasion of the xylem vessels (10). When infected plants become senescent and die, the fungus leaves the xylem, readily permeates the surrounding tissues, and MS are produced in large quantities (11). This is the primary method by which the population level of *Verticillium* increases in the soil (10). Population densities of 1,000 MS per gram of soil have been found by direct assay in fields cropped repeatedly to potatoes.

Reproduction starts when the crop starts maturing. Abiotic external factors such as temperature and humidity affect the reproduction of the pathogen (1,2,14). It may be expected that by manipulating dying plant debris the reproduction can be influenced. Several (modifications of) current techniques are considered such as mechanical haulm killing methods.

When the majority of the MS production takes place on the aerial plant parts, then a haulm treatment could affect the increase of the inoculum potential to control *V. dahliae*. For a proper assessment of the effect of haulm treatments one needs to know the ratio between the reproduction on the separate organs of the plant.

2. Materials and methods

2.1. INDUCTION OF MS GERMINATION

To study the effect of roots on the germination of MS, root observation boxes (24x18.5x5 cm) with one transparent side were constructed. At the inner side of this transparent plate MS were fixed in a water-agar layer. The boxes were filled with non-sterilised sandy soil and slantwise placed in a growth chamber at 22/15 °C day/night. Two weeks after planting a crop, the root system had covered the glass plate. Per box a zone of 1 mm around five root tips was examined. A microscope (magnification 50x) was used to count the percentage of MS germinated and the number of germination hyphae per MS and to measure the distance of a (non-)germinated MS from the root surface.

Four crops were grown in five replications: potato cv. Element (a susceptible and sensitive cultivar), potato cv. Mirka (a rather resistant and tolerant cultivar)(12), barley (a non-host, cv. Prisma), and field bean (a host, cv. Victor). A box without a crop was used as a control. In the control, five spots with a radius of one mm were chosen randomly. The experiment was carried out twice (Experiments 1 and 2).

A field trial on a sandy soil was carried out to compare the root length of crops grown under the same conditions (Experiment 3). The root length was estimated using the line-intersection method.

2.2. REPRODUCTION OF MS ON POTATO

To study their relative importance for reproduction leaf blades, petioles, aerial stems, subterranean stems, stolons, roots and tubers of cvs Element and Mirka were separately collected at two dates from plants grown in pots in forced draught air-conditioned glasshouses at 22/15°C day/night (Experiment 4). Before planting rooted sprouts were infested by immersing the roots in a spore suspension of *V. dahliae* and then planted in pure quartz sand. The first harvest date was 72 days after planting (DAP) (comparable to harvest dates of seed potatoes in the Netherlands) and the second one 113 DAP when plants were senescing. Plant parts were incubated for four weeks and the numbers of MS were counted in sub-samples.

2.3. EFFECTS OF HAULM TREATMENTS ON MS PRODUCTION

The potato haulm was cut into pieces of 1, 5, or 20 cm and the cut plant material was either covered with non-sterilised soil or left on the soil surface (Experiment 5). Potato haulm killed with a chemical was used as a reference to practice.

Before planting rooted sprouts of cv. Element were infested by immersing the roots in a spore suspension of *V. dahliae* and then planted in potting soil. The first harvest date was 72 days after planting (DAP) (comparable to harvest dates of seed potatoes in the Netherlands) and the second one 105 DAP when plants were senescing. Aerial plant parts were incubated for four weeks and the numbers of MS were counted in sub-samples.

2.4. COUNTING THE MS IN POTATO TISSUE

MS of *V. dahliae* are not only produced at (in) the outside of the plant material but also on surfaces of cavities in the material (for example hollow stem segments). To include all sclerotia dry plant material was ground and cooked with a sodium-hydroxide solution. During bleaching plant material discoloured but MS stayed black. After filtration the MS in a sample of 15 mg material were counted using an image analyzer. Results of this procedure showed high correlations with hand-countings and appeared to be very reproducible (L. Mol, unpublished).

3. Results and discussion

3.1. INDUCTION OF MS GERMINATION

Roots of host crops showed a stronger stimulating effect on germination of MS per root tip than crops that are known as non-hosts (Table 1). When non-hosts show a good score for activation of the MS this offers a possibility for the control of the pathogen. Therefore, it is necessary to know the part of the soil area influenced by the root system (Table 1).

TABLE 1. Effects of roots on the germination of microsclerotia (MS) in an area of 1 mm around the root tip, the percentage of the soil affected by the root system, and the percentage of the total population of MS in the soil induced to germinate by the crops.

| | Potato | | | | | |
	cv. Element	cv. Mirka	Barley	Field bean	Control	LSD (P=0.05)
MS germinated/ tip (%)[a]	44.6	35.2	33.4	43.5	16.4	6.7
Soil affected (%)[b]	4.9	6.2	13.2	6.9	-	
MS induced (%)[b]	2.2	2.2	4.4	3.0	-	

[a] based on Experiments 1 and 2
[b] values were obtained after multiplication of means of the Experiments 1, 2, and 3, so LSD values could not be calculated.

Combining the results of the experiment with root observation boxes with field observed root lengths shows that crops with a relatively low stimulating effect per root tip (barley) can give a higher reduction than crops with a relatively high stimulating effect per root tip (potato cv. Element) (Table 1). Especially for the potato crops the average number of germination hyphae decreased with an increasing distance of MS from the root tip (Figure 1). The concentration of exudates probably determines the intensity of germination. Germination of MS of *V. dahliae* seems to be rather unspecific, so that a certain percentage will respond to each stimulus even when there is no plant or no host to colonize. The chances of total elimination by spontaneous germination of MS may be low, because sclerotia have the ability to germinate more than once. However, the described experiments show clearly that crops can contribute to a reduction of the inoculum density in the soil.

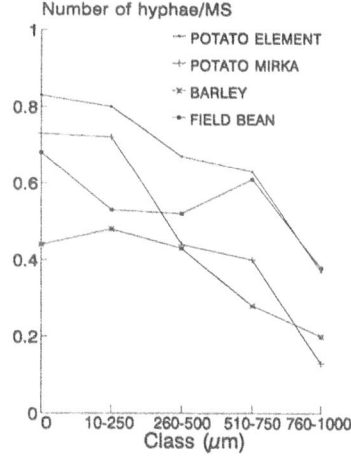

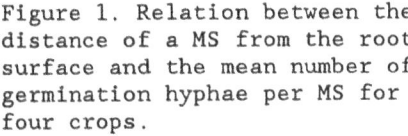

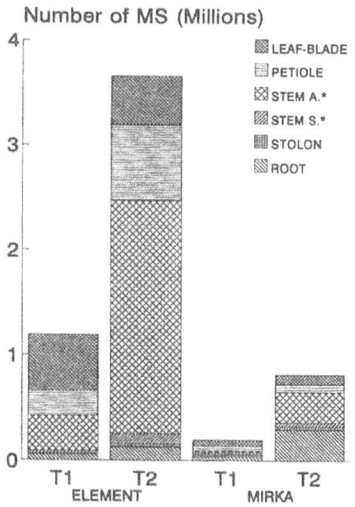

Figure 1. Relation between the distance of a MS from the root surface and the mean number of germination hyphae per MS for four crops.

Figure 2. Reproduction of MS on different parts of potato cvs Element and Mirka harvested 72 (T1) and 113 (T2) days after planting.
*A=aerial, S=subterranean.

Huisman (5) observed that the roots of four cotton cultivars differing in tolerance to *Verticillium* wilt had similar colonization frequencies for *V. dahliae*. In my experiments potato cultivars differed in their capacity to induce germination, indicating that part of the observed differences in susceptibility can be explained by differences in capacity to induce germination. Whether a plant is susceptible or immune to systemic invasion appears unrelated to the ability of the fungus to colonize the root cortex (4). Levy and Isaac (9) showed that there is more extensive growth of *V. dahliae* and MS formation over the root surface of a resistant than of a susceptible plant. This leads to the hypothesis that each crop can maintain a level of inoculum density in the soil. Krikun and Bernier (7,8) found that *V. dahliae* could systemically infect wheat under field conditions, although MS were never observed in above ground plant parts. MS were formed in large numbers in parts of the roots, with most of the roots being non-colonized (8). Taking into account root length density of wheat (13-77 $km.m^{-3}$ soil (13)) and the number of MS found, the inoculum potential from this source could be very meaningful (8). However, systemic invasion is of profound significance in the build-up of populations in the soil.

3.2. REPRODUCTION OF MS ON VARIOUS PLANT PARTS OF POTATO

Quantification of MS on different parts of the potato plant showed that aerial plant parts are by far most important for reproduction (Figure 2). More MS were found on plants harvested close to the maturing stage than on plants harvested in an earlier stage (Figure 2 and Table 2). Cv. Element yielded much more MS than cv. Mirka.

3.3. EFFECTS OF HAULM TREATMENTS ON MS PRODUCTION

Mechanical haulm killing methods inhibited reproduction of MS in comparison with chemical haulm killing with a herbicide up to 80 % (Table 2). Cutting the haulm into pieces of 1 cm was very effective in controlling MS formation at maturation when pieces were left on the soil surface. Increased rates of water loss by cutting is probably the cause of these effects, because cutting into small pieces was not effective when pieces were covered by soil. A slow rate of water loss also explains the very high numbers of MS when haulm is killed by chemicals because of the prolonged contact with a water supplying root system. At green harvesting cutting into pieces had no effect. Maybe the water loss for these viable pieces was too slow.

Covering the haulm with soil mostly tended to lower MS production, but this method was not very effective as control method.

TABLE 2. Effects of haulm treatments on the number of MS per aerial plant mass at 72 (T1) and 105 (T2) days after planting.

Haulm killing method	Mechanical						Chemical	
Haulm placement	On soil surface			In soil				LSD (P=0.05)
Stem piece (cm)	20	5	1	20	5	1		
T1	526	786	833	625	548	461	3516	449
T2	4558	4039	1905	4132	2938	3549	13457	962

Environmental conditions may affect the formation of MS on plant tissue. Ben-Yephet and Szmulewich (1) and Itoh et al. (6) found differences in MS production between autumn-grown and spring-grown crops. The difference between the two crops may be due to better reproduction of the fungus in plant tissue in cool and moist conditions in the autumn-winter season. The effects of humidity and temperature during MS-formation will be quantified in further research.

4. Acknowledgement

I thank Dr K. Scholte and Prof. P.C. Struik for their valuable comments.

5. References

1. Ben-Yephet, Y. and Szmulewich, Y. (1983) 'Inoculum levels of *Verticillium dahliae* in the soils of the hot semi-arid Negev region of Israel'. Phytoparasitica 13, 193-200.
2. Erwin, D.C., Tsai, S.D. and Khan, R.A. (1978) 'Reduced number of microsclerotia formed by *Verticillium dahliae* in cotton tissue exposed to systemic benzimidazole fungicides and desiccation'. Phytopathology 68(10), 1488-1494.
3. Gerik, J.S. and Huisman, O.C. (1988) 'Study of field-grown cotton roots infected with *Verticillium dahliae* using an immunoenzymatic staining technique'. Phytopathology 78(9), 1174-1178.
4. Huisman, O.C. (1988) 'Colonization of field-grown cotton roots by pathogenic and saprophytic soilborne fungi'. Phytopathology 78(6), 716-722.
5. Huisman, O.C. (1988) 'Seasonal colonization of roots of field-grown cotton by *Verticillium dahliae* and *V. tricorpus*'. Phytopathology 78(6), 708-716.
6. Itoh, S., Komoda, H., Monma, T. and Amano, T. (1989) 'Development of field diagnosis system (FDS) for preventing continuous cropping injury of crop'. 12. Study of factors related to the development of a prediction model of *Verticillium* yellows in Chinese cabbage. Bulletin of the National Agriculture Research Center, Japan; No. 16, 33-53.
7. Krikun, J. and Bernier, C.C. (1987) 'Infection of several crop species by two isolates of *Verticillium dahliae*'. Canadian Phytopathological Society. Sept. 1987 9(3), 241-245.
8. Krikun, J. and Bernier, C.C. (1990) 'Morphology of microsclerotia of *Verticillium dahliae* in roots of gramineous plants'. Canadian Journal of Plant Pathology 12(4), 439-441.
9. Levy, J. and Isaac, I. (1976) 'Colonization of host tissue of varying resistance to *Verticillium dahliae*'. Transactions of the British Mycological Society 67(1), 91-94.
10. Powelson, R.L. (1970) 'Significance of population level of *Verticillium* in soil', in T.A. Toussoun, R.V. Bega and P.E. Nelson (eds.), Root diseases and soil-borne pathogens. Univ. of California Press, Berkeley, CA, pp. 31-33.
11. Schnathorst, W.C. (1981) 'Life cycle and epidemiology of *Verticillium*', in M.E. Mace, A.A. Bell, and C.H. Beckman (eds.), Fungal wilt diseases of plants. Academic Press, NY, pp. 81-111.
12. Scholte, K. and s'Jacob, J.J. (1990) 'Effect of crop rotation, cultivar and nematicide on growth and yield of potato (*Solanum tuberosum* L.) in short rotations on a marine clay soil'. Potato Research 33, 191-200.

13. Vos, J. and Groenwold, J. (1987) 'The relation between root growth along observation tubes and in bulk soil', in H.M. Taylor (ed.), Minirhizotron Observation Tubes: Methods and Applications for Measuring Rhizosphere Dynamics. Special Publication 50. American Society of Agronomy, Crop Science Society of America, and Soil Science Society of America, Madison, WI, pp. 39-49.

14. Wilhelm, S. (1951) 'Effect of various soil amendments on the inoculum potential of the *Verticillium* wilt fungus'. Phytopathology 41, 684-690.

THE APPLICATION OF PHYSIOLOGICAL AND MOLECULAR UNDERSTANDING OF THE
EFFECTS OF THE ENVIRONMENT ON PHOTOSYNTHESIS IN THE SELECTION OF NOVEL
"FUEL" CROPS; WITH PARTICULAR REFERENCE TO C_4 PERENNIALS

STEPHEN P. LONG
Department of Biology
University of Essex
Colchester CO7 7DR
England, UK

ABSTRACT. Fuel crops offer a solution to two major problems
confronting W. Europe on the threshold of a new century: gainful use
of excess agricultural land and a decreased dependence on fossil
fuels. Success with both problems will depend on the choice, selection
and improvement of potential fuel crops. This will require a major
change in approaches to crop improvement. Key goals will be maximizing
efficiency of energy conversion but decreasing inputs of nitrogen and
pesticides. Because most of the plant represents economic yield, in a
fuel as opposed to a food crop, the benefits of increasing photosynthetic productivity in these crops are direct and represent the
major challenge in application of physiology to improvement. A specification for an "ideal" fuel crop for W. Europe is developed and shown
to correspond most closely to the characteristics of C_4 herbaceous
perennial plants. Despite the high potential photosynthetic capacity
attributed to C_4 species, yields are limited both by slow canopy
development and low temperature impairment of the photosynthetic
apparatus. A mechanistic model is used to assess the quantitative
significance of these two limitations, both are shown to account for
very significant losses in potential yield and suggest the areas in
which improvement could provide the greatest returns.

1. Fuel crops. A real role in W. Europe's agriculture of
   the 21st century?

1.1. IN THE CONTEXT OF ECONOMIC, ENVIRONMENTAL AND LAND USE CHANGE

The major problem facing W. Europe's, and much of the Developed
World's, agriculture on the threshold of a new century are large
surpluses of most arable crops. A partial solution has been to
encourage farms to take land out of production, through subsidies.
Furtherance of this policy though necessitates major changes in rural
employment and may prove unacceptable to trading blocks outside of the

P. C. Struik et al. (eds.), Plant Production on the Threshold of a New Century, 231–244.

EC. Simultaneously, the World looks to the developed countries to reduce their emissions of "greenhouse gases". Intensification of agriculture and particularly dependence on high nitrogen inputs over the past half century, particularly in W. Europe, are believed to have led to the rapid increases in CH_4 and N_2O in the atmosphere [1]. Simultaneously, continued applications of N fertilizers and improvements in the efficiency of N-fixing plants, promised by molecular manipulations, threaten to continue the progressive eutrophication of land and water in much of W. Europe. Over-production of food crops has and continues to depend on large inputs of pesticides, presenting growing problems for the provision of drinking water. At the same time much of W. Europe is committed to reduce its emissions of CO_2 from fossil fuels, in an attempt to slow the unprecedented rate of increase of this gas which has occurred over the past two decades. This increase in CO_2 concentration, together with increases in CH_4 and N_2O, are predicted to result in major climatic change through their greenhouse properties [1]. Fuel crops, plants grown to provide dry biomass or other combustibles, for the generation of heat and energy could promise a partial solution to all of these problems. Fuel crops would decrease dependence on fossil fuels, substitute for food crops in surplus and provide new opportunities to select plants with a lower dependence on nitrogen fertilizers or increased efficiency of use, and to select material or develop systems requiring lower inputs of pesticides.

1.2. THE WEALTH OF EXPERIENCE AND PRACTICE IN IMPROVING FOOD CROPS; HELP OR HINDERANCE?

The success of W. European agriculture over the past half century, which has resulted in excess production of many crops brings into question any further research into improved production. Research into improving the production of crops in excess can hardly be expected to remain a high priority, but can the research approaches and expertise developed for food crops be transferred to fuel crops? Three factors, above all else, have contributed to the marked increases in yields of the arable food crops of W. Europe: 1) Increased harvest index, i.e. improved partitioning of dry matter into the economic portion of the plant; 2) ability to show positive growth responses to increased levels of N applications; and 3) more effective pesticides [2]. Are these approaches desirable for the development of fuel crops?

1) Since the entire above-ground portion of a fuel crop can be used for combustion, there is little or no scope for improving harvest index. Improved allocation to the economic component, such as the ear of wheat, has predominated improvement of food crops - with outstanding success. This success has been achieved with no related increase in total biomass yield [2]. Thus, improved photosynthetic performance has rarely been a consideration. In fuel crops, where the total biomass also represents the bulk of the economic yield, total biomass and the photosynthetic performance which determines this quantity, are fundamental to economic yield. Thus, improvement of

these crops will depend little on allocation and very much on the efficiency with which they can capture the available light energy and convert this into chemically stored energy in the form of biomass; placing canopy structure and development, and photosynthetic efficiency at the fore of crop improvement.

2) A key consideration in fuel crop systems will be their "energy balance", i.e. the amount of energy obtained in the harvested crop vs. the inputs [3]. Nitrogen fertilizers require large inputs of energy in their production and further energy for their frequent application. This is coupled with the environmental desirability for reduced dependence on nitrogen fertilizers. In selecting for increased capacity to respond to increased nitrogen applications, modern arable crop cultivars are less efficient in their use of nitrogen at lower levels of application than their wild ancestors and older cultivars [2]. Further, a high nitrogen content of the harvested biomass is undesirable because of its implications for pollution on combustion and its use for direct combustion, e.g. as fluid particles, in turbines [3].

3) The narrow genetic base, high nitrogen contents and need to use monocultures, for modern cultivars, have all pre-disposed our food crops to epidemics of pests and pathogens, increasingly sophisticated mixtures of pesticides are an essential part of the cultivation of our food crops [2]. Pesticides similarly represent a substantial input of energy, used both in their production and application; again both economic and environmental considerations call for a reduced dependence.

1.3. SCIENTIFIC ADVANCES IN UNDERSTANDING OF LIMITATIONS TO PHOTO-SYNTHETIC PRODUCTIVITY

Whilst physiological improvement of food crops has centred on improved partitioning of biomass into the economic portion, to the neglect of increasing total biomass production, the scientific base for improving total dry matter production has expanded rapidly over the last two to three decades. Three advances in particular have obvious future importance to the selection and improvement of potential fuel crops. i) A detailed mechanistic understanding of controls on efficiency of the photosynthetic process and specific sites of impairment by specific environmental "stresses"; ii) Mechanistic mathematical models which allow quantitative assessment of limitations to production and can quantify the "benefits" or "costs" of change in specific aspects of the photosynthetic apparatus from changes in the properties of the carboxylating enzyme to manipulation of canopy size and architecture; iii) Advances in understanding of the molecular biology of photosynthesis and the potential to introduce DNA from unrelated species to alter the photosynthetic apparatus. These three areas of advance now provide the basis for designing and developing crops with increased potential photosynthetic productivity [2].

2. Requirements of a fuel crop; the case for C_4 perennials

2.1. WHAT QUALITIES SHOULD BE REQUIRED OF AN "IDEAL" FUEL CROP

The previous section highlights the contrasts between approaches to improving food crops and the requirements on future energy crops. A simple solution might be to use existing food crops already developed for W. Europe. For example, in wheat crops simply harvest the whole above-ground portion as dry biomass for combustion, rather than harvest the ears for grain. However, as outlined above the properties of many food crops are diametrically opposed to the qualities theoretically required of a fuel crop.

2.1.1. Maximum efficiency of light use

The economic yields and energy efficiency of fuel crops will be determined predominantly by the amount of biomass that can be formed per unit area and per unit of investment of other resources, notably nitrogen. The potential limit on biomass yield will be set by the amount of light available, its efficiency of interception and the efficiency with which intercepted light is converted into biomass:

$$P_n = S_{tot}.\varepsilon_i.\varepsilon_c.10^3/k_b \quad \ldots\ldots\ldots\ldots (1)$$

Where, for a given time interval (e.g. 1 day) P_n is the primary production (g m^{-2} d^{-1}), S_{tot} is incident solar radiation (MJ m^{-2} d^{-1}); ε_i the proportion of incident radiation which is absorbed by the crop canopy (MJ/MJ); ε_c the efficiency of transduction of absorbed radiant energy into chemical energy, in the form of plant biomass (MJ/MJ); and k_b the energy content of the biomass (MJ kg^{-1}). Interception efficiency depends on the duration, size and architecture of the canopy. A crop which can maintain a full canopy throughout the year, or at least through the period of maximum insolation (i.e. March - October) will clearly have the highest ε_i. In temperate regions the major factor determining ε_i is ability to develop leaves rapidly at the start of the growing season. Our winter annual crops (e.g. winter wheat, winter oil seed rape) and perennial crops (e.g. perennial ryegrass) meet this requirement effectively. The complete canopy cover needed to maximize ε_i also minimizes the ability of weeds to establish themselves so minimizing herbicide requirements. Conversion efficiency (ε_c) depends primarily on photosynthetic efficiency, i.e. the photosynthetic apparatus must be competent. Indeed whilst many evergreen plants maintain a high interception efficiency through the winter, the value of this to production is negated if the leaves are largely unable to photosynthesize. During the warmer months of the year photorespiration produces a significant decrease in ε_c thus, the ideal fuel crop would bear a complete canopy from March to October and would lack photorespiration.

2.1.2. Water content

Ideally the biomass harvested should be dry. Wet biomass will either require an input of energy for drying or if combusted will decrease the efficiency of sensible heat production. This conflicts with the need to maximize interception efficiency, since a requirement for a stand of dry shoots contradicts the need for maintaining a complete green canopy for light interception. A compromise would be for the shoots to die annually and dry-down in the winter, when the available solar radiation is small and decreased interception efficiency will be of least importance, i.e. a herbaceous perennial forming an annual crop of stems.

2.1.3. Nitrogen and nutrient use efficiency

Nitrogen use efficiency is determined at three levels. First, by maximizing the efficiency of energy transduction into biomass in photosynthesis per unit of nitrogen invested in the photosynthetic apparatus. Secondly, by maximizing the amount of N, and other nutrients, translocated out of the canopy components, on their senescence either into other leaves or storage organs; i.e. efficient internal recycling. Thirdly, by maximizing capture of nutrients from the soil. This property will also minimize both the quantities which need to be applied and which can be leached into drainage water.

2.1.4. Water use efficiency

Available soil water is a significant limitation to crop production over much of W. Europe and irrigation requires significant inputs of energy whilst placing a demand on diminishing water resources. Water use efficiency is therefore another important criterion in selecting fuel crops.

2.1.5. Minimal cultivation

Cultivation operations, ploughing, planting, and weeding all con-stitute energy inputs, fuel crops need therefore to have a form and life cycle which would minimize the need for these operations.

2.1.6. Pest and disease susceptibility

Energy efficiency and environmental acceptability will depend on selecting crops with a minimum need for pesticide, fungicide and herbicide applications. This could be achieved by selecting non-food crop species and maintaining the genetic diversity of the wild population. It would be further improved by using mixed rather than single species stands.

2.1.7. Minimizing changes in land use and farm machinery

Finally acceptability will be greatest and costs of conversion least, if the plants selected as fuel crops can be i) planted, cultivated and harvested with the machinery used for food crops; ii) easily inter-changed between fuel and food crop use, and iii) provide harvestable material in a short period of time.

2.1.8. A specification

The ideal crop would therefore show a maximum efficiency of light, water and nitrogen use. It would be similar in form to current arable crops and be capable of yielding dry biomass in the late autumn or winter.

2.2. C$_4$ PERENNIALS

The CO_2 concentrating mechanism of C$_4$ photosynthesis decreases the requirement for nitrogen, since lower carboxylase concentrations are needed when CO_2 saturated; increases water use efficiency by allowing the plant to maintain a higher external:internal CO_2 concentration gradient; and increases maximum light saturated rates of photo-synthesis by eliminating photorespiration. The cost is an increased requirement for light energy to drive the CO_2 concentrating mechanism, which decreases photosynthetic rates when light is limiting if temperatures are below ca. 25°C [2]. C$_4$ crops in Europe include maize and sweet sorghum. Both of these crops however have serious short-comings with regard to other specifications. Both have considerable cultivation and pesticide application requirements, and have a relatively short canopy duration leading to a poor ε_i [3]. As annuals they are unable to recycle nutrients from one year's growth to the next and at the beginning of the growing season their small root systems preclude efficient capture of nutrients, leading to signi-ficant leaching of applied fertilizer. Trees and shrubs, might provide an obvious alternative as fuel crops, wood providing a high density fuel. However, there are drawbacks. No temperate trees are C$_4$, wood at harvest contains large quantities of water, several years are required between planting and harvest, and once planted the land occupied cannot be converted easily back to arable use. Herbaceous rhizomatous perennials, i.e. perennial plants which produce an annual crop of shoots which die-back in the winter, lack these drawbacks. As perennials they can maintain a large root system providing efficient capture of nutrients, but should also retranslocate nutrients from the annual shoots to the perennating below-ground organs as winter approaches. They can provide dry standing biomass for a winter harvest. As perennials they require only a single cultivation and planting, but as herbaceous plants could be simply ploughed in for conversion of the land back to arable agriculture and could be harvested with conventional forage/herbage harvesters. By choosing a group of plants which have not been used as crops and have low leaf nitrogen contents, disease and pest problems should also be minimized [3].

2.3. LIMITATIONS OF HERBACEOUS C_4 PERENNIALS

Whilst in theory this group of plants comes the closest to the specified ideal for fuel crops, they have disadvantages in practice when cultivated in the cool temperate climates of the northern EC countries. Whilst there are a wide range of C_4 herbaceous perennials, the vast majority are tropical in origin, and in common with most plants of tropical origin, show a high temperature threshold for leaf growth and a susceptibility to low temperature dependent photo-inhibition. However, there are exceptions. In the New World, the genus *Spartina* includes several species with distribution extending into Canada. In the Old World, the genus *Miscanthus* includes species with distributions extending into the northern parts of Japan and China. The genera *Cyperus*, *Adropogon*, *Paspalum* and *Digitaria* include species which occur naturally in cool temperate climates or at high altitudes at the tropics with similar temperature regimes [4]. If C_4 herbaceous perennials are accepted as the potentially "ideal" fuel crops for W. Europe, then how should they be improved to achieve this potential? The utilization of physiological understanding within the mathematical framework now available for analyzing photosynthetic production from the biochemistry of CO_2 fixation to plant dry matter production, allows us to assess the relative "benefits" of hypothetical improvements, which in turn provide direction to further work. Thus, the desirability of specific improvements, e.g. via genetic engineering, can be quantified. The remainder of this paper demonstrates this approach by examining a simple application of mathematical modelling to quantify three issues: 1) the potential benefit of C_4 photosynthesis to dry matter yields in W. Europe; 2) the cost of late canopy development; and 3) the cost of low temperature impairment of photochemical efficiency.

3. Theoretical framework and analysis

3.1. A MECHANISTIC MODEL

3.1.1. Leaf photosynthesis

Farquhar *et al.* [5] provide a biochemically based mechanistic model of leaf photosynthesis, which has been widely tested and validated [6, 7] and encapsulates the primary mechanisms of response of the primary carboxylase to both CO_2 and O_2 concentrations, and temperature [8]. Equations developed from this model [8], incorporating the modifica-tions of Evans and Farquhar [9] provided the basic model of C_3 photo-synthesis used here. The C_4 mechanism has two consequences for photo-synthesis. First, it results in CO_2 saturation of the C_3 carboxylase,

Rubisco. This may be simulated by assuming that the CO_2 concentration at Rubisco is always saturating, with the resulting loss of any O_2 sensitivity. Secondly, the cost of this elimination of O_2 sensitivity and photorespiration is the increased energy cost of CO_2 uptake via the CO_2 concentrating mechanism; the C_4 photosynthetic dicarboxylic acid cycle [4]. This was simulated in the model by decreasing the maximum photon yield from 0.097 CO_2/photon to 0.064 CO_2/photon, to take account of this additional energy cost [10, 11, 12]. Figure 1 shows that this formulation mimics the well established differences in leaf photosynthesis between C_3 and C_4 types [2]. At 30°C leaf photosynthetic rates of C_3 leaves are inferior to those of C_4 at all photon fluxes, because photorespiratory losses more than offset the additional energy costs of the C_4 dicarboxylic acid cycle. At 10°C leaf photosynthetic rates in the C_4 are lower when light is limiting because at this temperature photorespiration is not favoured and thus losses are small, whilst the cost of metabolizing CO_2 through the C_4 pathway is a fixed cost, independent of temperature.

3.1.2. Canopy photosynthesis

To evaluate the significance of changes inferred at the leaf level to changes at the canopy level, a simplified model of canopy photosynthesis has been used. Norman [13] compared different approaches to predicting light distribution in canopies for the calculation of canopy photosynthetic rates from relationships of A (CO_2 uptake) to I (photon flux) determined at the leaf level. This comparison showed that by treating the canopy as two populations of leaves, sunlit and shaded, and by calculating the mean photon flux and in turn mean CO_2 uptake for each category, estimated canopy photosynthesis differed little from estimates made with more complex models, which treated the canopy as several component leaf populations. However, this division into sunlit and shaded leaves did provide a substantial improvement in prediction over models which simply assumed an exponential decline in I through homogeneously lit canopy layers. This approach of dividing the canopy into dynamically variable sunlit and shaded populations of leaves was employed here to extrapolate from the leaf photosynthesis model to a whole canopy.

The photon flux of the shaded (I_{shade}) and sunlit (I_{sun}) leaves, and the leaf area index in direct sunlight (F_{sun}) were calculated throughout the day, and for any given latitude [14]. These are functions of leaf area index (F), the ratio of the horizontally and vertically projected areas of the canopy (χ), the quantities of direct (I_{dir}) and diffuse (I_{diff}), and the angular distribution of the light entering the canopy [14]. Temperature was assumed to follow a daily sinusoidal pattern with a maximum offset 3h after solar noon, and an annual sinusoidal pattern with a peak in mid-July [15]. For these simulations, atmospheric transmissivity, mean annual temperature, mean annual temperature range and mean daily temperature range were the recorded averages for the given latitudes in W. Europe. Canopy parameters, unless stated otherwise were F = 3 and χ = 1. The daily and annual sums of net CO_2 uptake ($A_{c,tot}$) were obtained by integrating over 5 minute time steps using the Euler numerical method. Respiratory losses were linked only to canopy size, as described previously [16]. It was assumed that for each mol of CO_2 assimilated 30 g of dry matter were formed.

3.2. BENEFITS OF C_4 PHOTOSYNTHESIS

Figure 2 illustrates the annual cycle of dry matter production that could be achieved by both a C_3 and a C_4 canopy, assuming that a leaf area index of 3 is maintained throughout the year and that there is no environmental impairment of photosynthetic capacity, apart from the instantaneous effects of variation in light and leaf temperature. This demonstrates that, in W. Europe, a 21 %, 9 % and 0.5 % increase in net C uptake at latitudes 44°N, 52°N and 60°N, respectively, would be obtained by an equivalent canopy of C_4 leaves relative to C_3. Thus the potential of C_4 photosynthesis is greater, although this becomes almost insignificant at lat 60°N. Coincidentally, this is the highest latitude to which C_4 plants can be found in the natural environment [4].

3.3. SLOW CANOPY DEVELOPMENT

C_4 species, even of temperate origin, show a high temperature threshold for leaf growth [4]. *Zea mays*, the most widely grown C_4 food crop in W. Europe, at latitude 52°N does not form a closed canopy until mid-July, thus missing the annual peak of solar radiation. *Sorghum bicolor* (sweet sorghum) which has been suggested as a fuel crop for the EC shows an even slower rate of canopy formation (Figure 3) [3]. The C_4 perennial *Spartina cynosuroides* however shows an earlier canopy development with canopy closure by early June, but still 1-2 months behind many C_3 crops. What cost does this late canopy development impose? Figure 4 illustrates simulated daily rates of net dry matter gain, assuming for a canopy of F=6 from March - September. Throughout the equivalent C_4 canopy can be expected to provide a higher yield. If however, the canopy size observed for the *S. cynosuroides* canopy is used in place of a constant F=6, then dry matter gain by the C_4 canopy is inferior until June, removing all of the advantage of C_4 photosynthesis at this latitude. If the development described by *Z. mays* or *S. bicolor* are considered then this loss is even more substantial with the potential of these C_4 canopies being reduced by 35 and 52 % below that of a C_3 canopy.

240

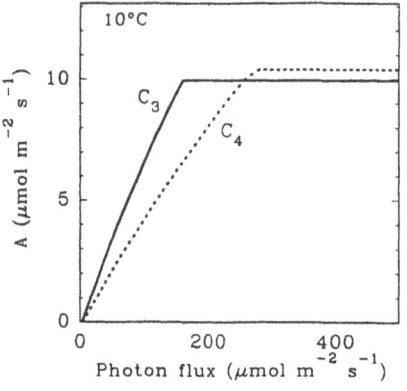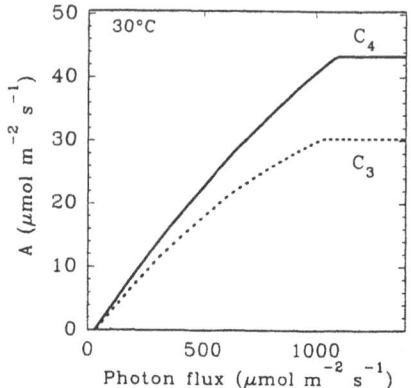

Figure 1. Simulated responses of leaf CO_2 uptake (A) to incident photon flux at leaf temperatures of 10°C and 30°C.

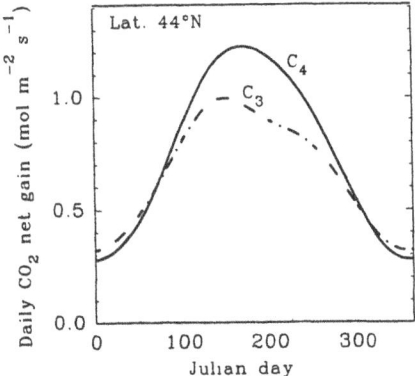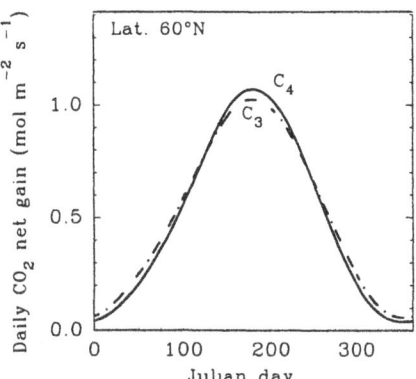

Figure 2. Simulated net canopy CO_2 uptake (leaf area index 3 and a random distribution of canopy elements) for a C_3 and a C_4 canopy, given mean temperature and light conditions at 44°N and 60°N in W. Europe.

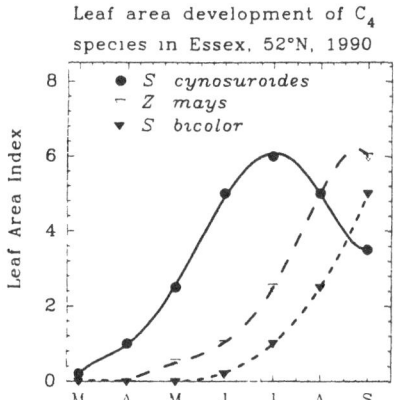

Figure 3. Leaf area index with time of year for pure stands of *Spartina cynosuroides*, *Zea mays* cv. LG11, *Sorghum bicolor* cv. Keller, grown at Wivenhoe Park in N.E. Essex, England.

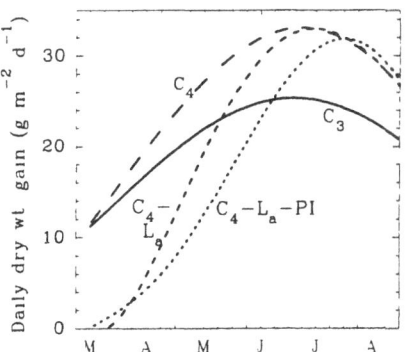

Figure 4. Simulated daily net canopy production, expressed as dry weight gain, for canopies of leaf area index 6 (C_4 and C_3), and for a C_4 canopy assuming the delayed leaf area development of *Spartina cynosuroides* illustrated in Figure 3 (C_4-L_a), and finally for the combined effects of a delayed leaf area development, plus the assumption of a 50 % decrease in photochemical efficiency from Mar.-June (C_4-L_a-PI).

3.4. IMPAIRMENT OF PHOTOCHEMICAL CAPACITY

C_4 species are prone to low temperature dependent photoinhibition, i.e. during the spring and early summer in temperate regions the efficiency of their photosynthetic apparatus is depressed. In Z. *mays* crops grown in S. England the maximum quantum yield of photosynthesis was depressed by ca. 50 % throughout May and June, and this corresponded to a significant decrease in the conversion of intercepted light into dry matter [17]. An impairment of maximum photochemical efficiency can be simulated in the model by decreasing the maximum quantum yield. Assuming a 50 % depression during early development (March-June), what additional decrease in dry matter gain could be expected? Figure 4 illustrates the additional decrease which this would impose on a canopy showing the leaf area development pattern of S. *cynosuroides*. This would delay by a month, the period in which the C_4 canopy can exceed the photosynthetic performance of the C_3 canopy, decreasing net carbon gain by 9 % below that of an equivalent and non-inhibited C_3 canopy. This rate of production would however yield an estimated 3.1 kg m^{-2} for S. *cynosuroides* and 2.2 kg m^{-2} for Z. *mays* by the end of August. Coincidentally, these values come close to the maximum dry matter yields of these two species grown in N.E. Essex, England, of 3.1 kg m^{-2} in 1988 for S. *cynosuroides* and of 2.4 kg m^{-2} for Z. *mays* [3].

4. Conclusion

Mechanistic models show that C_4 species, even disregarding their superior N and water use efficiencies, have the potential to provide higher yields of dry matter throughout W. Europe. This potential is negated by slow early canopy development and made inferior to C_3 canopies by the additional chronic effects of low temperatues on the efficiency of the photosynthetic apparatus. However, there is considerable variability in both of these properties among wild C_4 species [18, 19], and thus the prospect to engineer the "ideal" energy crop.

5. References

1. Watson, R.T. *et al*. (1990) 'Greenhouse gases and aerosols', in J.T. Houghton, G.J. Jenkins, and J.J. Ephraums (eds.), Climate Change: The IPCC Scientific Assessment, Cambridge University Press, Cambridge, pp. 1-40.
2. Beadle, C.B. *et al*. (1985) 'Photosynthesis in Relation to Plant Production in terrestrial Environments'. United Nations Environment Programme/Tycooly Publishing, London.

3. Long, S.P. *et al.* (1989) 'The potential of cord-grasses and galingale for low input biomass production in Europe - growth, photosynthesis and dry matter yields of stands in eastern England', in G. Grassi and W. Palz (eds.), Euroforum - New Energies. Stevens Associates, Bedford, pp. 231-236.

4. Long, S.P. (1983) 'C4 photosynthesis at low temperature'. Plant Cell Environ. 6, 345-363.

5. Farquhar, G.D., Caemmerer, S. von and Berry, J.A. (1980) 'A bio-chemical model of photosynthetic CO_2 assimilation in leaves of C_3 species'. Planta 149, 78-90.

6. Caemmerer, S. von and Farquhar, G.D. (1981) 'Some relationships between the biochemistry of photosynthesis and the gas exchange of leaves'. Planta 153, 376-387.

7. Long, S.P. (1985) 'Leaf gas exchange', in J. Barber and N.R. Baker (eds.), Photosynthetic Mechanisms and the Environment. Elsevier, Amsterdam, pp. 453-499.

8. Long, S.P. (1991) 'Modification of the response of photosynthetic productivity to rising temperature by atmospheric CO_2 concentrations: Has its importance been underestimated?' Plant Cell Env. 14, 729-740.

9. Evans, J.R. and Farquhar, G.D. (1991) 'Modeling canopy photosynthesis from the biochemistry of the C_3 chloroplast', in K.J. Boote and R.S. Loomis (eds.), Modeling Crop Photosynthesis - from Biochemistry to Canopy. Crop Science Society of America, Inc., Madison, WI, pp. 1-16.

10. Long, S.P., Postl, W.F. and Bolhár-Nordenkampf, H.R. (1993) 'Quantum yields for uptake of carbon dioxide in C_3 vascular plants of contrasting habitats and taxonomic groupings'. Planta 189, 226-234.

11. Björkman, O. and Demmig, B. (1987) 'Photon yield of O_2 evolution and chlorophyll fluorescence characteristics at 77K among vascular plants of diverse origins. Planta 170, 489-504.

12. Ehleringer, J. and Pearcy, R.W. (1983) 'Variation in quantum yield for CO_2 uptake among C_3 and C_4 plants'. Plant Physiology 73, 555-559.

13. Norman, J.M. (1980) 'Interfacing leaf and canopy light inter-ception models', in J.D. Hesketh and J.W. Jones (eds.), Predicting Photosynthesis for Ecosystem Models. CRC Press, Boca Raton, pp. 49-67.

14. Forseth, I.N. and Norman, J.M. (1993) 'Modelling of solar irradiance, leaf energy budget and canopy photosynthesis', in D.O. Hall *et al.* (eds.), Photosynthesis and Productivity in a Changing Environment. Chapman & Hall, London, pp. 207-219.

15. McMurtrie, R.E. (1993) 'Modelling of canopy carbon and water balance', in D.O. Hall *et al.* (eds.), Photosynthesis and Productivity in a Changing Environment. Chapman & Hall, London, pp. 220-231.

16. Long, S.P. and Drake, B.G. (1992) 'Photosynthetic CO_2 assimilation and rising atmospheric CO_2 concentrations', in N.R. Baker and H. Thomas (eds.), Crop Photosynthesis: Spatial and Temporal Determinants. Elsevier, Amsterdam, pp. 69-95.

17. Baker, N.R. *et al*. (1989) 'Measurements of the quantum yield of carbon assimilation and chlorophyll fluorescence for assessment of photosynthetic performance of crops in the field'. Phil. R. Soc. Lond. B. 323, 295-308.

18. Long, S.P. *et al*. (1986) 'Chilling dependent photoinhibition of photosynthetic CO_2 uptake', in W.J. Biggins (ed.), Proceedings VIIth International Congress on Photosynthesis Research. Martinus Nijhoff, Dordrecht, pp. 131-138.

19. Long, S.P. *et al*. (1990) 'Damage to photosynthesis during chilling and freezing, and its significance to the photosynthetic productivity of filed crops', in M. Baltscheffsky (ed.), Current Research in Photosynthesis. Kluwer Academic, Dordrecht, pp. 832-842.

DIAGNOSIS OF PRIMARY PRODUCTION OF PLANTS BY SPECTROSCOPIC ANALYSIS OF
THE PRIMARY EVENTS IN PHOTOSYNTHESIS[1]

J.F.H. SNEL AND W.J. VREDENBERG
Wageningen Agricultural University
Department of Plant Physiology
Arboretumlaan 4
6703 BD Wageningen
The Netherlands

ABSTRACT. Photosynthesis is the ultimate basis of plant productivity
and is therefore one of the crucial parameters available for analysis
of plant productivity. Photosynthetic energy conversion starts with
the absorption of light by the photosynthetic pigments in the two
photosystems. The energy of the resulting excited state of the pigment
molecule is transferred to the reaction centre where electron flow is
initiated. In these processes a fraction of the excitation energy is
lost, partly as chlorophyll fluorescence and partly as heat. These
losses decrease the efficiency of photosynthesis. Some of these
reactions are associated with optical changes of intrinsic pigments.
Chlorophyll fluorescence, heat emission and absorbance changes can be
used as phenomena to analyse the light reactions and estimate the
efficiency of photosynthesis. The principles of the methods and some
recent applications will be presented.

1. Introduction

The light reactions in photosynthesis provide the plant with reducing
power and energy which are utilized for growth and development. Plant
productivity is dependent on the availability of reducing equivalents
and chemical energy and therefore the light reactions are of crucial
importance in the analysis of plant productivity. Most plants are
hardly able to avoid large fluctuations in their light environment. On
the other hand the use of photosynthesis products is limited by the
availability of nutrients. The efficiency of the light reactions can

[1]Acknowledgements. Experimental contributions of Wilma Groen-Versluis,
Marc Polm and Hans Boumans are greatly acknowledged. The authors wish
to thank Jack van Rensen, Margreet Bossen and Bert van Hove for
stimulating discussions.

P C Struik et al (eds), Plant Production on the Threshold of a New Century, 245–262
© 1994 Kluwer Academic Publishers

be down-regulated in order to match the supply and use of NADPH and
ATP. The efficiency of the light reactions thus reflects the use of
NADPH and ATP in the mesophyll cell. In the last decade a number of
spectroscopic techniques have emerged which are based on the
biophysics of the primary events in photosynthesis. First the light
reactions of photosynthesis will be briefly introduced to appreciate
the advantages and drawbacks of these methods.

1.1. PHOTOSYNTHESIS

The light reactions start with the absorption of light by the
photosynthetic pigments in the antennae of the photosystems II and I.
The energy of the resulting excited state of the pigment molecules, in
higher plants mainly chlorophyll a and b, is transferred to the
primary electron donors P680 and P700 in the reaction centres of
photosystems II and I, respectively, where a charge separation
initiates vectorial electron flow. The ultimate electron donor water
is split into molecular oxygen, protons and electrons which are
transferred to $NADP^+$ via the electron transport chain (see Andréasson
and Vänngård, 1989 for a review). The protons liberated in the lumen
during water splitting and the protons translocated across the
thylakoid membrane during electron flow from Q_A to cyt f, result in an
accumulation of protons in the inner compartment, the thylakoid lumen.

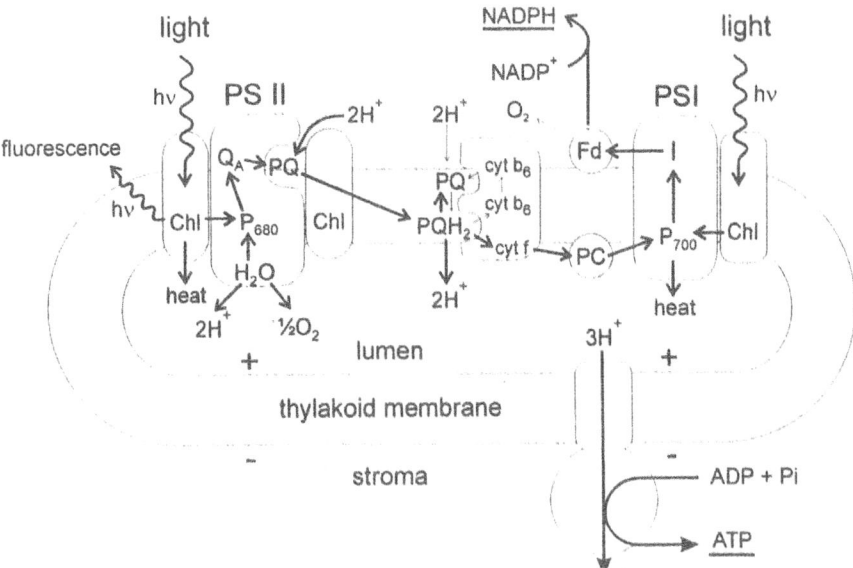

Figure 1. Scheme of the light reactions in plant photosynthesis.
Abbreviations: Chl - antennae chlorophyll, PS - photosystem, P680 -
primary donor of PSII, Q_A - primary pastoquinone acceptor of PSII, PQ
- plastoquinone, cyt - cytochrome, PC - plastocyanin, P700 - primary
donor of PSI, I - primary acceptor of PSI, Fd - ferredoxin.

The proton gradient and the electrical potential across the thylakoid membrane, generated by vectorial electron flow, constitute the driving force for ATP synthesis via ADP-phosphorylation by the ATP-ase.

1.2. PRIMARY REACTIONS

The efficiency of excitation energy transfer from the antenna to the reaction centre is less than 100 %. A fraction of the excitation energy is lost, partly as chlorophyll fluorescence and partly as heat. Figure 2 shows a simplified scheme of the primary reactions in PS II as proposed by Butler and Kitajima (1975). According to this scheme absorption of a photon results in excitation of a chlorophyll molecule. This excited state remains until the excitation energy is either 1) transferred to a primary donor P; 2) dissipated by radiationless conversion into heat; or 3) emitted as fluorescence.

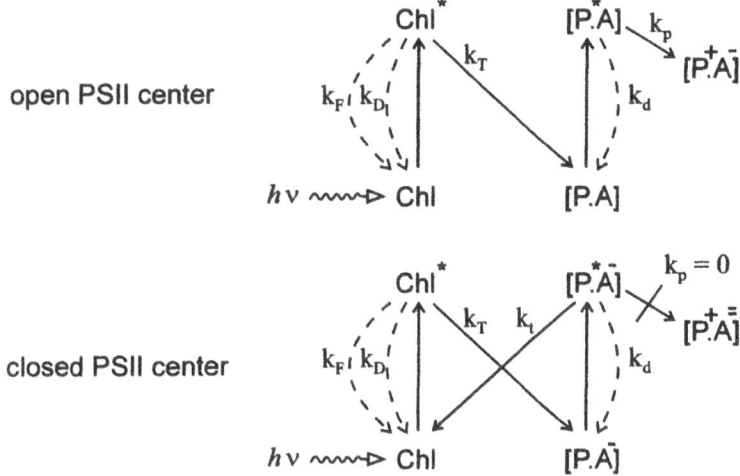

Figure 2. Simplified model of the primary reactions in open and closed photosystem II centres. Abbreviations: Chl, chlorophyll, P - primary donor of photosystem II, k_T, k_D, k_F - rate constants for de-excitation via energy transfer to P, radiationless conversion into heat and fluorescence, respectively; k_p, k_t, k_d - rate constants for de-excitation of P^* via photochemistry, backtransfer to Chl and heat emission, respectively.

Transfer of excitation energy to the primary donor P leads to excitation of P. The energy of P^* can be used to drive photochemistry or can be transferred back to Chl, depending on the redox state of the primary acceptor A. When the primary acceptor is in the oxidized form (A) the probability of photochemistry is close to 1. The fluorescence yield is minimal. When the primary acceptor is in the reduced form (A⁻) the probability of electron transfer to A⁻ is very small. As the probability of de-excitation by heat emission is negligible, the excitation energy of P* is transferred back causing re-excitation of Chl. This leads to an enhanced PSII fluorescence yield. In other words: the fluorescence yield of chlorophyll in the antenna is lower, or quenched, by photochemistry in the reaction centre. This quenching of fluorescence is therefore called photochemical quenching. This form of quenching is associated with photochemistry.

In contrast to PSII, PSI does not show variable fluorescence; moreover the fluorescence yield is much lower at physiological temperatures. This means that *in vivo* virtually all of the observed chlorophyll fluorescence is from PSII.

1.3. REGULATION OF THE EFFICIENCY OF PHOTOSYNTHETIC ELECTRON FLOW

The transfer of excitation energy from the antennae to the primary donor of PSII is regulated by the proton concentration in the lumen. At a low pH in the lumen the efficiency of energy transfer from the antennae to the primary donor is lowered due to an increase in heat emission. Although the mechanism of this phenomenon has not been fully resolved yet, the increase in heat emission causes the fluorescence yield to be lower, a phenomenon called non-photochemical quenching. Non-photochemical fluorescence quenching is not a well-defined process as several mechanisms of non-photochemical quenching have been described (Krause and Weis, 1989). Quantitatively the most important one is the energy-dependent quenching related to the acidification of the lumen as described above. The pH of the lumen is also involved in the regulation of the efficiency of PSI electron flow. The rate of oxidation of PQH_2 is slowed down at low pH. This means that at low pH the donation of electrons to $P700^+$, the oxidized primary donor of PSI, is too slow to keep P700 in the reduced state. $P700^+$ is an efficient trap for excitation energy, but the energy is not used for photo-chemistry but lost as heat.

2. Diagnostic tools for assessment of photosynthetic performance

2.1. CHLOROPHYLL FLUORESCENCE

For more than half a century after the discovery of fluorescence transients in plants by Kautsky in 1931, the relation between chlorophyll fluorescence yield and photosynthesis in the intact leaf could not be quantified. This was mostly due to the occurrence of another fluorescence quenching process: non-photochemical quenching.

2.1.1. Quenching analysis

The 'light-doubling' method introduced by Bradbury and Baker (1981) was modified to become known as the 'saturating pulse' method (Quick and Horton, 1984; Dietz *et al.*, 1985). A pulse-modulated fluorometer

(Schreiber, 1986) enabled registration of chlorophyll fluorescence quenching components (Schreiber et al., 1986). It was shown that even at the highest light intensity used it takes about 300-500 ms to reach the maximal fluorescence. Figure 3 shows a typical measurement protocol illustrating the concept of quenching analysis. In a dark-adapted leaf all photosystems II are open and the fluorescence yield is minimal (F_0). During a saturating light pulse Q_A is reduced and the fluorescence yield of photosystem II is maximal (F_M). In the light-adapted state the saturating light pulse also reduces Q_A. The maximal fluorescence, denoted by F_M' in the light, is smaller than the F_M. This decrease of fluorescence is due to non-photochemical quenching. Schreiber et al. (1986) first described the coefficients of photo-chemical and non-photochemical fluorescence quenching. The nomen-clature and calculation of fluorescence quenching has been slightly modified (Van Kooten and Snel, 1990) following new insights in the quenching mechanisms. Photochemical quenching reflects the redox state of Q_A. Non-photochemical quenching is an indicator of dissipative processes, i.e. a decrease of the efficiency of photosystem II. In many conditions energy-dependent quenching, which is related to the proton concentration in the lumen, is the major component of non-photochemical quenching.

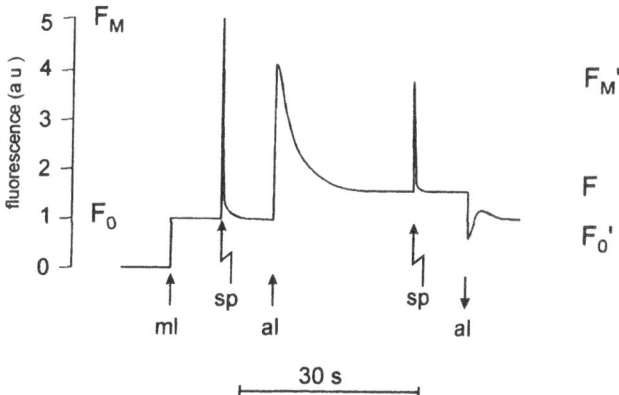

$$q_N = 1 - \frac{(F_M'-F_0')}{(F_M-F_0)} \qquad q_P = \frac{(F_M'-F)}{(F_M'-F_0')} \qquad \phi_{PSII} = \frac{F_M'-F}{F_M'}$$

Figure 3. Fluorescence induction transient demonstrating the saturating pulse method (Schreiber et al., 1986) and the calculation of the quenching coefficients according to Van Kooten and Snel (1990). Abbreviations: ml - measuring light; sp - saturating light pulse; al actinic light. Up- and downwards arrows indicate the switching on and off, respectively, of measuring and actinic light.

2.1.2. The intrinsic photochemical yield of photosystem II

The yield of photochemistry in open photosystems II, expressed as the probability of photochemistry after absorption of a photon, equals F_V/F_M (Butler and Kitajima, 1975). Although the sample has to be dark-adapted for at least 15 min., this parameter has proven very useful to assess photoinhibition, i.e., light-dependent damage to the primary reactions in photosystem II. The absolute value of F_V/F_M is dependent on the excitation and detection wavelength (Genty et al., 1990) which makes it difficult to compare measurements obtained with different fluorescence equipment.

In addition to the saturating light pulse method the pump and probe technique (Falkowski and Kolber, 1990) has been used to determine F_V/F_M from marine phytoplankton at very low chlorophyll concentrations. Due to the fact that the maximal fluorescence is not reached in a single-turnover light flash (Schreiber, 1986), this method yields lower figures than the saturating pulse method.

2.1.3. The photochemical yield of photosystem II electron flow (ϕ_{PSII})

The photochemical yield, or quantum efficiency, of photosystem II electron flow is defined as the yield of photochemistry in a population of open and closed photosystems II per absorbed photon. The efficiency of photosystem II is determined by F_V'/F_M', the intrinsic photochemical yield, and qP, the number of open photosystems II. The product of these two parameters yields $(F_M'-F)/F_M'$ (Genty et al., 1989). This parameter can be determined in less than 2 s since only two measurements have to be made: F and F_M'. The efficiency of photosystem II can be used to estimate the steady state rate of photosynthetic electron flow. In steady state the electron fluxes through PSI and PSII are identical. If the number of photons absorbed by photosystem II per unit of leaf area is known, the rate of PSII electron flow can be expressed per unit of leaf area. The number of photons absorbed by PSII can be determined by measuring the absorption cross section (see Mauzerall and Greenbaum, 1989). The chlorophyll concentration has also been used as a measure for the PSII concentration (Snel et al., 1991).

2.1.4. Fluorescence equipment

Several more or less specialized chlorophyll fluorometers have become commercially available. The non-modulated types (e.g., Plant Stress Meter, Biomonitor, Umeå, Sweden; Plant Productivity Fluorometer, Richard Branker Research Ltd., Ottawa, Canada; Plant Efficiency Analyser, Hansatech, King's Lynn, UK) rely on a complete shielding from stray light and cannot be used to perform quenching analysis. The modulated types allow for measurements in daylight; several also are capable of quenching analysis (MFM, Hansatech, King's Lynn, UK; PAM-2000, Walz, Effeltrich, FRG) and/or measurement of ϕ_P (Plant

Productivity Meter, EARS, Delft, Holland) using the saturating light pulse. For measurement of primary production in aquatic systems submersible fluorometers using the sensitive pump and probe method have become available (Ecomonitor, Russia).

2.1.5. Field measurements

The modulated fluorometers are in principle suitable for field use under ambient light; not all types, however, are able to operate under all weather conditions. The fluorescence yield of the devices employing solid state technology (LED's, photodiodes), however, is dependent on the ambient temperature which can lead to errors in absolute fluorescence yield measurements. Another limitation of the saturating pulse method can be the intensity of the saturating light pulse. When the ambient light intensity and the photosynthesis rate are high, it can be difficult to obtain the maximal fluorescence F_M. (Markgraf and Berry, 1990). On the other hand the minimal fluorescence F_0 may be difficult to measure from plants grown under low light, as even a small amount of measuring light can induce a fluorescence induction.

2.2. LIGHT-INDUCED ABSORBANCE CHANGES

2.2.1. ΔA820

Chlorophyll cation radicals absorb weakly at 820 nm. The only chlorophyll cation observed *in vivo* is $P700^+$. This means that the redox state of P700 can be simply measured using modulated spectrophotometers (Harbinson and Woodward, 1987; Weis et al., 1987; Schreiber et al., 1988). As PSI electron flow is proportional to the fraction reduced P700, this method forms a relatively simple method to measure the rate of PSI electron flow. The technique has been success-fully applied in research on the relation between the efficiency of CO_2 assimilation and the efficiency of PSI electron transport (Harbinson et al., 1990). The rate of PSI electron transport can be obtained by multiplying the fraction P700, as determined from ΔA820 measurements, with the rate of photon absorption by PSI.

2.2.2. P515

The thylakoid membrane contains pigments, mainly β-carotene and chlorophyll, which respond to the electrical field across the thylakoid membrane with a red shift of the absorption band. This so-called electrochromic bandshift results in an absorbance change at 515 nm which is proportional to the membrane potential across the thylakoid membrane. The mechanism and analysis of the P515 change and the involved instrumentation have been reviewed by Vredenberg (1981) and Vredenberg (1986), respectively, and will not be repeated here.

When a single-turnover saturating light flash is applied, the amplitude of the P515 change is a measure for the number of open photosystems that have performed a charge separation. This parameter has been employed to identify photosystems in which the rate of electron transport to the plastoquinone pool apparently is extremely slow (Chylla and Whitmarsh, 1989; Snel *et al.*, 1992; Vredenberg *et al.*, 1992). The dark-decay of the P515-signal is complex and contains information on electrogenic events related to the Q-cycle (Ooms *et al.*, 1989) and on the overall conductance of the thylakoid membrane (Ooms *et al.*, 1991).

2.2.3. Equipment

The $\Delta A820$ measurements can be carried out using the fluorometers equipped with optional 820 nm emitter-detector units (Hansatech, Walz). The measurements can be performed under field conditions, but the measurements are sensitive to mechanical disturbances as the absorbance changes are relatively small. Commercial spectrophotometers capable of measuring the P515 changes under field conditions have not yet been described.

2.3. THE PHOTOACOUSTIC EFFECT

The principle of photoacoustic effect is based on periodically heating a sample in a gas-filled, closed PA-cell by illuminating the sample with intensity-modulated light. The heat waves diffuse to the sample surface where periodical fluctuations of the surface temperature are generated. These temperature fluctuations are transferred to the surrounding gas in the PA-cell where the resulting pressure fluctuations in the gas are detected with, *e.g.*, a microphone. See Rosencwaig (1990) for a general introduction to the photoacoustic effect.

In a photochemically active sample the absorbed light is partly converted into chemical energy and partly lost as heat. The photoacoustic signal is maximal when photochemistry is blocked. The decrease of the photoacoustic signal by photochemistry is a direct measure for the rate of energy storage by photochemistry. The relative decrease of the photoacoustic signal normalized to the maximal photoacoustic signal is referred to as 'photochemical loss' and is a measure of the efficiency of energy storage by photochemistry (Malkin and Cahen, 1978).

In plant photosynthesis oxygen is produced and carbon dioxide consumed. The concentration changes of these gases contribute to the observed PA signal when the oxygen and/or carbon dioxide concentrations contain a modulated component. A contribution of carbon dioxide has not been observed. The pressure changes related to carbon dioxide uptake must be small due to 1) the large number of steps between light absorption and carbon dioxide fixation; 2) the large pool sizes of the intermediates involved in carbon dioxide fixation;

and 3) the large pool size of inorganic carbon in the cell. Oxygen and heat diffuse from their site of production, the thylakoid membrane in the chloroplast, to the intercellular airspaces where the pressure changes are generated. The diffusion constants of heat and oxygen are sufficiently different to cause a measurable delay in the propagation of the oxygen signal with respect to the photothermal signal. Moreover the finite rate constant of oxygen evolution ($t_{\frac{1}{2}} \approx 1 ms$) causes an additional delay. These properties cause the oxygen signal to be much more damped at high modulation frequencies.

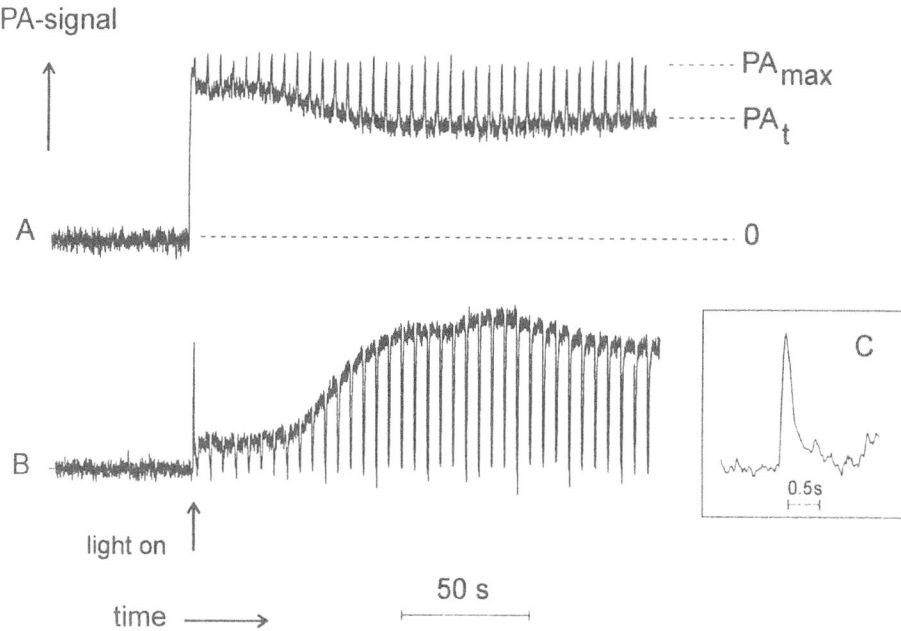

Figure 4. Photoacoustic signal transients measured simultaneously from a spinach leaf-disc as in Snel *et al*. (1992). A: Thermal signal measured at 600 Hz. B: Oxygen signal measured at 134 Hz. C: Oxygen signal as in B, timescale expanded 100x, showing the high time resolution of the PA method.

2.3.1. Photochemical energy storage

At high modulation frequencies (100 - 1000 Hz, depending on leaf morphology) the oxygen component is strongly damped because of the finite rate constant of oxygen evolution. The photoacoustic signal at high modulation frequency is a direct measure for heat emission by the leaf. At very high irradiances photosynthesis is saturated and the efficiency of photosynthesis is extremely low causing nearly all the absorbed (modulated) light to be converted into (modulated) heat. The photochemical loss, or photochemical energy storage, can thus be measured conveniently at high modulation frequency using a non-modulated light pulse to saturate photosynthesis (Bults *et al.*, 1982). The photochemical energy storage (PES) can then be expressed as the relative decrease of the maximal photoacoustic signal due to photochemistry: PES = $(PA_{max}-PA_t)/PA_{max}$.

2.3.2. Photosynthetic oxygen evolution

The oxygen component of the photoacoustic signal can be separated from the photothermal component in the phase domain using phase-sensitive analyzers (Bults *et al.*, 1982; Poulet *et al.*, 1983; Snel *et al.*, 1992) or in the time-domain (Kolbowski *et al.*, 1990). The photoacoustic signal is dependent on several instrumental and sample properties and the signals from different instruments or samples are not comparable. The oxygen signal has been normalized to the maximal photothermal signal (Poulet *et al.*, 1983) to overcome this problem. At high CO_2 concentrations the oxygen signal can contain a negative component indicative of oxygen uptake (Malkin, 1987; Kolbowski *et al.*, 1990).

2.3.3. Equipment

Although photoacoustic cells are made by several manufacturers, complete instruments for field use are not yet available. Figure 4 shows an example of a PA measurement with a dark-adapted spinach leaf-disc using a home-built PA set-up (Vredenberg *et al.*, 1992). In this measurement a dual modulation technique was employed to measure both the oxygen yield and the photochemical energy storage simultaneously (Snel *et al.*, 1992). One modulated light beam was used to measure the PA signal at 138 Hz. The second light beam was modulated at 600 Hz to measure the photothermal signal. The photoacoustic method thus can monitor energy utilization and PSII electron flow in leaf-discs with a time resolution in the sub-ms range.

3. Applications

3.1. MEASUREMENT OF PRIMARY PRODUCTION IN MARINE ECOSYSTEMS

The Tidal Waters Division of the Ministry of Public Works and Transportation manages the Dutch coastal waters. One of the main problems is the occurrence of phytoplankton blooms, some of which may involve toxic species or species which produce toxic products. These blooms can lead to negative economic effects on tourism and/or fishery). The ecological and physiological factors which lead to the occurrence of these hazardous species are still largely unknown. One of the key parameters is primary production. The commonly used methods (^{14}C-incorporation, oxygen production) are time consuming or laborious or both. Estimation of primary production by means of measurement of the efficiency of PSII with chlorophyll fluorescence as described above might be an alternative or supplemental tool.

In a pilot study (Hofstraat *et al.*, 1993) a positive correlation was found between the estimated rate of PSII electron flow and the relative growth rate in batch cultures of *Dunaliella tertiolecta*. One of the main conclusions of the project was that the sensitivity of the PAM fluorometer was at least 3 orders of magnitude too low. As a result of this project a new, highly sensitive, modulated chlorophyll fluorometer was developed in collaboration with dr. U. Schreiber (Schreiber *et al.*, 1993). With this fluorometer *in situ* measurements are feasible and its versatile measuring light-source, a Xenon flashlamp, will be used to investigate the problem of how to measure the efficiency of photosynthesis in a population of different phytoplankton classes.

3.2. EFFECTS OF AIR POLLUTANTS ON PHOTOSYNTHESIS OF DOUGLAS FIR

In highly populated areas, as in large parts of the Netherlands, air pollution has become a major problem. Experiments involving long-term exposure of plants to realistic concentrations of air pollutants are of crucial importance to understand the effects of air pollutants on the physiology of the plant. In a series of experiments Douglas fir (*Pseudotsuga menziesii*) seedlings were grown in laboratory growth chambers exposed to a number of air pollutants in concentrations typical for some areas in Holland. The plants were exposed to either NH_3, SO_2 or O_3 or to the combinations NH_3+SO_2 or NH_3+O_3. The effects were assessed by measuring CO_2 assimilation, chlorophyll content, the rate of PSII electron transport as described before with plants exposed to filtered air as a control (Van Hove *et al.*, 1992, Snel *et al.*, 1991).

One of the problems addressed was how to assess these effects at the canopy or even vegetation level. Remote sensing of chlorophyll fluorescence might be one of the tools which could prove useful in this respect. Therefore we measured a number of fluorescence parameters which are "accessible" to active remote sensing fluorometers with a laboratory-built simulator (Schroote and Snel, 1992) and related these results to the observed effects. One of the major barriers for a more widespread application of remote sensing of chlorophyll fluorescence is the lack of an equivalent to the saturating pulse method. The major reason is the amount of energy involved in a saturating light pulse; in non-restricted areas these energies are hazardous to man and do not comply with safety regulations. Therefore the information has to be extracted from simple steady-state fluorescence yield measurements. This problem can best be illustrated by the following experiment. A spinach leaf-disc is kept in the dark for 30 min. Then the rate of oxygen evolution and the fluorescence yield are measured upon a dark-to-light transition for 5 min. Figure 5 shows the fluorescence yield as a function of the rate of oxygen evolution. There are three distinct phases in the relation between fluorescence and oxygen evolution. The first phase (1) after application of light is characterized by a decrease of oxygen evolution and a simultaneous increase in fluorescence; this phase is completed in about 1 s. The following phase (2) lasts about 30 s and is characterized by a large decrease in fluorescence yield and small changes in the rate of oxygen evolution. In the final phase (3), from about 30 s until steady-state at the end of the experiment, the situation is just the opposite: the fluorescence yield is relatively constant and the large changes occur in the rate of oxygen evolution.

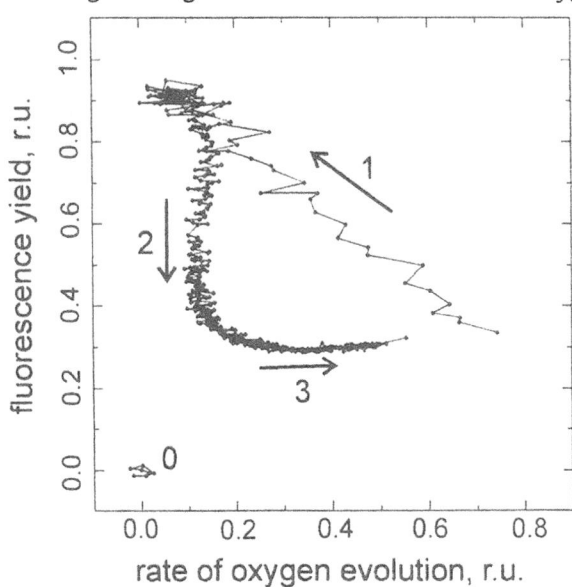

rate of oxygen evolution, r.u.

Figure 5. Relation between chlorophyll fluorescence yield and the rate of oxygen evolution in a dark-adapted spinach leaf-disc. Fluorescence yield and the rate of oxygen evolution, as detected by photoacoustics, were measured as in Vredenberg et al. (1992). The numbers in the figure denote the different phases in the transition from the dark-adapted to the light-adapted state.

While the relation between fluorescence and oxygen evolution is more or less linear in phase 1, it is obvious that the relation between fluorescence and the rate of oxygen evolution is very complex in phase 3. As explained in 2.1 this absence of a linear relation is the result of heat emission related as reflected in non-photochemical quenching. Unfortunately most measurements in situ in daylight will be performed somewhere in the phase 3, depending on the kinetics of the fluctuations in the local light environment.

Preliminary results show that, in spite of this complex relation between chlorophyll fluorescence and electron flow, the effect of the treatments on photosynthetic electron flow can be observed using steady-state fluorescence yield measurements (Schroote et al., 1993).

3.3. DIAGNOSIS OF HERBICIDE RESISTANCE

In the last decade resistance against herbicides has developed in many weeds and in many countries over the world. Resistance against photosynthesis-inhibiting herbicides is nowadays widespread. The mechanism of resistance against atrazine is studied in detail (see Van Rensen, 1993 for a review). This resistance is specific for all triazine herbicides; the concentration causing 50 % inhibition of photosystem II electron transport in isolated chloroplasts is often about 1000 times higher than in chloroplasts of the susceptible plants. There is little difference in sensitivity for urea herbicides, like diuron, while resistant biotypes are more sensitive to phenol-type herbicides. The resistance is caused by lack of binding of triazine herbicides (Naber and Van Rensen, 1991) and could be correlated with an alteration of the D1 protein of the photosystem II reaction centre: a substitution of serine for glycine at position 264 of the D1 protein. This alteration causes also a decrease in the rate of electron flow from Q_A to Q_B, which can be measured by fluorescence techniques.

Resistance against herbicides can be monitored and studied in detail using fluorescence of intact leaves (Van Oorschot and Van Leeuwen, 1992; Curwiel et al., 1993). Leaves may be infiltrated with a herbicide solution and fluorescence induction curves are measured. In the susceptible leaves the fluorescence remains at maximum (F_M) level, because electron flow from Q_A^- to Q_B is inhibited by the herbicide; in the resistant leaves the fluorescence decreases from F_M to the steady state terminal level T. Also the inflection level, I, between F_O and F_M, is higher in the resistant leaves and can be used for diagnosis of resistance against photosynthesis-inhibiting herbicides (Van Oorschot and Van Leeuwen, 1992). Using fluorescence techniques Curwiel et al. (1993) showed that photochemical quenching, photochemical quantum yield and the rate of photosystem II electron transport are lower in the triazine-resistant biotype of Chenopodium album.

3.4. CHARACTERIZATION OF PHOTOSYSTEM II HETEROGENEITY

Another example of the use of spectroscopic methods is the use of the light-induced P515 response in the characterization of inactive PSII centres. It has been known for several years that the PSII pool appears to contain a number of functionally different PSII species (see Melis, 1991 for a review). One of these PSII species, referred to as 'inactive' PSII centres (Chylla and Whitmarsh 1989; Snel et al., 1992), is characterized by a remarkably slow electron transfer from Q_A to the plastoquinone pool. The rate of electron flow in this step is reported to be slowed down by almost three orders of magnitude. These 'inactive' PSII centres, the fraction of which can vary from < 5 % to 40 % (Snel et al., 1992), would hardly contribute to steady-state electron flow. This raises questions about the function of these 'inactive' centres. The flash-induced P515 method can be applied to the intact leaf avoiding the elaborate procedure of isolating chloroplasts and the 'inactive' PSII centres are now subject of further study. One of the aspects we are currently investigating is the mechanism of the conversion of the active form to the inactive form. It appears that most of the inactive PSII centres have disappeared after prolonged dark adaptation (Snel et al., 1992). It is not yet clear whether the 'inactive' PSII centres were somehow 'activated' or whether the inactive centres were disassembled and replaced by active centres.

3.5. ESTIMATION OF DAMAGE BY PHOTOINHIBITION

Photoinhibition of photosynthesis occurs when a plant is exposed to irradiance higher than that which it can convert or dissipate without harm. This is generally the case when a plant is exposed to a higher irradiance than experienced during growth. Since photosystem II is the major target of the photoinhibitory effect, this damage can be estimated by fluorescence.

The F_V/F_M value is widely used as a measure for the photoinhibitory damage. Using F_V/F_M values Curwiel et al. (1993) demonstrated that the observed lower production in triazine-resistant plants was due to a higher sensitivity to photoinhibition.

Schansker et al. (1992) studied the effect of a photoinhibitory treatment of pea leaves on the induction kinetics of chlorophyll fluorescence and oxygen evolution by simultaneous measurement of the fluorescence and photoacoustic signal. They found that the 'inactive PSII centres' (described before in 3.3) are considerably less sensitive to a photoinhibitory treatment than the 'active centres'.

4. Concluding remarks

The methods and results described here are based on the primary events in photosynthesis and give a measure of solar energy utilization at the leaf level. It is important to keep in mind that these methods do not give information concerning the 'fate' of the products of the light reactions in photosynthesis.

Due to the non-invasive character and the high sensitivity of some of the methods and the robustness and the potential for further miniaturization of the hardware, these methods could find applications in, e.g., plant breeding or in horticulture in the next millennium for (semi-)continuous monitoring of photosynthesis at multiple small spots on the plant. This will provide the investigator with more detailed information on the actual energy utilization by the plant.

5. References

Andréasson, L.-E. and Vänngård, T. (1988) 'Electron transport in photosystems I and II', Ann. Rev. Plant Physiol. 39, 347-411.

Bradbury, M. and Baker, N.R. (1981) 'Analysis of the slow phases of the *in vivo* chlorophyll fluorescence induction curve. Changes in the redox state of photosystem II electron acceptors and fluorescence emission from photosystems I and II'. Biochim. et Biophys. Acta 635, 542-551.

Bults, G., Horwitz, B.A., Malkin, S., Cahen, D. (1982) 'Photoacoustic Measurements of Photosynthetic Activities in whole Leaves'. Photochemistry and Gas Exchange. Biochim. Biophys. Acta 679, 452-465.

Butler, W. L. and Kitajima, M. (1975) 'Fluorescence quenching in photosystem II of chloroplasts', Biochim. et Biophys. Acta 367, 116-125.

Chylla, R.A. and Whitmarsh, J. (1989) 'Inactive Photosystem II Complexes in Leaves. Turnover rate and quantitation', Plant Physiol. 90, 765-772.

Curwiel, V.B., Schansker, G., Vos, O.J. de and Rensen, J.J.S. van (1993) 'Comparison of photosynthetic activities in triazine-resistant and susceptible biotypes of *Chenopodium album*'. Z. Naturforsch. 48c, 278-282.

Dietz, K.J., Schreiber, U. and Heber, U. (1985) 'The relation between the redox state of Q_A and photosynthesis in leaves at various carbon dioxide, oxygen and light regimes'. Planta 166, 19-26.

Evans, L.T. (1992) 'From leaf photosynthesis to plant productivity', in N. Murata (ed.), Research in Photosynthesis. Kluwer Acad. Publ., Dordrecht, Vol. IV, pp. 587-594.

Falkowski, P.G. and Kolber, Z. (1990) 'Phytoplankton photosynthesis in the atlantic ocean as measured from a submersible pump and probe fluorometer *in situ*', in M. Baltscheffsky (ed.), Current Research in Photosynthesis. Vol. IV Kluwer Academic Publishers, Dordrecht, pp. 923-926.

Genty, B., Briantais, J.M., Baker, N.R. (1989) 'The Relationship between the quantum yield of photosynthetic electron transport and quenching of chlorophyll fluorescence'. Biochim. Biophys. Acta 990, 87-92.

Genty, B., Wonders, J. and Baker, N.E. (1990) 'Non-photochemical quenching of F_0 in leaves is emission wavelength dependent. Consequences for quenching analysis and its interpretation'. Photosynth. Res. 26, 133-139.

Harbinson, J. and Woodward, F.I. (1987) 'The use of light-induced absorbance changes at 820 nm to monitor the oxidation state of P-700 in leaves'. Plant Cell & Envir. 10, 131-140.

Harbinson, J., Genty, B. and Baker, N.R. (1990) 'The relationship between CO_2 assimilation and electron transport in leaves'. Photosynthesis Research 25, 213-224.

Hofstraat, J.W., Peeters, J.C.H., Snel, J.F.H. and Geel, C. (1993) 'Simple determination of photosynthetic efficiency and photo-inhibition of *Dunaliella tertiolecta* by saturating pulse fluorescence measurements'. Marine Ecology Progress Series, in press.

Horton, P. and Bowyer, J.R. (1990) 'Chlorophyll fluorescence transients', in J.L. Harwood and J.R. Bowyer (eds.), Methods in plant biochemistry. Academic Press, New York, pp. 259-296.

Hove, L.W.A. van, Bossen, M.E., Mensink, M.G.J. and Kooten, O. van (1992) 'Physiological effects of a long term exposure to low concentrations of NH_3, NO_2 and SO_2 on Douglas fir (*Pseudotsuga menziesii*)'. Physiologia Plantarum 86, 559-567.

Kolbowski. J., Reising, H. and Schreiber, U. (1990) 'Computer-controlled pulse modulation system for analysis of photoacoustic signals in the time-domain'. Photosynth. Res. 25, 309-316.

Kooten, O. van and Snel, J.F.H. (1990) 'The use of chlorophyll fluorescence nomenclature in plant stress physiology', Photosynth. Res. 25, 147-150.

Krause, G.H. and Weis, E. (1989) in H.K. Lichtenthaler (ed.), 'Applications of chlorophyll fluorescence in photosynthesis research, stress physiology, hydrobiology and remote sensing'. Kluwer Acad. Publ., Dordrecht, pp. 3-12.

Malkin, S. and Cahen, D. (1978) 'Photoacoustic spectroscopy and radiant energy conversion: Theory of the effect with special emphasis on photosynthesis'. Photochem. Photobiol. 29, 803-813.

Malkin, S. (1987) 'Fast photoacoustic transients from dark-adapted intact leaves: oxygen evolution and uptake pulses during photosynthetic induction - a phenomenology record'. Planta 171, 65-72.

Markgraf, T. and Berry, J.A. (1990) 'Measurement of photochemical and non-photochemical quenching: correction for turnover of PS2 during steady-state photosynthesis', in M. Baltscheffsky (ed.), Current Research in Photosynthesis, Vol. IV. Kluwer Academic Publishers, Dordrecht, pp. 279-282.

Mauzerall, D. and Greenbaum, N.L. (1989) 'The absolute size of a photosynthetic unit'. Biochim. Biophys. Acta 974, 119-140.

Melis, A. (1985) 'Functional properties of photosystem II (beta) in spinach chloroplasts', Biochim. Biophys. Acta 808, 334-342.

Melis, A. (1991) 'Dynamics of photosynthetic membrane composition and function'. Biochim. Biophys. Acta 1058, 87-106.

Naber, J.D. and Rensen, J.J.S. van (1991) 'Activity of photosystem II herbicides is related with their residence times at the D1 protein'. Z. Naturforsch. 46c, 575-578.

Ooms, J.J.J., Vredenberg, W.J. and Buurmeijer, W.F. (1989) 'Evidence for an electrogenic and a non-electrogenic component in the slow phase of the P515 response in chloroplasts', Photosynth. Res. 20, 119-128.

Ooms, J.J.J., Versluis, W., Vliet, P.H. van and Vredenberg W.J. (1991) 'The flash-induced P515 shift in relation to ATPase activity in chloroplasts'. Biochim. Biophys. Acta 1056, 293-300.

Oorschot, J.L.P. van and Leeuwen, P.H. van (1992) 'Use of fluorescence induction to diagnose resistance of *Alopecurus myosuroides* Huds. (black-grass) to chlorotoluron'. Weed Res. 32, 473-482.

Poulet, P., Cahen, D. and Malkin, S. (1983) 'Photoacoustic detection of photosynthetic oxygen evolution from leaves. Quantitative analysis by phase and amplitude measurements'. Biochim. Biophys. Acta 724, 433-446.

Quick, W.P. and Horton P. (1984) 'Studies on the induction of chlorophyll fluorescence quenching by redox state and transthylakoid pH gradient'. Proc. R. Soc. London, series B, 217, 405-416.

Rensen, J.J.S. van (1993) 'Regulation of electron transport at the acceptor side of photosystem II by herbicides, bicarbonate and formate', in Y.P. Abrol, P. Mohanty and Govindjee (eds.), Photosysnthesis: Photoreactions to plant productivity. Oxford & IBH Publishing CO. PVT. LTD., New Delhi, pp. 157-180.

Rosencwaig, A. (1990) 'Photoacoustics and photoacoustic spectroscopy'. R.E. Krieger Publishing Company, Malabar, Florida.

Schansker, G., Snel, J.F.H. and Rensen, J.J.S. van (1992) 'Analysis of the induction kinetics of chlorophyll fluorescence and oxygen evolution obtained by simultaneous measurement of the fluorescence and photoacoustic signal: effect of a photoinhibitory treatment', in N. Murata (ed.), Research in photosynthesis, Vol IV. Kluwer Academic Publishers, Dordrecht, pp. 475-478.

Schreiber, U. (1986) 'Detection of rapid induction kinetics with a new type of high-frequency modulated chlorophyll fluorometer', Photosynth. Res. 9, 261-272.

Schreiber, U, Schliwa, U. and Bilger, W. (1986) 'Continuous recording of photochemical and non-photochemical quenching with a new type of fluorometer'. Photosynth. Res. 10, 51-62.

Schreiber U., Klughammer, C. and Neubauer, C. (1988) 'Measuring P700 absorbance changes around 830 nm with a new type of pulse modulation system'. Z. Naturforsch. 43c, 686-698.

Schreiber, U., Neubauer C. and Schliwa, U. (1993) 'PAM fluorometer based on a medium-frequency pulsed Xe-flash measuring light: A highly sensitive new tool in basic and applied photosynthesis research'. Photosynthesis Research, in press.

Schroote, J.J. and Snel, J.F.H. (1991) '(L)EAF: A remote sensing fluoresensor', in Proceedings of the 5th international Colloquium 'Physical Measurements and Signatures in Remote Sensing', Vol.2, ESA Publication Division, ESTEC, Noordwijk.

Schroote, J.J., Snel, J.F.H., Bossen, M.E. and Hove, L.W.A. van (1993) 'On the use of chlorophyll fluorescence parameters in the assessment of plant stress by means of remote sensing'. BCRS report, in press.

Snel, J., Boumans, H. and Vredenberg, W.J. (1992) 'Formation of inactive PSII centers during light adaptation in spinach leaves', in N. Murata (ed.), Research in Photosynthesis. Kluwer Acad. Publ., Dordrecht, Vol. IV, pp. 615-618.

Snel, J.F.H., Kooijman, M. and Vredenberg, W.J. (1990) 'Correlation between chlorophyll fluorescence and photoacoustic signal transients in spinach leaves'. Photosynth. Res. 25, 259-268.

Snel, J.F.H., Kooten, O. van and Hove, L.W.A. van (1991) 'Assessment of plant stress by means of analysis of photosynthetic performance'. Trends in Analytical Sciences 10, 26-30.

Snel, J.F.H., Polm, M.W., Buurmeijer, W.F. and Vredenberg, W.J. (1992) 'Deconvolution of photobaric and photothermal signals from spinach leaves', in D. Bićanić (ed.), Photoacoustic and Photo-thermal Phenomena III. Springer Series in Optical Sciences 69, 65-68.

Vredenberg, W.J. (1981) 'P515. A monitor of photosynthetic energization in chloroplast membranes', Physiol. Plant. 53, 598-602.

Vredenberg, W.J. (1986) 'Fluorescence and absorbance measurements in leaves: sensors of photosynthetic performance', in W. Gensler (ed.) Advanced agricultural instrumentation. Design and use. NATO ASI series E no. 111, Martinus Nijhoff Publishers, Dordrecht, pp. 107-132.

Vredenberg, W.J. Snel, J.F.H., Buurmeijer, W.F. and Boumans, H. (1992) 'Application of non-invasive spectroscopic and photoacoustic techniques in research on photosynthetic performance of intact leaves'. Photosynthetica 27, 207-215.

Weis, E., Ball, J.T. and Berry, J.A. (1987) 'Photosynthetic control of electron transport in leaves of *Phaseolus vulgaris*: Evidence for regulation of photosystem 2 by the proton gradient', in J. Biggins (ed.) Proc. 7th Int. Congr. Photosynth., Vol. 2. M. Nijhoff Publ., Dordrecht, The Netherlands, pp. 553-556.

Weis, E. and Berry, J.A. (1987) 'Quantum efficiency of photosystem II in relation to 'energy'-dependent quenching of chlorophyll fluorescence', Biochim. Biophys. Acta 894, 198-208.

REMOTE SENSING FOR SCREENING ECOSYSTEM QUALITY

A. ROSEMA
EARS Remote Sensing Consultants
Kanaalweg 1
2628 EB Delft
The Netherlands

ABSTRACT. A brief introduction on remote sensing is given with some technical and historical aspects. Thereafter the development and present stage of three remote sensing techniques, that are considered relevant in relation to ecosystem quality, are discussed. They are: forest monitoring using LANDSAT, evapotranspiration monitoring with METEOSAT and laser induced chlorophyll fluorescence. The role of physical models in this kind of research is emphasized.

1. Introduction

Probably because the sun is the most abundant radiator in our very small part of the universe, living beings evolved with eyes. For several millions of years we have used reflected solar radiation to gather information on our environment. In 1800 W.F. Herschell found by means of a thermometer that outside the solar colour spectrum created with a prism, invisible radiation is present. Just two years later T. Young discovered the wave nature of light. Since then it became clear that light represents only a small portion of a more general phenomenon: electromagnetic radiation. It includes an almost infinite range of wavelengths and extends far beyond the visual range. These discoveries constituted an enormous challenge to man and can be considered the impetus for "remote sensing"· to extend our vision beyond its natural limitations.

2. Remote sensing techniques

Sun and earth are the two major natural radiators in our environment and remote sensing techniques based on them are named "passive". In all cases in which we use an artificial source of radiation, we speak about "active" remote sensing techniques. Most well known is radar, which is an active microwave technique. In the visual and infrared range lasers may be used as a source of radiation.

P C Struik et al (eds), Plant Production on the Threshold of a New Century, 263–279

Extending our ability to "see" is not only a question of extension beyond the visible part of the spectrum. Our eyes mix wavelengths. We cannot see the difference between green and a mixture of blue and yellow. In addition we have other limitations that can be overcome with remote sensing techniques: we are subjective, we must sleep now and then and we cannot be everywhere at the same time. In this respect remote sensing systems have some great advantages. Remote sensing data are:

- objective
- synoptic
- multi-spectral
- multi-temporal

But there is one main problem: the interpretation. In many respects remote sensing data cannot call on our visual experience. For this reason detailed research of the interaction between object and radiation is required in order to draw full advantage of this kind of information. In addition a multi-disciplinary approach, involving both fundamental and applied sciences, is required to develop the applications.

2.1. PHOTOGRAPHY

In the widest meaning of remote sensing, photography is the oldest technique. The first photographs were reported by Daguerre and Niepce in 1839. Modern colour photography is a wonderful remote sensing technique because of its high spatial detail and relatively low costs. Its main limitation, however, is its spectral performance. Photographic film has three colour layers, which produce the red, green and blue tones in the photograph. The sensitivity of these layers shows a considerable spectral overlap. The reproduction of eye vision is very good, but the ability to discriminate fine spectral features is low. An advantage is that the sensitivity of photographic film extends beyond the range of the eye until 1.0 μm. In this range, named the photographic infrared (0.7-1.0 μm) green vegetation has a very high reflectance, much higher than in the visual range. This feature was used to develop an infrared film, by which it became possible to discriminate between vegetation and camouflage painting. Modern colour infrared film has three colour producing layers with a shifted sensitivity:

Main sensitivity:	Reproduced as:
Infrared	Red
Red	Green
Green	Blue

For this reason this film is also called "false colour" film. It is possible by means of filters to measure the densities of the three colour layers separately and in this way a semi-multi-spectral approach is possible. However, complete spectral separation is not possible, whereas the red producing layer is not only sensitive to infrared but also to red and green. For a better spectral separation one might rely on a multi-camera system with suitable film-filter combinations. Visual interpretation, however, will then require colour additive viewing facilities. A detailed treatment on the quantitative application of multi-spectral aerial photography in agriculture may be found in the thesis of Clevers (1986).

2.2. SCANNING

An important instrument in remote sensing, that created a whole new range of observation possibilities, is the scanner. A scanner is an imaging instrument that measures radiation point by point and line by line. The first air-borne scanners were developed in the 1960's. With these instruments a line perpendicular to the flight line was scanned by means of a small rotating mirror. Due to the forward movement of the aircraft, subsequent lines could be used to reconstruct an image. By means of a prism or grating, to separate the various wavelengths, a multi-spectral scanner was created. Modern scanners have no rotating parts but use detector arrays.

The possibility to use scanning techniques for imaging is spectrally limited by the functional characteristics of the optics and detectors used. Using gold mirror coatings, germanium lenses and liquid nitrogen detector cooling, it was possible to develop thermal infrared imaging systems.

2.3. RANGING

In the microwave region, however, these scanning systems do not work. Relatively large antennae are required to collect microwave radiation. Air-borne imaging systems, however, were developed as well, but based on a different principle: ranging. A radar antenna attached to the airplane is sending a radar pulse perpendicular to the flight line. The radar pulse will scatter against objects at the ground surface and the backscatter is received by the antenna. By measuring the time delay between the radar pulse and the backscatter, the distance to the scattering object may be calculated. Again by means of the forward movement of the airplane, a complete radar image of the ground may be reconstructed.

2.4. SATELLITES

Today these kinds of instruments are working on a routine basis on board of various satellite systems. In terms of orbit there are two kind of satellites: those in a polar sun-synchronous orbit and those in an equatorial geostationary orbit. Polar earth observation satellites have an orbit (nearly) across the poles at a height of say 1000 km. Examples are LANDSAT, NOAA (both USA), SPOT (French), ERS

(European). Such satellites have a relatively high spatial resolution, but a low temporal resolution or "repeat coverage".

At our side of the earth the only equatorial remote sensing satellite is METEOSAT. Its orbit is 35,000 km over the equator and its orbital velocity corresponds exactly to the rotation velocity of the earth. For this reason its position is stationary relative to the earth surface. The spatial resolution in this case is relatively low (5 km), but the repeat coverage is very high (every half hour).

This brief overview of remote sensing systems is not intended to be complete. Moreover, new techniques and new satellite systems are under development. In this respect technology tends to run faster than the development of its applications. It seems sometimes easier to find money for the development and launching of new satellite systems than to get a substantial fraction of that amount for developing the applications.

In the Netherlands government programmes have supported the research and the development of remote sensing techniques and applications. The first programme (NIWARS) was carried out during the early years of remote sensing: 1971-1977. The "National Remote Sensing Programme" (NRSP) started in 1986 and got a follow-up in 1991 for a period of 10 years. The main problem in remote sensing is to understand the data and to translate the data into information that the user understands. From the very beginning our national remote sensing efforts have been marked by a fundamental and multi-disciplinary approach to this problem. This has been called the "Dutch school". It is characterized by research into the interaction of electromagnetic radiation with soil and vegetation, by the development of models to describe and study this interaction and by the use of related algorithms to extract information from the remotely sensed data.

Three cases of this approach will be discussed, which in the author's opinion have value in relation to the quality and condition of ecosystems, and which he is familiar with. He will not try to give an overview of the whole remote sensing field. To this end the reader is referred to the various textbooks. In Dutch an excellent textbook was published (Buiten and Clevers, eds., 1990).

3. Multi-spectral satellite data to extract forest closure and LAI

It was noted a long time ago that the multi-spectral reflection of bare soil is completely different from that of vegetation, as shown in Figure 1. The fact that vegetation strongly absorbs red (R) and reflects near infrared (NIR), while for soils there is not such a difference, has lead to the notion that a combination of these two bands may provide an estimate of vegetation ground coverage, or even of "leaf area index" (LAI) and biomass. Most widely used is the "normalised difference vegetation index" (NDVI), which is defined as:

$$NDVI = (NIR - R)/(NIR + R)$$

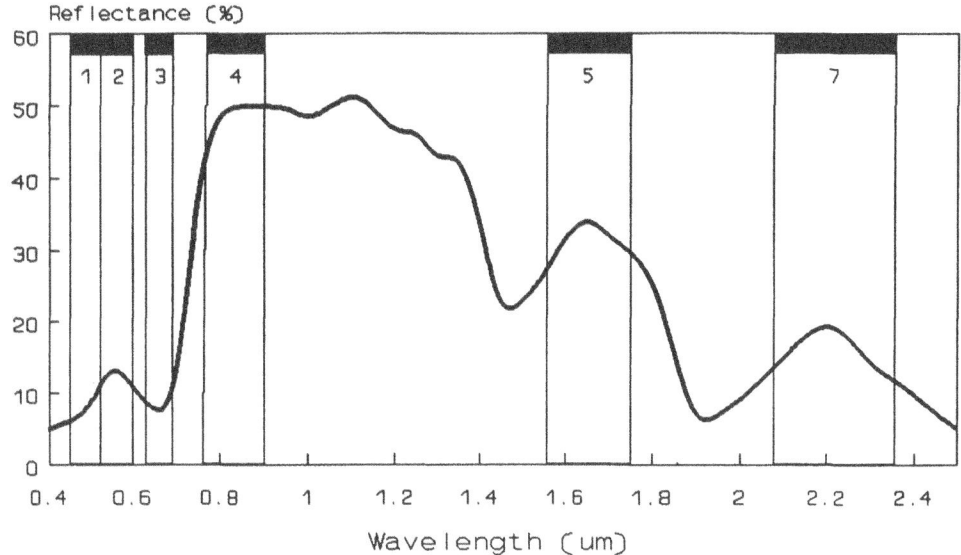

Figure 1. Spectral reflection of vegetation canopy and LANDSAT-TM spectral bands. Band 3 = red (R), band 4 = near infrared (NIR), band 5 = mid infrared.

The American polar weather satellites (NOAA) have been equipped with these two bands for this special purpose. Many people are using this index and are trying to calibrate and recalibrate it without questioning its validity. There is, however, a fundamental failing of this index: it applies only to a flattened earth surface, where shadows do not exist. And this is far from reality.

In the period 1988-1991 the project "Monitoring of Dutch Forest with Landsat-TM" has been carried out in the framework of our National Remote Sensing Programme and in cooperation with the National Forestry Institute "De Dorschkamp" (now IBN) in Wageningen and the National Aerospace Laboratory. The thematic mapper (TM) on board of the LANDSAT satellite is a 7-band multi-spectral scanner (Figure 1) with a "footprint" of 30 m. The objective was to study the possibilities of this tool for forest management and forest vitality monitoring. It was initially believed that by literature study one would be able to select indices based on combinations of spectral bands that would be suitable for this purpose. A large literature search was done and finally about 35 titles were studied in depth. The results were disappointing. They were often contradictory and many studies relied on multiple regressions between forest data and TM bands. Regressions, however, are not causal, they do not explain and have no general validity.

LANDSAT colour composites of the state forestry "Kootwijk" were studied by eye. It strikes that from space forest looks so dark. This is not because the soil is dark. When visiting the forest, which consists mainly of Douglas fir, it became clear that almost everywhere a nearly complete green ground cover existed of grass and herbs. It appeared also that the most open forest stands were darker on the satellite image than the dense stands and clear cuts. Gradually the awareness grew that the differences in appearance of the forest parcels was mainly effected by forest structure and that this had to be explained in terms of crown shadows casted on the ground. The idea grew that by a kind of geometrical model of crowns and shadowing, we could learn more about this phenomenon, and that we might turn the problem into an advantage.

3.1. FOREST MODELLING

In terms of vegetation modelling quite some work had already been done. The Dutch SAIL model (Verhoef, 1984) had been developed to simulate the reflectance of a homogeneous transparent canopy layer using the multi-spectral reflectances and transmittances of a leaf as starting point. This requires a lot of input data. Strahler and Li (1981) had developed a geometrical optical model which considers trees as opaque geometrical bodies. Although dealing with shadows, it was clear that this model was not applicable to the Dutch forest. On the contrary, the crowns transmit light. The needle density is variable and can be considered a characteristic of tree vitality.

It was decided to develop a new model, the "Forest Light Interaction Model" (FLIM) that combines the effects of shadowing and crown transmission and that uses a minimum of input data (Rosema et al., 1992). In this model the forest is considered as a discontinuous canopy layer with crowns and gaps. Then, given the position of the sun, the reflectance of the model forest was expressed in terms of the probabilities to view the following categories:

- tree crown with shadowed background
- tree crown with sun-lit background
- open space that is shadowed
- open space that is sun-lit

In remote sensing, simulation models often suffer from the problem that they have too many parameters. Such models are difficult to use, because of lack of information, and they cannot be inverted for the purpose of feature extraction. We have deliberately restricted the number of parameters by which the forest is characterised. They are:

- crown closure (horizontal crown projection) in %
- crown leaf area index (LAI)
- crown "yellowness"

3.2. SATELLITE DATA

Three LANDSAT-TM bands: TM3 (red), TM4 (near infrared) and TM5 (mid infrared) are used to derive the afore-mentioned parameters (Figure 1). A fully automatic method has been developed to extract this information from the LANDSAT imagery and to present the results in image format. An important and most practical element in this approach is that very little additional input is required. Only the LANDSAT spectral values of two types of reference plot are required:

- clear-cuts with herb vegetation (crown closure = 0 %)
- "infinitely dense" forest (crown closure=100 %, LAI=high)

This information is extracted from the LANDSAT imagery itself. The clear-cuts are easily identified on the LANDSAT imagery. However "infinitely dense" forest does not really exist. For this purpose the most dense forest parcel in the area may be chosen as a reference to which an estimated crown closure is assigned. From here the spectral values of the "infinite dense" forest are extrapolated with the model. So, the LANDSAT spectral values of two reference parcels are used for input and on this basis the complete relation between LANDSAT spectral

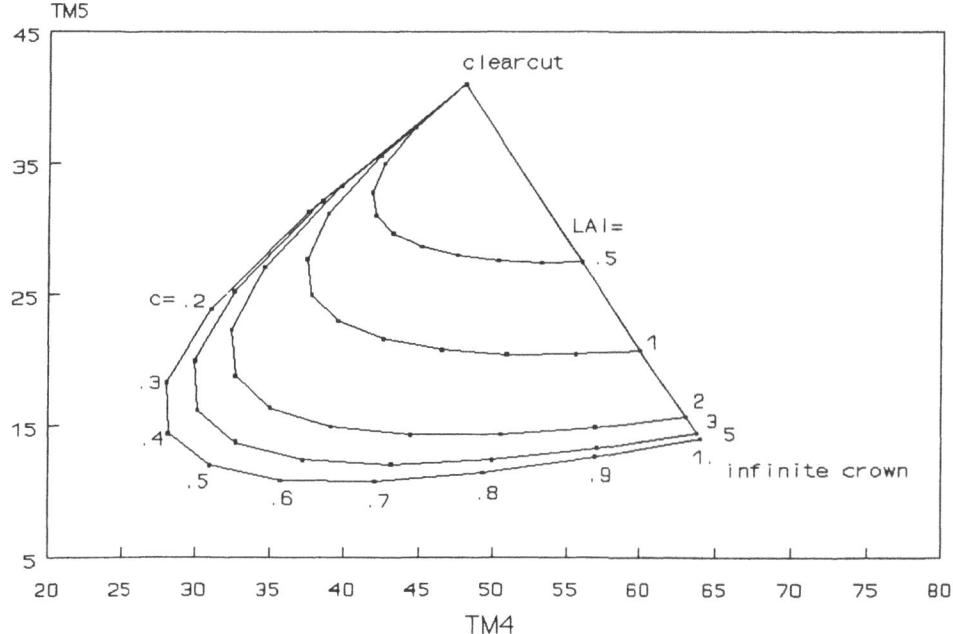

Figure 2. FLIM nomogram of TM4 versus TM5 for various combinations of forest closure (c) and leaf area index (LAI). TM values are in units of 0.4 % reflectance. Without shadowing all points would fall on the straight line.

values and above-mentioned forest parameters is established. Figure 2 shows a nomogram of TM band 4 (near infrared) against band 5 (mid infrared). The two "anchor points" named "clearcut" and "infinite crown" are shown. The nomogram shows the forest reflectance path for increasing crown closure and various values of the crown LAI. It is clear that the pixels are darkest for a crown coverage of 40 % and that the "colour" on the low and high coverage side of the curve is notably different. We can do the same for band 3 (red) and band 4 (near infrared) from which the NDVI is derived. Figure 3 shows a scatter plot of the NDVI against the forest green biomass (crown cover x crown LAI). From this figure it is clear that the NDVI and the forest green biomass are not uniquely and not linearly related. Biomass estimation errors up to 100 % are possible. This does not only apply to forest. The forest can be scaled down to bush or crop, i.e. any vegetation in which shadowing is significant.

The "Forest Light Interaction Model" (FLIM) has opened a new perspective for better interpretation and exploitation of LANDSAT and SPOT satellite data. This perspective is closely related to "ecosystem quality". Every year the vitality and timber stock of the Dutch forest is assessed by means of large field surveys. From our research it has appeared that timber stock is fairly well related to crown closure and could thus be assessed on the basis of LANDSAT. Forest vitality is

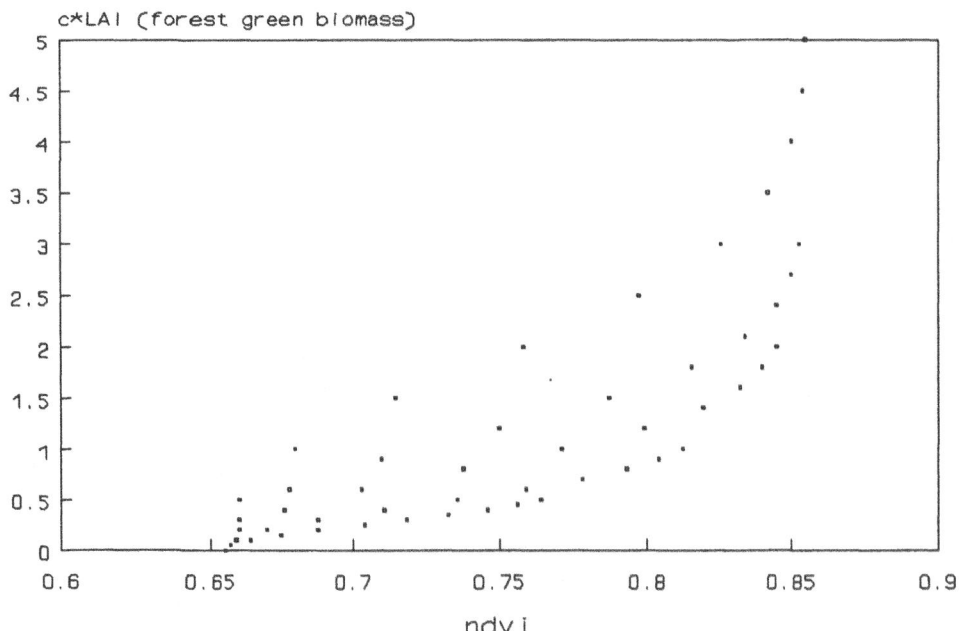

Figure 3. FLIM scattergram of forest green biomass (c*LAI) versus NDVI. Biomass and NDVI have no unique relation. Biomass estimates on the basis of NDVI may show large errors.

closely related to crown LAI and yellowness and could also be assessed on the basis of LANDSAT. With this new approach to forest assessment on the basis of satellite data, the development of an operational forest monitoring system can be envisioned.

4. Thermal and visible satellite data to monitor evapotranspiration and growth

From the very beginning we have used the physics and mathematics of heat flow to create models of the surface temperature behaviour in response to the daily radiation cycle. Such models were based on differential equations describing transient heat (and moisture) flow in the ground, which are solved for certain boundary conditions, in particular the heat (and moisture) budget at the ground surface. The heat budget at the surface plays an important part. It states that the net radiation absorbed by the surface (I_n) is used to heat the soil (G), to heat the air (H) and to evaporate water (LE) and reads:

$$I_n = G + H + LE$$

4.1. THERMAL MODELLING

The first thermal models were developed in the 1960's for the moon and were later applied to the earth by American geologists. In these models the atmosphere was lacking (H=0, LE=0). The next category had an atmosphere but neglected the presence of water and consequently evaporation (LE=0). During the first Dutch national remote sensing program (NIWARS, 1971-77) a considerable modelling effort was carried out and this resulted in a very detailed thermal model that included coupled heat, water and vapour transport in the soil, and heat and water exchange at the ground-atmosphere interface. It also included (in)stability effects on atmospheric heat and vapour transport (RADMOD model; Rosema, 1975). With this numerical model it was possible to simulate the development of drying and wetting fronts in bare soil for various soil types. A similar model was developed for the case of a crop cover (TERGRA model; Soer, 1980). These models allowed to study the surface temperature effects of parameter changes and allowed the development of strategies for thermal scanning flights and methods of data interpretation. There was also a disadvantage in these large, complicated models. They had many parameters and could not be inverted so as to derive useful information from surface temperatures observed with thermal scanning.

In 1978 NASA launched the experimental "Heat Capacity Mapping Mission" (HCMM) satellite. This gave rise to a European research effort, coordinated by the EC Joint Research Centre in Ispra: the TELLUS-project. This was the first time that noon and night thermal infrared satellite data came available in a more or less regular way and with a fair spatial resolution (400 m). EARS was requested to develop an operational HCMM interpretation algorithm. From this

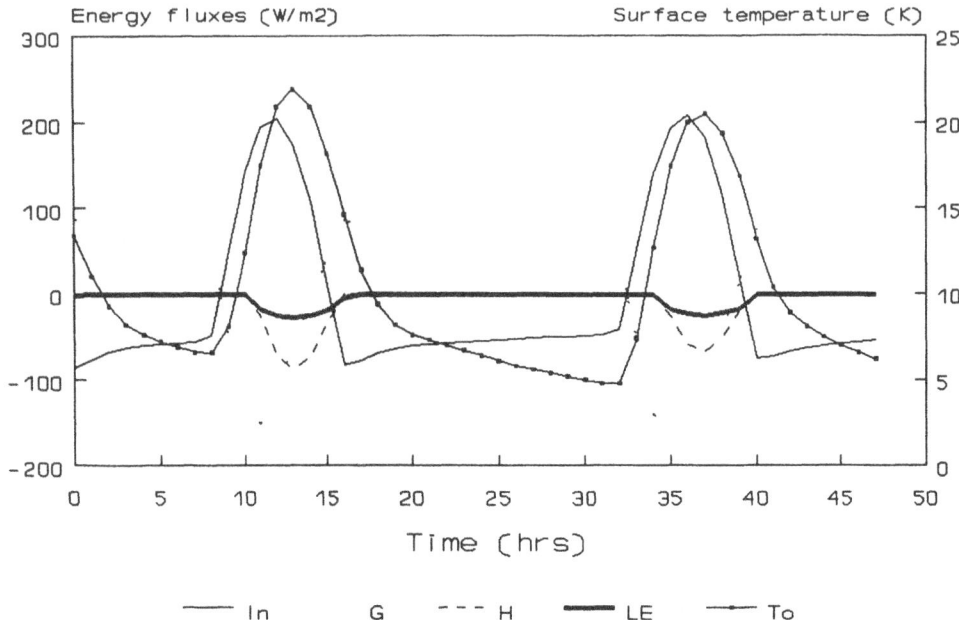

Figure 4. Example of TELL-US model simulation of the course of the daily energy budget components and the surface temperature.

satellite we could obtain regularly noon and night surface temperatures. We used our experience with the RADMOD model to develop a simplified model. With this model we could construct two dimensional look-up graphs and tables from which for each combination of a noon and a night surface temperature we could read the thermal inertia of the ground (square root of thermal conductivity and heat capacity), the evaporation resistance and also the daily evaporation (Rosema *et al.*, 1978). It was believed that the model applies as well to a canopy cover, in which case the evaporation resistance is identified as the canopy stomatal resistance. The algorithm was called "TELL-US". However, in the framework of the TELLUS project it was never used for actual monitoring or mapping purposes.

4.2. SATELLITE DATA

A new opportunity came in 1980 when Working Group 7 of the European Association of Remote Sensing Laboratories (EARSeL) obtained a contract from the European Space Agency to carry out the "Group Agromet Monitoring Project" (GAMP). The objective was to investigate the agrometeorological monitoring applications of METEOSAT in the semi-arid zone of Africa. METEOSAT provides a visible and thermal

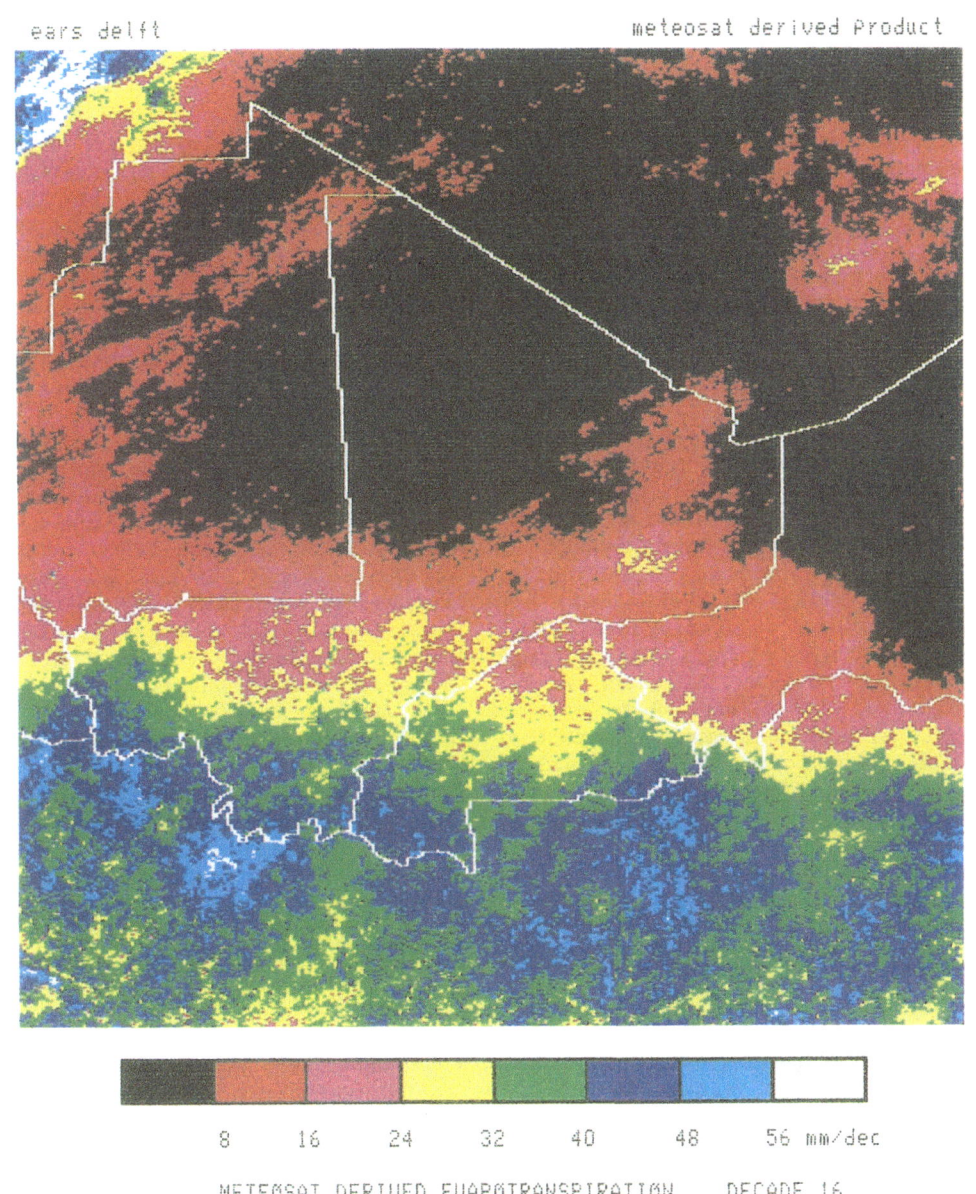

Figure 5. METEOSAT derived evapotranspiration in the Western Sahel during the first decade of June, 1990.

overview of the whole earth disk every half hour, but at a limited
spatial resolution of 5 km. This satellite is very useful for hydro-
meteorological monitoring on national and regional scales. Until
recently METEOSAT was only used to look at clouds and in the
meteorological user community this is still the case. In our
application we look particularly at the ground surface.

By means of Fourier series analysis of the daily cycle it could
be shown that 12.00 and 24.00 hrs are the most suitable observation
times because they allow a relatively simple extraction of the
evapotranspiration and thermal inertia information from the satellite
data. In GAMP for the first time evapotranspiration and thermal
inertia were actually monitored, mapped and analysed during a growing
season in an area of 260x340 km, that included the Niger delta. A
growth model was added that simulated the development of the Savannah
vegetation throughout the growing season on the basis of the METEOSAT
derived evapotranspiration resistance, with favourable results
(Rosema, 1986a,b).

Later new projects were carried out in the framework of the
Netherlands Remote Sensing Programme. The methodology was further
developed and the applications were investigated. For example: the use
of thermal inertia for monitoring flooding in the Niger Delta and the
Chari/Lagone/Lake Chad system (Rosema and Fiselier, 1990). The
capabilities for vegetation monitoring were studied in Burkina Faso
(Rosema, 1993). In 1990 a real time monitoring experiment was carried
out, that was commissioned by FAO in Rome. During the complete growing
season (June-October) evapotranspiration and growth conditions were
followed in the Western Sahel. Decadely evapotranspiration and
relative growth maps were produced and mailed to FAO immediately,
together with a summary report that included warnings for unfavourable
growing conditions.

Today EARS (Delft) is operating the first operational evapotranspiration and thermal inertia monitoring system in the world.
Besides evapotranspiration, also rainfall is estimated on the basis of
cloud temperature indices. The system is applied for regional energy
and water balance studies in Botswana. But presently its main task is
the monitoring and assessment of desertification in the Mediterranean
region. This is the ASMODE project, a part of the EC Environment
programme.

Satellite thermal scanning has developed to the point that
routine monitoring of evapotranspiration and thermal inertia is
possible. So we have an operational tool available to detect and map
water stress in vegetation and to estimate growth and crop production
at national and regional scales.

5. Chlorophyll fluorescence techniques to detect photosynthetic stress

During the 1980's environmental pollution became "the" issue and the
concern about ecosystem degradation, was strongly growing.
Particularly forest decline received a lot of attention. It seemed a
creeping mechanism, that was only observed in its final stage. We felt

that this was a real challenge for remote sensing: to detect the effects of soil and air pollution on plants in a very early stage, when no visible signs are present yet. We also believed that in this respect false colour photography and multi-spectral scanning techniques would not perform much better than the eye.

In the 1970's an instrument called "Fraunhofer Line Discriminator" (FLD) had been developed in the US. The FLD instrument was able to measure fluorescence in these dark lines. With this instrument enhanced fluorescence of Pinus growing in copper and zinc rich soil had been observed. Later on, also an effect of moisture stress on the fluorescence of Citrus trees was reported. With a grant of the Netherlands Ministry of Education we were able to start an explorative study in 1982. From the plant physiological literature we learned that chlorophyll fluorescence is not only depending on the amount of chlorophyll, but that the fluorescence yield is variable and inversely related to the photosynthetic electron transport. When the photosynthetic electron transport is blocked, fluorescence may increase 5 or 6 fold. Our initial laboratory measurements showed some effect of soil pollution on the fluorescence level of grass. Also a leaf fluorescence model was developed and from this model we could conclude that stimulated fluorescence is present in the solar reflection spectrum to a measurable level. Thus there seemed to be a perspective for chlorophyll fluorescence techniques.

5.1. LASER-INDUCED FLUORESCENCE

In the first half of the 1980's several laser induced fluorescence ("lif") systems had been developed, and were mainly used for measurements on algae and oil pollution of the sea. In 1985 contact was made with the IROE-CNR research group in Florence. They had developed a most advanced and manageable laser induced fluorescence system, the FLIDAR. A joint project was devised that would combine this unique instrument with the experimental facilities in Wageningen, the Netherlands. To this end cooperation was established with the Forestry Institute "De Dorschkamp" (now IBN), the institute of Plant Disease Research and the Department of Plant Physiological Research of Wageningen University. This project "Laser Induced Fluorescence in Trees" (LIFT) was financed by the Netherlands Remote Sensing Board and the Italian Special Programme on Improvement of Agricultural Production. Pioneering laboratory and field measurements were carried out. The field measurements were done on Douglas fir at the "acid deposition" research site in the forest near Garderen. At that time the FLIDAR was still a laboratory instrument and numerous problems caused that only 36 hours of continuous measurements on Douglas fir came available. Nevertheless, these measurements were very exciting since a 16 fold change in the fluorescence signal was observed instead of the 5 to 6 fold maximum range reported in the plant physiological literature (Figure 7). Moreover the variations in tree fluorescence during the day were large and seemed to be related to the NO_x and O_3 levels (Figure 6). No such correlation was observed with solar

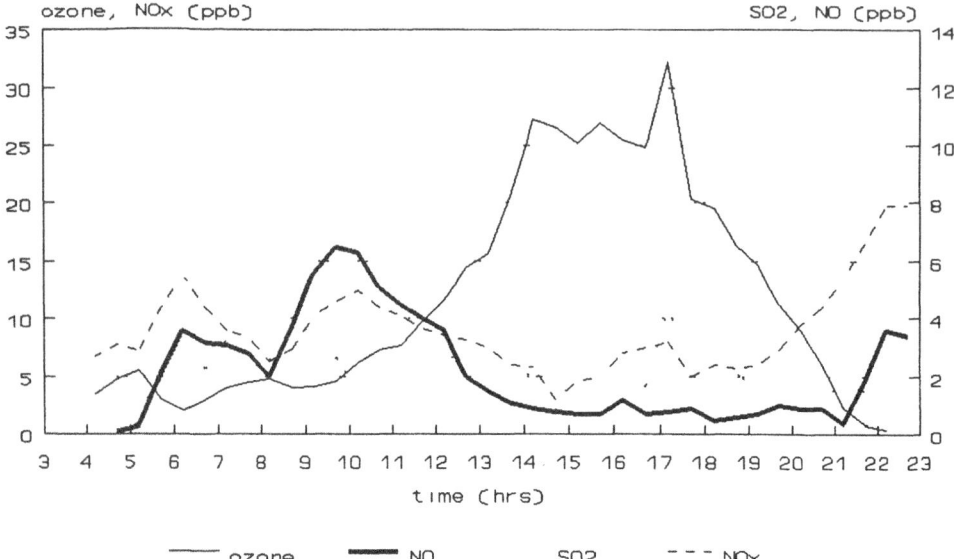

Figure 6. Concentration of some air pollutants during a 24 hr-period measured at the first site in Garderen (1987).

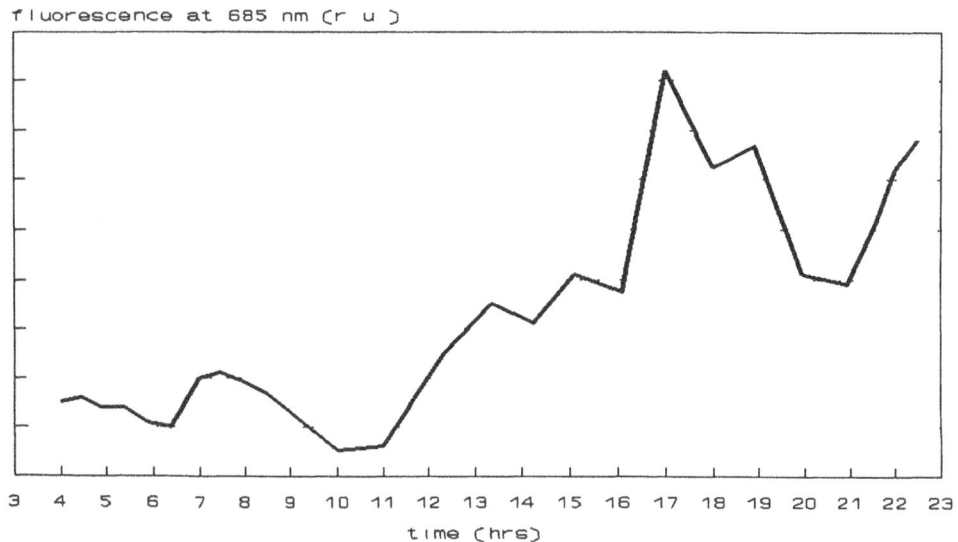

Figure 7. Laser induced fluorescence of Douglas fir at the forest site in Garderen (1987). Very high fluorescence after 16.00 hrs suggests severely hampered photosynthesis.

radiation and air temperature and we believe that the fluorescence changes observed were caused by the influence of these air pollutants on leaf photosynthesis. The extremely high fluorescence values in the late afternoon were interpreted as an almost complete blocking of the photosynthesis due to ozone (Rosema et al., 1988).

These preliminary results pointed to the exiting perspective of obtaining information on photosynthesis and stress in plants in a stage before any visual signs show up. It has led to a follow-up project in which a "lif"-measuring system has been designed. The prototype of this system was developed: the "Laser Environmental Active Fluorosensor" (LEAF). A laboratory simulator of LEAF was developed and with this instrument (EAF) measurements were carried out on young Douglas fir (*Pseudotsuga menziesii*) that had been treated with ozone, ammonia and SO_2 and mixtures of these gases (Schroote and Snel, 1991). Notable effects of these gases on the fluorescence were indeed observed, particularly after longer treatments, but these effects were always small compared to what had been observed in the field.

5.2. FLUORESCENCE MODELLING

When measuring the fluorescence of a leaf or a canopy, the signal does not only depend on the fluorescence yield, which is related to photo- synthesis, but also on the degree of reabsorption within the crop and consequently on its structural properties. As long as we keep the position of the "lif"-system fixed, the amount of chlorophyll and the structural properties of the canopy within the field of view are not expected to change and consequently changes in the signal can be attributed to changes in photosynthesis. However, if we move the "lif"-system, chlorophyll and canopy texture will play a role. Since we are mainly interested in photosynthesis related information, this information must be extracted selectively. This has been investigated in detail by means of canopy fluorescence light interaction modelling. Building on earlier work of Bunnik (1978), a multispectral canopy reflection model (SAIL) has been created (Verhoef, 1984). We had already been working on a Kubelka-Munk Fluorescence (KMF) model of plant leaf reflection and fluorescence. In a joint effort both models have been combined into a canopy model that is able to simulate the multi-spectral reflection ànd fluorescence of a canopy: the FLSAIL model. With this model we have been able to carry out a detailed sensitivity study of the various parameters involved (Rosema et al., 1991). In this way we have also formulated a possible strategy to extract the photosynthesis information from the LEAF measurements (Rosema and Verhoef, 1991).

The LEAF instrument came available april 1993 and new measuring campaigns are carried out at test sites in the forest. Besides the LEAF instrument, we are using the EARS-PPM, a field instrument to measure photosynthetic electron transport. In addition, parallel CO_2 and H_2O exchange measurements are done on the same tree. Also air pollution and meteorological data are measured at the site. In this way we hope to obtain better insight in the relation between "lif" and photosynthesis.

The first results of the LEAF field campaigns show interesting changes and differences in the two main bands of fluorescence emission: 685 and 730 nm, that are closely related to the photosynthetic light reactions but are not yet fully understood.

It is believed that in terms of ecosystem quality (laser induced) chlorophyll fluorescence is a most promising technique, whereas the signal is directly related to the photosynthetic light reaction. However more research is needed to bring this technique to the stage that it is fully understood and can be applied.

6. Conclusion

Various techniques in remote sensing have grown to the stage that they can be used successfully. Among these are the more traditional false colour and multi-band photography in combination with mainly visual interpretation. Scanning techniques have made considerable progress. New canopy-light interaction modelling has enabled the automatic extraction of LAI, crown closure and yellowness information from LANDSAT-TM data with 30 m resolution. Thermal and visual image data of meteorological satellites is operationally used for evapotranspiration and growth monitoring on national and regional scales. Finally, chlorophyll fluorescence techniques bear a promise for future monitoring of photosynthetic stress in an early stage.

7. References

Buiten, H.J. and Clevers, J.G.P.W., eds. (1990) 'Remote sensing, theorie en toepassingen van landobservatie', Pudoc, Wageningen, 515 pp.

Bunnik, N.J.J. (1978) 'The multispectral reflectance of shortwave radiation by agricultural crops in relation with their morphological and optical properties' PhD thesis, Wageningen Agricultural University, H. Veenman & Zonen B.V., Wageningen.

Clevers, J.G.P.W. (1986) 'Application of remote sensing to agricultural field trials', PhD Thesis, Agricultural University Wageningen Papers 86-4, 227 pp.

Rosema, A. (1975) 'A mathematical model for simulation of the thermal behaviour of bare soils, based on heat and moisture transfer', NIWARS publication nr. 11, Delft, The Netherlands, 92 pp.

Rosema, A. (1986a) 'Results of the Group Agromet Monitoring Project (GAMP)', ESA Journal 10(1), 17-41.

Rosema, A. (1986b) 'GAMP methodology; integrated mapping of rainfall, evapotranspiration, germination, biomass development and thermal inertia, based on METEOSAT and conventional meteorological data', in Proc. ISLSCP Conference, Rome, 2-6 December 1985, ESA SP-248, May 1986, pp. 549-557.

Rosema, A. (1993) 'Using METEOSAT for operational evapotranspiration and biomass monitoring in the Sahel region', Remote Sens. Environ. 45, 1-25.

Rosema, A., Bijleveld, J.H., Reiniger, P., Tassone, G., Blyth, K. and Gurney, R.J. (1978) 'A combined surface temperature, soil moisture and evaporation mapping approach, in Proc. 12th Int. Symp. on Remote Sensing of the Environment, Manila, Philippines, April, pp. 2267-2275.

Rosema, A. and Fiselier, J.L. (1990) 'METEOSAT-based evapotranspiration and thermal inertia mapping for monitoring transgression in the Lake Chad Region and Niger Delta', Int. J. Remote Sens. 11(5), 741-752.

Rosema, A. and Verhoef, W. (1991) 'Modeling of fluorescence light-canopy interaction', in Proc. 5th Int. Coll. Physical Measurements and Signatures in Remote Sensing, Courchevel, France, 14-18 January, 1991, EAS SP-319, pp. 743-748.

Rosema, A., Verhoef, W., Schroote, J. and Snel, J.F.H. (1991) 'Simulating fluorescence light-canopy interaction in support of laser-induced fluorescence measurements', Remote Sens. Environ. 37, 117-130.

Rosema, A., Cecchi, G., Pantani, L., Radicatti, B., Romuli, M., Mazzinghi, P., Kooten, O. van, and Kliffen, C. (1992) 'Monitoring photosynthetic activity and ozone stress by laser induced fluorescence in trees', Int. J. Remote Sensing 13 (4), 737-751.

Rosema, A., Verhoef, W., Noorbergen, and Borgesius, J.J. (1992) 'A new Forest Light Interaction Model in support of forest monitoring', Remote Sens. Environ. 42, 23-41.

Schroote, J.J. and Snel, J.F.H. (1991) '(L)EAF: A remote sensing fluoresensor' in Proc. 5th International Symposium 'Physical Measurements and Signatures in Remote Sensing', vol. 2, ESA Publ. Div., ESTEC, Noordwijk, the Netherlands.

Soer, G.J.R. (1980) 'Estimation of regional evapotranspiration and soil moisture conditions using remotely sensed crop surface temperature', Remote Sens. Environ. 5, 137-145.

Strahler, A.H. and Li, X. (1981) 'An invertible forest canopy reflectance model', Proc. 15th Int. Symp. Remote Sensing of Environment, Environmental Research Institute of Michigan, Ann Arbor.

Verhoef, W. (1984) 'Light scattering by leaf layers with application to canopy reflectance modeling: the SAIL model', Remote Sens. Environ. 16, 125-141.

BREEDING FOR ABIOTIC STRESS TOLERANCE

J.E. PARLEVLIET
Plant Breeding Department
Wageningen Agricultural University
P.O. Box 386
6700 AJ Wageningen
The Netherlands

ABSTRACT. The combined effect of all abiotic stress factors is world-wide seen enormous and the deterioration of many soils due to poor farm management is aggravating this. Especially stress factors that reduce and retard growth and development (drought, salinity, acidity, frost, heat, water logging, iron shortage etc.) are very important. These stresses tend to occur highly heterogeneously in time and space making selection in the field very inefficient. If the tolerance to the stress factor is simply inherited (tolerance to acid soil and Al-toxicity) it is possible to develop an efficient greenhouse or laboratory screening test. Is the tolerance a complex trait no efficient screening test is yet available and breeding for it is a tedious one (drought tolerance) or may not even be advisable at all (salt tolerance).

1. Introduction

1.1. STRESS

Nearly all organisms have to live in an environment that is heterogeneous in time (weather, climate) and space (soil, topography). An organism therefore cannot be optimally adapted all the time; it must be under stress at least part of the time.

Stress can be described as the direct effect on the organism when it is exposed to a detrimental factor for some time. Because of this stress the organism cannot function optimally and damage develops as a result.

Individuals that experience on average the least stress have the greatest chance to survive. The survival of an individual is a function of the ability to **avoid**, to **resist** or to **tolerate** the stress factors or to **recover** from the consequences of the stress factors.

Plants are exposed to a variety of variables that may hamper growth and/or development. Variables of a non-biological nature such as frost, heat, drought, acid soil, wind, hail and water logging are

P. C. Struik et al. (eds.), Plant Production on the Threshold of a New Century, 281–294.

collectively classified as abiotic stress. In agriculture stress too is always present. Boyer showed this convincingly (Boyer, 1982). For each of a number of crops he compared the highest farm yields ever realized in the U.S.A. with the mean yields. The highest yield is an estimate of the genetic potential of the crop. For maize the highest farm yield recorded was 19.3 ton/ha, the country wide average 4.6 ton/ha. Averaged over the main crops the mean yields were only 25 % of the highest yields. This reduction of 75 %, he estimated, was due to biotic stress for only 11 % and to abiotic constraints for 64 %. Due to resistance breeding, pesticides and agronomic measures the biotic stresses (parasites, weeds) are reasonably under control. With abiotic stress the situation is far less favourable.

1.2. STRESS DAMAGE

The stress factors can be classified into two groups; those that induce an undesirable reaction and those that suppress a desirable reaction.

Undesirable reactions are for instance sprouting of the maturing grains in the ears during humid conditions, lodging of cereals and legumes as a result of rain and wind, and bolting of sugar beet plants in the first year, when they should remain vegetative and form a beet. Resistance to these stress factors is generally not too difficult to obtain.

Desirable reactions that are suppressed usually pertain to the suppression of growth, leading to a reduced biomass and yield. Through various measures the farmer tries to maximize yield. This asks for an unhindered growth of his crop all the time. Stress factors such as drought, acid soil, salinity, frost, heat, shallow soil and water logging may hamper this growth process for shorter or longer periods with reduced yields as a consequence.

The damage shown by Boyer (see 1.1) is to a large extent caused by factors of this second group (Boyer, 1982). To solve such stress problems by breeding asks for a diversified approach as each factor has its own peculiarities.

1.3. STRESS TOLERANCE

In adapting to abiotic stress factors plants employ a wide range of mechanisms of which tolerance mechanisms are probably the most important. Because of this and because the real mechanism of adaptation is often not known the term tolerance is used in a wider context to indicate the adaptation to abiotic stress.

Agronomy and breeding both try to reduce stress to which the crop is exposed. The former does so by adapting the environment to the crop (irrigation, fertilization, drainage) while the latter aims at adapting the crop genetically to the environmental constraints (winter hardiness in winter cereals, drought tolerance).

Whether a breeder is successful in selecting tolerant cultivars depends on the uniformity of exposure in time and space of the breeding populations to the stress factor and on the heritability (genetic part of the total variance) of the stress tolerance. Both the homogeneity of exposure and the heritability vary greatly among the stress factors and crops. Table 1 shows that the combination of a sufficiently homogeneous exposure and a fairly high heritability in the field just does not occur, making breeding for abiotic stress tolerance often a difficult job. However, if the breeder can develop a screening procedure in the greenhouse or laboratory that gives a response representative for the response in the field under stress conditions selection may become much more efficient. Tolerance to acid soils and aluminium toxicity are good examples.

For the breeder tolerance to a given stress factor is one of the many traits he wants to improve. The easiest traits to improve are those that are simply inherited and are expressed homogeneously across the testing site. Then the heritability is high. The other extreme is formed by complex traits that are heterogeneously expressed over the testing situation. A complex trait is the accumulated or totalized result of a number of other traits, each with its own inheritance. Tolerance to the most important abiotic stress factors are highly complex (drought tolerance, salt tolerance, winter hardiness) and the exposure to these factors is very heterogeneous.

TABLE 1. Homogeneity of occurrence of stress factors and ease of assessment (heritability) of tolerance to the stress factor in the field or in a screening test (Scr.) on a scale of 1 to 9, 1 being extremely poor and 9 extremely good.

Trait	Homogeneity of stress factor in		Heritability	
	Time[1]	Space[2]	Field	Scr.
Drought tolerance (cereals)	3	3	2	_[3]
Salt tolerance (cereals)	6	1	2	5
Winter hardiness (wheat and barley)	3	5	4	4
Tolerance to low temperatures, W. Eur. (maize)	3	7	3	5
Tolerance to acid soils (wheat, barley)	7	3	5	8
Tolerance to Al-toxicity (wheat)	7	2	4	8

[1]within and between seasons.
[2]within and between experimental field plots.
[3]no suitable screening test available.

Since the inheritance of tolerance to abiotic stress and the exposure to the stress vary greatly with the stress factor the selection procedures for tolerance to abiotic stress vary greatly from each other. To show the diversification in approaches three examples will be discussed; breeding for tolerance to acid soil and aluminum toxicity, to drought and to saline soils.

2. Tolerance to acid soil and Al-toxicity

Worldwide there are large areas with low pH values with and without excess of Al. The root growth of sensitive crops or cultivars is reduced and so is the total biomass and yield. An excess of Al in acid soils intensifies the stress considerably (Al-toxicity). Crops differ in tolerance to acidity and Al-toxicity and within a crop the variation in tolerance to both factors is even greater. Drought increases the acidity and Al-stress significantly. The pH value varies during the season and across fields. Especially the variation within fields can be large and is associated with variations in water holding capacity, actual water content and differences in soil structure and type. This within field variation often causes a large experimental error when selecting.

Fortunately a good screening test is developed (Mesdag and Balkema-Boomsma, 1984) for wheat and barley. Acid soil, peat and some H_2SO_4 are mixed to give a soil with a pH-KCl of about 3.6. For wheat screening slightly more H_2SO_4 is added than for barley. Flats filled with this soil are sown with the entries to be screened. In each flat two check cultivars are sown; for barley Bavaria (very tolerant) and Alfor (very sensitive) and for wheat Colonais (very tolerant) and Thatcher (very sensitive) can be used. Ten (barley) and 15 (wheat) days from sowing the root system is assessed in relation to the controls on a scale of 1 (Alfor, Thatcher) to 9 (Bavaria, Colonais). The test conditions are considerably more severe than reality in the field, but the exposure is much shorter. The screening results agreed very well with those from the field and with the experience gained in practice.

In barley and wheat the inheritance of tolerance to low pH is fairly simple. Only one or a few genes seem to be involved (Mesdag and Balkema, 1984).

Screening for acidic, Al-rich soils can be done even in the laboratory. Germinated seeds are placed on a construction floating on a nutrient solution. After some time they are transferred to a nutrient solution containing Al for a given time, after which the seedlings are washed and returned to the nutrient solution without Al. Tolerant entries show a clear regrowth of the roots, sensitive ones do not resume root growth. Table 2 shows some data from a similar test. The test assessment agrees quite well with the field results. All cultivars that react sensitively in the field show no regrowth at 6 mg/l Al. All tolerant cultivars show a variable regrowth at 6 mg/l Al. But it is not always possible to distinguish slightly tolerant cultivars from sensitive ones or moderately tolerant from tolerant cultivars. Through the described tests above the tolerant and moderately tolerant entries can be selected. These should be field tested to ensure that the tolerance is of the required level and is embedded in genotypes of a suitable agronomic performance.

TABLE 2. Root length regrowth (mm) of 9 Brazilian wheat cultivars after the seedlings were grown in a nutrient solution for 72 hrs followed by 48 hrs in solution with different Al-concentrations. After Camargo and Felicio (1988).

Cultivar	Suitable for soils, which are:	Al^{+++} concentration in mg/l			
		0	2	6	10
BH 1146	acidic	95	63	40	27
IAC 5	acidic	52	31	23	3
IAC 28	acidic	69	39	31	17
IAC 24	acidic	51	34	29	4
IAC 161	acidic	63	41	10	0
IAC 22	moderately acidic	54	30	13	3
Anahuac 75	slightly acidic	62	0	0	0
IAC 162	non-acidic	66	31	0	0
Paraguay 281	non-acidic	51	12	0	0

3. Drought tolerance

For a continued growth and development of the crop there should be sufficient water available all the time. This is often not the case unless man interferes by e.g. irrigation. The availability of water depends on various factors. The amount and distribution of the precipitation are of prime importance, but the water holding capacity and the run off of the soil (soil type and soil depth) too are of considerable effect. Evaporation is another factor of importance. These factors together determine the intensity and duration of the drought stress experienced by the crop.

If the breeder wishes to select for differences in tolerance the stress factor should occur homogeneously over the field at a time considered representative for the occurrence of the drought stress. This is unfortunately very difficult to realize as the precipitation in many regions of the world is very erratic both between and within seasons.

The start, the intensity and the duration of the drought stress therefore vary from season to season and from location to location. The drought stress may occur at or just after germination, during the seedling stages, during tillering, during flowering or during the grain filling period. The sensitivity of the crop to the drought stress varies with the development stage, but is also dependent on the growth history of the crop preceding the drought stress (acclimatization). Even within a field, where the erratic rainfall can be considered the same, the drought stress may and often does vary considerably due to differences in soil depth and run off due to differences in slope.

As drought stress varies from year to year and from location to location one cannot select for tolerance to a given drought stress. There is no clear representative period and intensity of the drought stress. A cultivar adapted to a certain region must be yielding well relative to other cultivars at various levels of drought stress.

Apart from the great difficulties in obtaining a representative and uniform drought stress exposure the breeder has the problem of measuring the drought stress. In cereals drought stress is measured through grain yield, itself a complex trait. This only increases the problems for a breeder who wishes to select efficiently for drought tolerance.

Table 3 illustrates some of the problems in measuring the drought stress and the tolerance to drought. In this assumed situation a large number of cultivars are yield tested at a very dry location over several years. Each year five treatments are applied, full irrigation (I), no irrigation (V) and partial irrigation (II to IV). The mean yields over the years of the five treatments averaged over all cultivars could be seen as representing the drought stress, zero in treatment I and most severe in treatment V. The yield reductions due to drought stress range from 24 % to 65 % for the treatments II to V. The tolerance of a cultivar can be taken as the deviation from this average pattern. Cultivars A and B are less tolerant than the average cultivar, their yields drop faster with increased drought stress than the mean does. Cultivars C and D show an above average tolerance, their yields drop considerably slower. D is the most tolerant cultivar. But how can one express the level of tolerance in a simple figure? This can be done by taking the ratio of yield under stress to yield without stress. If environments V and I are used A and B are equally sensitive to drought, the ratio being 0.30 and C and D are considerably more tolerant, D with a ratio of 0.50 slightly more so than C with 0.45. In this case the information of only one stress environment is used. By using the regression coefficient of the cultivar effects on the environment index (the yields averaged over all cultivars per environment) one uses all the available information. The higher the regression coefficient the stronger the cultivar reacts to the drought stress. The ranking order from sensitive to tolerant is A, B, C and D. The regression coefficient, using more information, seems the more appropriate parameter to represent the drought tolerance provided the drought stress and the tolerance to the drought stress are linearly related.

However the data of cultivar E create a problem. According to its regression coefficient it is less drought tolerant than D, but it outyields this cultivar at all environments with 24 % including the most severely stressed environment. If it outyields cultivar D even in the most drought stressed environment it cannot be less drought tolerant than D. This problem was recognized by Fisher and Maurer (1978), who observed that the higher yielding cultivars tend to have a higher regression coefficient. This is easily explained if one realizes that biological yield should be viewed within a multiplicative system rather than in an additive one and the regression coefficient as shown in Table 3 is based on additive effects.

To meet this problem Fisher and Maurer (1978) developed the **drought susceptibility index, S**.

$$S = \frac{Y_p - Y_d}{Y_p} \times \frac{X_p}{X_p - X_d}$$

where Y is the yield of the cultivar of which S is calculated and X is the mean yield of all cultivars present in the same yield experiment. The number of cultivars should not be too small and preferably a random sample. Y_p and X_p represent the yields under non-drought stress conditions (the potential yield) and Y_d and X_d the yields under drought stress.

When the S-values are compared a better assessment of the true drought tolerance is obtained than with the regression coefficient (Table 3).

TABLE 3. Yield in ton per ha of five barley cultivars in five environments (I to V), the regression coefficient, b, of cultivar yields on the environments index and the mean drought susceptibility index, S.

Cultivar	I	II	III	IV	V	Mean II-V	b	S
A	5.0	3.8	3.0	2.2	1.5	2.63	1.13	1.04
B	4.6	3.5	2.7	2.0	1.4	2.40	1.05	1.05
C	4.4	3.5	2.9	2.4	2.0	2.70	0.81	0.85
D	4.2	3.4	2.9	2.4	2.1	2.70	0.71	0.78
E	5.2	4.2	3.6	3.0	2.6	3.34	0.89	0.78
Mean*	4.6	3.5	2.8	2.1	1.6	2.50	1.00	1.00

*Mean over a large number of cultivars. In each environment this mean represents the environment index.

In this assumed situation cultivar E is not a very realistic example, but the other four do represent a realistic situation. Rosielle and Hamblin (1981) concluded that yield potential and stress tolerance can be expected to be negatively associated. Parlevliet (1988) indicated that this is the case for oats in The Netherlands for yield potential and drought tolerance. The breeder therefore must try to find a compromise between the yield potential and the drought tolerance. If he has to choose between the cultivars A to D of Table 3 his choice depends on the average and range in drought stress that occurs in the region for which he is breeding. If the environments II and III tend to represent the region A is the obvious choice, if IV and V are more representative C and D are the likely candidates.

The data from Ehdaie and Waines (1989) clearly show the above mentioned problems. They compared seven spring wheats insensitive to photoperiod. Three genotypes were lines selected from South-West Iranian land races adapted to terminal drought stress conditions. The other four were recently bred cultivars, three were selected in South-West Iran, the fourth was the Californian cultivar Anza. The seven genotypes were tested in six environments with different drought stress levels, ranging from no stress to severe. These environments were obtained through differential irrigation and different sowing dates in two years at Moreno, California, a location with terminal drought as in Iran. They applied the Finlay and Wilkinson (1963) model, without a logarithmic transformation though. With these untransformed data the three Iranian landrace lines had a regression coefficient ranging from 0.99 to 1.23. Table 4 shows the data for the two highest yielding cultivars and of two of the landrace lines.

The authors concluded that the landrace lines were more drought tolerant because of the lower regression coefficient. A second conclusion was that Anza and Sholeh are the more desirable cultivars in even the severely stressed environments because of their high yielding ability. The latter conclusion is difficult to refute, but the first conclusion is less certain. By not using a logarithmic transformation one assumes implicitly additivity of yield effects, or in other words a change from 1.0 to 2.0 ton/ha is the same as from 6.0 to 7.0 ton/ha. Biologically this makes little sense. If the yields of the four cultivars/lines in this experiment had been ln-transformed before the regression coefficients had been calculated another conclusion would have emerged (Table 4, columns 6, 7 and 8); the Iranian landrace lines would not have appeared more drought tolerant, and Anza would certainly not have been classified as drought sensitive.

The drought susceptibility index S, gives a ranking similar to the regression coefficient after ln-transformation, but its discriminative power seems somewhat less.

The best parameter for expressing the drought tolerance therefore seems to be the drought susceptibility index S, the lower it is the higher the tolerance. A regression coefficient based on ln-transformed yield data would even be better because of its greater discriminative power but one needs data from a fair number of different drought stress environments, which are not always available.

Yield testing, especially under drought stress, is very inaccurate and so assessment of drought tolerance is equally inaccurate. This cannot be done with thousands of entries. And it would help the breeder if he could apply indirect selection. For this purpose one or more traits well correlated with drought tolerance, but with a higher heritability are needed. Much research in various crops has been carried out but the results up till now are not encouraging. There is no single trait that can be used for this purpose. In a few cases there are such traits that may help to remove the most sensitive entries. In cereals rapid leaf rolling and leaf firing are such indirect indications for drought sensitivity.

TABLE 4. Regression coefficient (r) of cultivar yields on environment index (yields averaged over all cultivars per environment), mean yields over six environments in ton/ha, yields in the highest (H) and lowest (L) yielding environment, these yields ln-transformed, the regression coefficients of cultivar yields on environment index after ln-transformation and the mean drought susceptibility index, S (adapted from Ehdaie and Waines, 1989).

Cultivar or line	r	Mean yield	Yield* H	Yield* L	Yields, ln-transformed H	Yields, ln-transformed L	r	S
Anza	1.20	2.67	4.40	0.91	1.48	-0.09	0.86	0.95
Sholeh	1.23	2.56	4.33	0.75	1.47	-0.29	0.97	0.99
Line 14	0.82	1.86	3.04	0.66	1.11	-0.42	0.84	0.93
Line 25	0.82	1.66	2.84	0.46	1.04	-0.78	1.00	1.00
Mean of 7 genotypes	1.00	2.03	3.47	0.56	1.24	-0.58	1.00	1.00

* These yields have been obtained from the mean yields and the regression coefficient because the actual yields were not given in Ehdaie and Waines' publication. As the genotype x environment interaction was largely of a linear nature the estimated data presented here must be close to the actual ones.

4. Salt tolerance

Injury due to salt in the soil is a widespread and old problem especially in arid and semi-arid regions. High salt concentrations in the soil water create high osmotic pressures, reducing the availability of water to the plants. At the same time specific ions such as sodium and chloride may prove toxic at higher concentrations.

Of the cereals barley is the most tolerant to salty conditions. The importance of having salt tolerant cultivars has been discussed already for several decades. The fact that, taken over all crops, no truly tolerant cultivars have been issued until now suggests that it must be very difficult at least to produce cultivars with an agronomic desirable performance together with a high salt tolerance.

There are several reasons why progress has been close to zero.
i) Saline soils are exceedingly variable in both the kinds of salts present and their concentration. The cation exchange properties of soils add more complexity, as do spatial and temporal variations in all these features (Epstein et al., 1980).
ii) Tolerance to salinity is almost certainly complexly inherited.

iii) Tolerance has to be measured either as a reduction in loss of biomass or as a reduction in loss of yield. Both biomass and yield are not easy to assess accurately.

iv) The damage caused by some salinity is extremely large. At salinity levels of 40 % of that of seawater the biomass in wheat was reduced to about 9 % for the most tolerant and to about 2 % for the sensitive entries (Kingsbury and Epstein, 1984). Although the differences in tolerance are highly significant, the cultivar effects are small compared to the total damage. So one needs unusually high levels of tolerance to prevent such a large damage to a significant extent.

4.1. HETEROGENEITY OF SALINITY

Epstein *et al*. (1980) stated that the nature and the intensity of the stress varies enormously both in space and in time. Which means that what is selected in one salty environment may not be adapted to another salty environment. But even within one field the heterogeneity is very high as shown in Table 5. So screening for salt tolerance in the field is rather inefficient. To screen for salt tolerance the entries to be screened have to be exposed uniformly over a given period of time to a predetermined salinity stress. This is in principle possible through salinized solution cultures (hydroponics) or through sand cultures watered with a standardized salinized solution. As salt tolerance probably is a complex trait it is unlikely that seedling screening only will give a representative result. So the exposure should be during most of the life cycle of the plant, which asks for considerable areas with such facilities. Other screening methods have been tested as well without encouraging results. Ray (1988) summarized it by concluding that at present there is no screening method available that indicates salt tolerance quickly and accurately.

Even if such a method would be developed only the tolerance to salt will be assessed. About its agronomic performance, especially yield under non-saline conditions no information is obtained. And it is essential that salt-tolerant cultivars yield also well at non-saline conditions (Shannon and Qualset, 1984; Richards, 1983) as the salinity within salty fields varies greatly, from almost zero to levels where no crop can grow. This means that if it is decided to select for salt tolerance the breeder has to select for salt tolerance and yield potential under non-stress conditions both. A preliminary tolerance test to separate the salt tolerant from the salt-sensitive genotypes must be followed by extensive yield testing under non-stress conditions of the tolerant entries.

4.2. TOLERANCE TO SALINITY

Within crops, and also within barley and wheat, genotypes clearly differ in degree of tolerance to salinity (Epstein *et al*., 1980; Kingsbury and Epstein, 1984; Richards *et al*., 1987). So, provided a good screening test is available, selection for salt tolerance is possible. Such a screening method has not yet been developed as mentioned above. The tolerance cannot be measured as with acidic soils through the effect on root growth or another trait directly affected by the saline conditions. It is the reduction in grain yield which is indicative of the tolerance. The higher the tolerance the smaller the reduction compared to yield in non-saline conditions. This makes it automatically a complex trait, whereby the exposure to the saline conditions has to last over the whole growth cycle of the plant or crop.

In some crops cell cultures are exposed to saline conditions, an attractively simple test. Apart from the fact that regeneration still forms a problem in small cereals it is not yet clear how useful cell culture testing will become. It is not yet clear whether salt tolerance at the cell culture level and salt tolerance of the plant or crop are identical or similar traits. The reports are as yet conflicting (Stavarek and Rains, 1983). Also the somaclonal variation going together with regenerated cell cultures may cause problems. And here too the tolerant lines have to be field tested extensively to assess the agronomic performance under non-stress conditions.

4.3. TOLERANCE TO SALINITY; IS IT NEEDED?

The extreme heterogeneity of the salinity in saline fields and the very rapid loss of yield performance with increased salinity create a special situation. Crops grown on saline fields are exposed to levels of salinity varying from close to zero to too high for crop growth (Shannon and Qualset, 1984). The best genotype for such conditions may not be a cultivar with the highest salt tolerance level, but a genotype that yields very good at non-saline conditions and reasonably well at the high saline patches. Richards (1983) recognized this. He estimated the percentage of land in seven salinity classes, as shown in Table 5, of an extremely saline, a saline and a non-saline field. From another saline field carrying a yield experiment with 16 barley cultivars and lines with varying levels of salt tolerance in six replicates he obtained grain yields in relation to the salinity. Per genotype about 20 quadrates of 0.37 m² were chosen ranging from areas with visually no observable damage to areas where grain yield was greatly affected. Of these quadrates the salinity in dS/m and the grain yield in g/m² were measured. From the strong negative linear relationship between grain yield and salinity the grain yield per salinity class (Table 5) was estimated averaged over cultivars. These yields were used to estimate barley yields on the extremely salty, salty and non-salty fields to give the data in Table 5. The saline fields appeared very heterogeneous, forming a mosaic with a considerable part not being salty, while at neighbouring patches no crop could grow. The areas in the fields above 20 dS/m did not contribute to the yield. The greater part of the yield (over 80 %) was produced on the non-saline or slightly saline patches. The yields used here represent the yield of barley genotypes that on average can be considered moderately tolerant to saline conditions.

What would be the best breeding strategy to improve the barley yields on such saline soils? Table 6 depicts four improved cultivars assumed to be derived from breeding for higher yield under non-saline conditions or under uniform very saline conditions. An increase of yielding potential of 10 % without change of the level of salt tolerance would give yields as shown in Table 6 (high yielding-1). At all levels of salinity the yields increase with 10 %. This is not very likely to occur. Rosielle and Hamblin (1981) stated that selection for stress tolerance is expected to produce a negatively correlated response in mean yields in non-stressed environments. The reverse is equally probable. Starting from the on average moderately tolerant population selection for high yielding potential per se is likely to lead to some loss in tolerance to salty conditions (high yielding-2 in Table 6). Selection for a high level of salt tolerance without losing yield potential (very tolerant-1) is therefore also not likely to occur. More realistic in such a case will be the very tolerant-2 type of cultivar in Table 6. Comparing the four improvement situations it is quite clear that only increasing the tolerance level, even if it is very considerable (two fold at the 16-20 dS/m level of salinity) is inferior to increasing the yield under non-saline conditions. This is primarily due to the fact that by far the greater part of the yield is produced on the non-saline or slightly saline patches of saline fields.

TABLE 5. Grain yield of barley in g/m^2 on soil of diverse saltiness classes in fields varying in levels of saltiness. The salinity is measured by measuring the conductivity of soil saturated with water and is expressed in decisiemens/m (dS/m). Seawater diluted to 40 % has a dS/m value of about 20. (After Richards, 1983).

	Salinity class in dS/m						
	0-4	4-8	8-12	12-16	16-20	20-24	>24
Grain yield in g/m^2	485	431	321	211	102	5	0
	% of field in salinity class						
Field very salty	23	26	7	6	11	6	21
Field salty	52	19	10	7	6	1	5
Field not salty	100	0	0	0	0	0	0
	% of total yield of the field						Yield g/m^2
Field very salty	41	41	8	5	4	0	0	270
Field salty	65	21	8	4	2	0	0	387
Field not salty	100	-	-	-	-	-	-	485

TABLE 6. Expected grain yield in g/m^2 at five levels of salinity and in fields of three salinity classes for barley cultivars selected for high yields under non-salty conditions and under very salty conditions. High yielding-1 represents a cultivar with a yield potential 10 % higher than the cultivars on which the experiments were based (starting level). High yielding-2 represents the same yield increase at non-salty conditions but associated with a slight loss in salt tolerance. Very tolerant-1 represents a highly tolerant cultivar, without loss in yield under non-salty conditions, while tolerant-2 shows a loss of yield potential under non-salty conditions of 5 %.

Selected for	Salinity in dS/m					Field		
	0-4	4-8	8-12	12-16	16-20	very salty	salty	not salty
Starting level	485	431	321	211	102	270	387	485
High yielding-1	534	474	353	232	112	297	426	534
High yielding-2	534	470	343	219	100	293	423	534
Very tolerant-1	485	450	390	310	205	297	411	485
Very tolerant-2	460	440	390	315	220	291	397	460

This poses the important question, as brought forward by Richards (1983), whether it is useful to select for salt tolerance even if a suitable screening method would be available. Richards' data clearly indicate that selection under non-stress conditions is to be preferred. The progress in time will be better especially because selection for yield under saline conditions is far more tedious than under non-stress conditions. The h^2 for yield under severe stress is much lower than under optimal growth conditions. But even if selection for tolerance would be not so tedious selection should be directed at the non- or slightly saline conditions because the greater part of the yield is produced on such patches.

The decision not to select for salt tolerance but for high yield potential per se with possibly a removal of only the really salt sensitive entries is of such an importance that this aspect should be investigated more thoroughly.

5. References

Boyer, J.S. (1982) 'Plant productivity and environment', Science 218, 443-448.
Camargo, C.E. de O. and Felicio, J.C. (1988) 'Wheat breeding at the Campinas agronomic institute', in M.M. Kohli and S. Rajaram (eds.), Wheat breeding for acid soils, CIMMYT, Mexico D.F., pp. 39-49.

Ceccarelli, S. (1987) 'Yield potential and drought tolerance of segregating populations of barley in contrasting environments', Euphytica 36, 265-273.

Edhaie, B. and Waines, J.G. (1989) 'Adaptation of landrace and improved spring wheat genotypes to stress environments', J. Genet. & Breed. 43, 151-156.

Epstein, E., Norlyn, J.D., Rush, D.W., Kingsbury, R.W., Kelly, D.B., Cunningham, G.A. and Wrona, A.F. (1980) 'Saline culture of crops: A genetic approach', Science 210, 399-404.

Finlay, K.W. and Wilkinson, G.W. (1963) 'The analysis of adaptation in a plant breeding program', Austr. J. Agric. Res. 14, 742-754.

Fisher, R.A. and Maurer, R. (1978) 'Drought resistance in spring wheat cultivars. I. Grain yield responses', Austr. J. Agric. Res. 29, 897-912.

Kingsbury, R.W. and Epstein, E. (1984) 'Selection for salt-resistant spring wheat', Crop Sci. 24, 310-315.

Mesdag, J. and Balkema-Boomstra, A.G. (1984) 'Varietal differences for reaction to high soil acidity and to trace elements; a survey of research in The Netherlands', Fert. Res. 5, 213-233.

Parlevliet, J.E. (1988) 'Problems with selection for drought tolerance', in M.L. Jorna and L.A.J. Slootmaker (eds.), Cereal breeding related to integrated cereal production, PUDOC, Wageningen, pp. 78-81.

Ray, N. (1988) 'Parameters for salt tolerance in crop plants. A review', Agric. Reviews 9, 37-43.

Richards, R.A. (1983) 'Should selection for yield in saline regions be made on saline or non-saline soils?' Euphytica 32, 431-438.

Richards, R.A., Dennett, C.W., Qualset, C.O., Epstein, E., Norlyn, J.D. and Winslow, M.D. (1987) 'Variation in yield of grain and biomass in wheat, barley and triticale in a salt-affected field', Field Crops Res. 15, 277-287.

Rosielle, A.A. and Hamblin, J. (1981) 'Theoretical aspects of selection for yield in stress and non-stress environments', Crop Sci. 21, 943-946.

Shannon, M.C. and Qualset, C.O. (1984) 'Benefits and limitations in breeding salt tolerant crops', Calif. Agric. 38(10), 33-34.

Stavarek, S.J. and Rains, D.W. (1983) 'Mechanisms for salinity tolerance in plants', Iowa State J. of Res. 57, 457-476.

CAN CHLOROPHYLL FLUORESCENCE AND P700 ABSORPTION CHANGES DETECT ENVIRONMENTAL STRESS?

H.R. BOLHAR-NORDENKAMPF[1], CH. CRITCHLEY[2], J. HAUMANN[1], M.M. LUDLOW[3], W. POSTL[1] AND A.J. SYME[2]
[1] Institute of Plant Physiology, University of Vienna, Austria
[2] Department of Botany, The University of Queensland, Australia
[3] Division of Tropical Crops & Pastures, CSIRO, Brisbane, Australia

ABSTRACT. At a particular location several natural and anthropogenic stress factors cause a pattern of stress which shows seasonal variations. With measurements of chlorophyll fluorescence, modifications in photochemical efficiency (Fv/Fm,) which often shows the earliest signs of stress, were determined on a daily as well as an annual basis. By the comparison of field measurements with laboratory measurements after recovery short term, long term and permanent stresses can be distinguished.

We have shown that stress of short duration is induced daily but can be species dependent in terms of time taken for recovery. This stress response may require the function of the xanthophyll cycle and may be based on a photoprotective, down regulating mechanism operating in PSII. Longer lasting stress such as occurs in winter (frost, photochilling) seems to reduce the amount of PSII but not of PSI, while any remaining PSII appears to function normally. Recovery from photochilling exhibited a fast phase, perhaps also based on pigment and reaction centre rearrangements, and a slow phase which may require more extensive metabolic and structural adjustments. All these stress responses have a photoinhibition component, whose mechanism remains to be explained. No straightforward relationship was seen between stress-induced changes in Fv/Fm and the velocity of D1 protein degradation, indicating that D1 turnover is not a stress-related metabolic response.

1. Introduction

Stress in plants is induced by natural and anthropogenic stress factors which cause changes in photosynthetic rates and by this growth and development of plants are influenced. On a site, stress factors have daily and seasonal patterns. Forestry, agriculture and horti-

P C Struik et al (eds), Plant Production on the Threshold of a New Century, 295–302
© 1994 Kluwer Academic Publishers

culture, by managing a site, will modify the pattern of stress factors. All environmental factors at a location have the capacity to induce stress. Therefore, depending on synergistic effects of these stresses and the physiological status of the plants, they will have different effects at different times of the year.

The capacity of a plant to cope with environmental stress varies during ontogeny. Disturbances in metabolism and consequent damage of functional structures will occur if the stress load exceeds the capacity for adaptation or repair (Heath, 1980; Larcher, 1987). The metabolism of plants will be modified to achieve an adaptation to any arising pattern of stress factors. This strategy requires regulatory work and always an additional energy input to run the maintenance of this regulation.

With respect to the factor, time different stress situations occur: i. diurnal short-term stress, e.g. excessive light; ii. long-term stress, e.g. frost, photochilling; iii. permanent stress, e.g. nutrient deficiency, air pollution. The level of stress induced can be defined by the time needed to complete recovery. For description of the stressing impact of environmental factors, stress-index-values can be worked out. To detect the stress level induced in the plant, measurements of physiological parameters and their dependence on ontogenesis, seasons and other time-dependent modifications are required.

Determination of the photochemical efficiency by means of chlorophyll-a-fluorescence measurements is a proper method to record not only permanent but also slight temporary impairment of the photosynthetic apparatus (Bolhar-Nordenkampf and Öquist, 1993), which is often induced by climatic as well as anthropogenic stress factors because of their considerable influence on different photosynthetic processes.

In this paper a number of short, medium and long-term measurements were made on several plant species subjected to stress and their recovery and other responses were assessed.

2. Materials and methods

Plants of Norway spruce (*Picea abies* (L.) Karst.); sunflower (*Helianthus annuus* L.); bush bean (*Phaseolus vulgaris* var. *nanus* L.); maize (*Zea mays* L.); sorghum (*Sorghum bicolor* L.); *Schefflera polybotrya* (syn. *arboricola*) Forst.; pea (*Pisum sativum* L.) and sugar cane (*Saccharum officinarum* L.) were found in natural stands, were growing in the field or were cultivated in controlled environments. Chlorophyll fluorescence of dark adapted leaves was measured with portable, time resolving fluorometers (PSM, Biomonitor AB, Umea, S; PEA, Hansatech, King's Lynn, UK; Bolhar-Nordenkampf *et al.*, 1989) or in the laboratory with a modulated system (MFMS, dual channel, Hansatech, King's Lynn, UK). The MFMS system was also used to determine P700 absorption changes. Leaf temperature was recorded by attached fine thermocouples (Cu-Const.). Leaves were heated by an infrared lamp and illuminated with halogen or high pressure mercury lamps. Irradiation was recorded by quantum and energy sensors simultaneously (SKP 215, SKS 1110, Skye Instruments, Powys, UK). D1 degradation was studied in pulse chase-type experiments with ^{35}S-methionine. Thylakoid membrane proteins were separated on LiDS polyacrylamide gradient gels. Wet gels were scanned for β-radiation with an AMBIS scanner (Syme et al., 1992).

3. Results and discussion

On a sunny day in the field, plants were exposed to non-saturating light levels in the morning, whilst at noon time they had to tolerate a surplus of light. The adaptation to light stress resulted in photo-inhibition, an irradiation-dependent loss of photochemical efficiency manifested as a reduction in quantum yield (Powles, 1984).

The comparison of artificially shaded plants with those exposed to full sunlight showed large differences in the photochemical efficiency (Fv/Fm, Figure 1). At noon the leaves of sun exposed plants exhibited a well developed reduction in photochemical efficiency, which recovered to differing degrees during the afternoon: Helianthus (Figure 1c) was by far less sensitive to photoinhibition than Phaseolus (Figure 1a) or even maize (Figure 1b; Bolhar-Nordenkampf et al., 1991) when considering the relative decrease in Fv/Fm in the sun compared to the shade. It did not recover at all in the afternoon, while Phaseolus recovered fully from noon depression.

Measurements of Fv/Fm on the upper (adaxial) and lower (abaxial) side of pea leaves demonstrated clearly that photoinhibition was only developed in mesophyll layers near the upper surface (Figure 2a). According to the shade characteristics of chloroplasts in the abaxial mesophyll illumination from below gave rise to a more pronounced photoinhibition.

Shade and sun leaves of Schefflera exhibited varying degrees of photoinhibition corresponding to the light intensity they were exposed to (Figure 2b). In shade leaves a photon flux density of 1000 μmol.m^{-2}.s^{-1} resulted in a strong loss of photochemical efficiency in the adaxial mesophyll after 12 hours of exposure, whereas the reaction of sun leaves was moderate, comparable with the response to 300 μmol.m^{-2}.s^{-1} of both sun and shade leaves.

When using chlorophyll fluorescence in light stress studies, it is recommended to compare data from the upper, adaxial and the lower, abaxial surface of a leaf. In addition the light environment of the leaves under investigation together with their anatomical features should be taken into consideration when interpreting the data.

In droughted Sorghum, high light together with high temperature can cause more damage to photosynthesis, than high temperature alone (Ludlow, 1987). Sorghum leaves with a water potential of -2 MPa were exposed to a photo flux density of 2200 μmol.m^{-2}.s^{-1} for 20 minutes (Figure 3). Measurements of gas exchange (data not shown) and fluores-cence showed that the photochemical efficiency (Fv/Fm) began to fall at 49°C for high temperature alone and at 46.4°C when it is accompanied by high light.

298

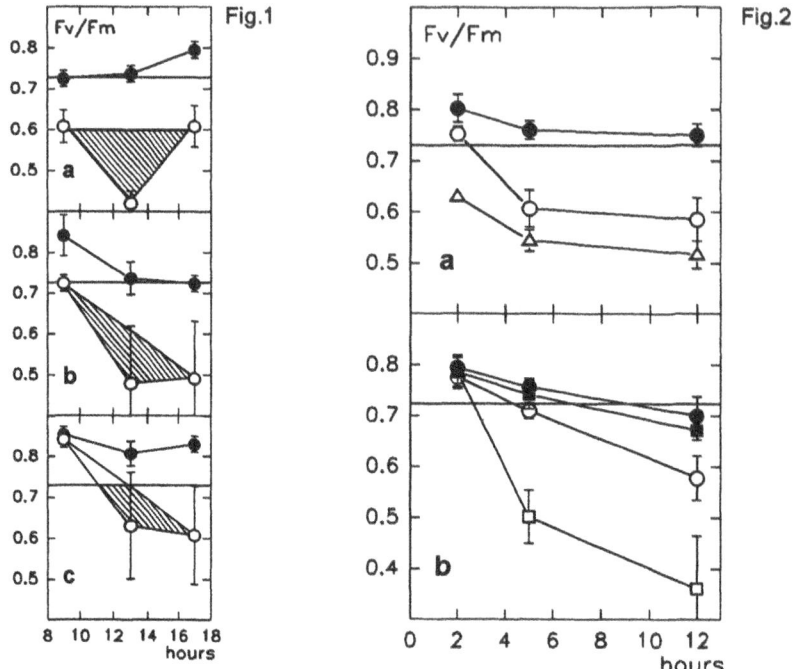

Figure 1. Changes in the photochemical efficiency (Fv/Fm) in artificially shaded (at noon 310 μmol.m^{-2}.s^{-1}) - • - and sun exposed (at noon 1350 μmol.m^{-2}.s^{-1}) - o - plants. Threshold for severe disturbances in the photosynthetic performance at Fv/Fm = 0.725. Cross hatched area indicates fast relaxing photoinhibition. a) *Phaseolus vulgaris* bush bean; b) *Zea mays*, corn, maize; c) *Helianthus annuus*, sunflower.

Figure 2. Changes in the photochemical efficiency (Fv/Fm) in floating leaf disks exposed to 1000 μmol.m^{-2}.s^{-1} up to 12 hours. a) *Pisum sativum*, pea; - o - upper, adaxial mesophyll; - • - lower, abaxial mesophyll, - △ - abaxial mesophyll illuminated from below; b) *Schefflera polybotrya*; sun leaves = circles; shade leaves = squares; full symbols = 300 μmol.m^{-2}.s^{-1}; open symbols = 1000 μmol.m^{-2}.s^{-1}.

Photochilling events are known to reduce dramatically the photochemical efficiency of evergreen needles of coniferous trees. Spruce trees growing near the timber line (1700 m a.s.l.) were nearly permanently subjected to climatic and ozone stresses (Figure 4). Especially in March the combination of chilling temperatures with high irradiances gave rise to strong impairment of the photosynthetic apparatus. Changes in the "ground fluorescence" (F0) indicated an influence of ozone (Bolhar-Nordenkampf and Lechner, 1988).

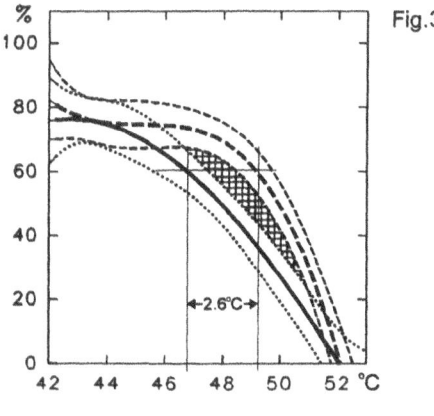 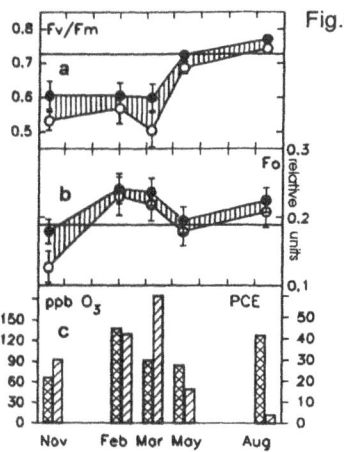

Figure 3. Changes in the photochemical efficiency (% of control) in leaves from *Sorghum bicolor* after drought stress (water potential - 2 MPa), strong illumination (2200 μmol.m^{-2}.s^{-1}) and heat treatment. ----- upper side illuminated, - - - - lower side shaded; regression lines 3rd order, 95 % confidence interval. Comparable reduction in Fv/Fm at 49°C on the shaded side occurs at 46.4°C on the illuminated side of the leaves.

Figure 4: Spruce needles (*Picea abies*) were measured from trees at the timber line (1700 m a.s.l.). a) changes in the photochemical efficiency (Fv/Fm) - o - at the site, - • - after 8 hours recovery; b) changes in ground fluorescence Fo, relative units, - o - at the site, - • - after 8 hours of recovery; c) cross hatched bars = ppb O_3, hatched bars = PCE = photochilling events.

During winter stress the PSII fluorescence signal is strongly reduced. In February Fm - values reached only 35 % of those in April (Figure 5), which represented a substantial reduction in reaction centres with a lesser effect on the functionality of those that remained (Fv/Fm values much less reduced than Fm). Measurements of P700 absorption changes showed no differences in the amount of P700 reaction centres whereas the oxidation status changed from 74 % to 17 % (Bolhar-Nordenkampf et al., 1993).

Field studies with chilled sugar cane (-2°C at 8 a.m.) demonstrated that photochemical efficiency of only the sun exposed tips of the leaves were reduced ("photochilling"), whereas the shaded leaf bases were unaffected (Figure 6). The increasing photon flux density during the day induced further photoinhibition in both parts of the leaf. When leaves were put under favourable conditions (25°C, 90 % RH, 10 μmol.m^{-2}.s^{-1}), there were two phases of recovery, a fast rate followed by a slower rate, which probably involved repair (Krause, 1988; Van Wijk and Krause, 1991). The fast recovery was

300

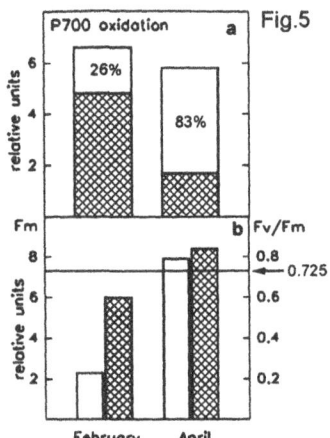

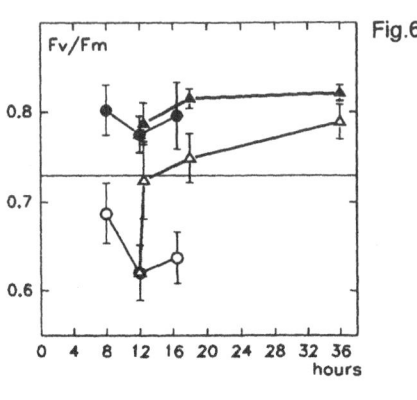

Figure 5. Changes in the amount of PSII and PSI reaction centres in needles from spruce trees. a) bar = amount of PSI centres, which can be oxidized, cross hatched area = amount of PSI centres oxidized by white light; b) bar = amount of PSII reaction centres present, cross hatched bar = Fv/Fm.

Figure 6. Changes in photochemical efficiency (Fv/Fm) in sugar cane (*Saccharum officinarum*) leaves measured in the field. - • - shaded base, - o - sun exposed tip, recovery of detached leaves (25°C, 90 % RH, 10 μmol.m^{-2}.s^{-1}): -▲ - base, - △ - tip.

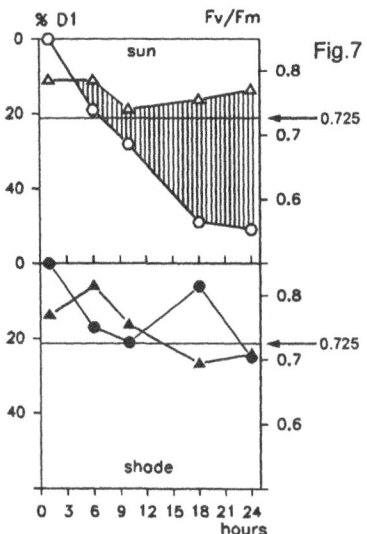

Figure 7. Changes in photochemical efficiency (Fv/Fm) and in D1-protein degradation. *Schefflera polybotrya*: circles = D1-degradation in % labelled D1, triangles = Fv/Fm, open symbols = sun leaf, closed symbols = shade leaf.

completed during 30 minutes. It is thought to correlate with the rearrangement of Photosystem II reaction centres (Garab *et al.*, 1991; Walters and Horton, 1993), In contrast, the slow recovery took 6-8 hours, which probably corresponds with repair of PSII reaction centres.

Studies with sun and shade leaves clearly demonstrated that Dl protein turnover was high in sun leaves and low in shade leaves and showing no correlation with photoinhibition (Figure 7; Syme *et al.*, 1992). These results may lead to the conclusion that arrangement of PSII reaction centres and the xanthophyll cycle are strongly involved in the recovery process after a reduction in photochemical efficiency.

The response of the photochemical efficiency to different stress factors is unspecific. The normally observed reduction in the value F_v/F_m can be the result of a short-term, a long-term or a permanent disturbance in the photosynthetic performance. A causal relationship can only be found if additional information as gas exchange data and/or climatic and air pollution data are available.

4. References

Bolhar-Nordenkampf, H.R. and Lechner, E.G. (1988) 'Temperature and light dependent modifications of chlorophyll fluorescence kinetics in Spruce needles during Winter'. Photosyn. Res. 18, 287-298.

Bolhar-Nordenkampf, H.R., Long, S.P., Baker N.R., Öquist G., Schreiber, U. and Lechner, E.G. (1989) 'Chlorophyll fluorescence as a probe of the photosynthetic competence of leaves in the field: a review of current instrumentation'. Funct. Ecology 3, 497-514.

Bolhar-Nordenkampf, H.R., Hofer, M. and Lechner, E.G. (1991) 'Analysis of light-induced reduction of the photochemical capacity in field-grown plants. Evidence for photoinhibition?' Photosyn. Res. 27, 31-39.

Bolhar-Nordenkampf, H.R. and Öquist, G. (1993) 'Chlorophyll fluorescence as a tool in photosynthesis research', in Hall *et al.* (eds.), Photosynthesis and productivity in a changing environment, a field and laboratory manual. Chapman and Hall, pp. 193-206.

Bolhar-Nordenkampf, H.R., Haumann, J., Lechner, E.G., Postl, W.F. and Schreier, V. (1993) 'Seasonal changes in photochemical capacity, quantum yield, P700 - absorbance and carboxylation efficiency in needles from Norway spruce', in H. Yamamoto and C. Smith (eds.), Photosynthetic Responses to the Environment. Am. Soc. Plant Phys., in press.

Garab, G., Kieleczawa, --, Sutherland, J.C., Bustamante, C. and Hind, G. (1991) 'Organisation of pigment-protein complexes into macrodomains in the thylakoid membranes of wild-type and chlorophyll *b*-less mutant of barley as revealed by circular dichroism. Phytochem. and Phytobiol. 54, 273-281.

Heath, R.L. (1980) 'Initial events in injury to plants by air pollutants'. Ann. Rev. Plant Physiol. 31, 395-431.

Krause, G.H. (1988) 'Photoinhibition of photosynthesis. An evaluation of damaging and protective mechanisms'. Physiol. Plant. 74, 566-574.

Larcher, W. (1987) 'Streß bei Pflanzen'. Naturwiss. 84, 158-167.

Ludlow, M.M. (1987) 'Light stress and high temperature', in D.J. Kyle, C.B. Osmond and C.L. Arntzen (eds.), Photoinhibition. Elsevier Sci. Pub., pp. 89-109.

Powles, S.B. (1984) 'Photoinhibition of photosynthesis induced by visible light'. Ann. Rev. Plant. Physiol. 35, 15-44.

Syme, A.J., Bolhar-Nordenkampf, H.R. and Critchley, C. (1992) 'Light-induced Dl protein degradation and photosynthesis in sun and shade leaves', in N. Murata (ed.), Research in Photosynthesis. Kluwer Acad. Pub., pp. 337-340.

Walters, R.G. and Horton, P. (1993) 'Theoretical assessment of alternative mechanisms for non-photochemical quenching of PSII fluorescence in barley leaves'. Photosyn. Res. 36, 119-139.

Wijk, K.J. van, and Krause, G.H. (1991) 'Oxygen dependence of photo-inhibition at low temperature in intact protoplasts of Valerianella locusta L.'. Planta 186, 135-142.

INDEX OF FREEZING BASED ON TIME SERIES

ANA M. TARQUIS[1], ANTONIO SAA[2] AND MAITE CASTELLANOS[1]
[1] *Dept. of Applied Mathematics*
[2] *Dept. of Edafology and Climatology*
*E.T.S Ing. Agrónomos (U.P.M) Ciudad Universitaria s.n.
Madrid 28040, SPAIN*

ABSTRACT. A time series of monthly minimum temperatures and their monthly averages (1961-84) from Guadalajara (Spain) have been analyzed by the Box-Jenkins method. The ARIMA model obtained was identical for both temperature series: $(1\ 0\ 0)\ (0\ 1\ 1)_{12}$ N. The aim of our work has been to study a method of calculating an index of freezing based on time series analysis, and try to predict the probability of this situation.

1. Introduction

Crop establishment and development can be studied as an investment of time, work and resources that always involves a high risk. This is mainly because climatological factors such as low temperatures cannot be controlled (6).

These variables have been studied to develop models to simulate and quantify several ecological processes (8). The monthly averages of the most relevant variables (temperature, rainfall, etc.) have been used as input data in the model to relate them to physiological responses (3) and development (9).

On the other hand, other studies have shown that meteorological data sets are time series data that can be fitted into ARIMA models (7). This means that a datum point of a series is the sum of a tendency, stationary oscillations and a white noise (1). This type of analysis has been widely applied in economics, mainly because it allows forecasting (4).

We will try to apply these concepts to agriculture and offer a tool that can be used in crop decision and planning as related to low temperatures.

2. Materials and methods

The observations of minimum temperatures represent the absolute minimum of the daily temperatures during each month, and their monthly

P C Struik et al (eds), Plant Production on the Threshold of a New Century, 303–309
© 1994 *Kluwer Academic Publishers*

averages are the average of the daily minimum temperatures of each month. The data have been supplied by the Spanish National Meteorological Institute and all temperatures are expressed in centigrades (°C).

Both series are complete, beginning in January 1961 and ending in September 1985. The last year (1985) was not used to complement the model, but was used to verify it. Also, the number of days at or below 0°C during an agricultural year (from September to September of the next year) have been recorded.

All the calculations have been done on a 386/33 PC using a statistical software package (STATGRAPH version 6.0).

2.1. MODEL FITTING

The Box and Jenkins method has been applied to analyze univariate time series data (1). The objective of this method is to obtain a simple parametric model that can explain the relationship between observed data points. It is useful to forecast future values, too.

The models introduced by Box and Jenkins can be classified into two types: autoregressive models (AR) and moving average models (MA). Both can be combined giving a mixed model which, in this case, verifies some conditions that allow the application of an integrated autoregressive moving-average model.

Three steps have been taken to fit an ARIMA model:

a) Identification. In this study the original series, the simple autocorrelation function (AF) and the partial autocorrelation function (PAF) have been observed. Based on this, it is possible to see if a series is stationary or not, and to determine the type of the model and its order of differentiation.

b) Estimation of parameters. The estimation has been made with the backcasting method (1) that the statistical package uses.

c) Validation. Three cases were considered to validate the model. First, residual series were studied to assure that they were white noise following an adopted method (2). Second, the model was over-adjusted to get another model of the same type and a higher order to see if any improvement was achieved (5). And finally the model was applied to calculate the values for 1985 which were compared with the observed data.

Once the model was validated the temperatures for the next six months were forecasted.

2.2. INDEX OF FREEZING

The number of days at or below freezing was chosen as the index of freezing. This is defined as the total number of days during an agricultural year that the minimum temperature is equal or below 0°C.

To estimate the value of this index it was assumed that the distribution of temperatures follows a normal density function, based on results from other meteorological stations (data not published).

So, it is possible to determine it by considering the monthly minimum temperature average as the average of the normal distribution, and to calculate the standard deviation proportional to $(t_{abs} - t_{avg})$. The constant of this relation is estimated by the least square method based on the data recorded from 1961 to 1985.

2.3. CROP PLANNING

Two crops were chosen as examples to calculate the possibility of establishment and harvest date: tomato (*Solanum lycopersicum*) and cabbage (*Brassica var. capitata* L.F. Alba D.C.).

Based on the critical temperatures for both crops (Table 1) the percentage of days per month below or over a certain value are estimated to analyze the protection they need to grow.

TABLE 1. Critical temperatures of freezing resistance, and growth for tomato (*Solanum lycopersicum*) and cabbage (*Brassica var. capitata* L.F. Alba D.C.).

Crop	Temperatures (°C)				
	Freezing	Growth			
		Critical	Minimum	Optimum	Maximum
Tomato	-3	+6	+7	+15,+18	+30
Cabbage	-9	+4	+5	+16,+18	+30

3. Results and discussion

3.1. MODEL FITTING

The resulting ARIMA model was identical for both time series, showing a seasonal difference of order 1 with a seasonal length of 12. The parameters estimated are shown in Table 2. The estimated white noise in monthly minimum temperatures has a variance of 2.32803 with 272 degrees of freedom and a Chi^2 test on the first 20 residual auto-correlations of 25.5879 with probability of a larger value given white noise of 0.109561. In the minimum temperatures the variance of the estimated white noise is double, but the Chi^2 test on the first 20 residual autocorrelations has a value of 32.6267. The partial auto-correlation function of the series differenced once is shown in Figure 1.

The forecast for 1985 was compared with the measured data, resulting in a R^2 of 86 % and 96 % respectively for the monthly minimum temperature and the monthly minimum temperature average (Figure 2). It is reasonable that the predicted average fits better because the minimum values are more extreme.

TABLE 2. Estimated parameters of ARIMA model $(1,0,0)$ $(0,1,1)_{12}$ N for both temperatures series time, including standard error, T-value and P-value.

Temp.	Parameters	Estimated Value	Standard Error	T-value	P-value
Average					
	AR(1)	.30244	.05855	5.16524	.00000
	SMA(12)	.71483	.04642	15.39906	.00000
Minimum					
	AR(1)	.21275	.06009	3.54082	.00047
	SMA(12)	.73663	.04327	17.02381	.00000

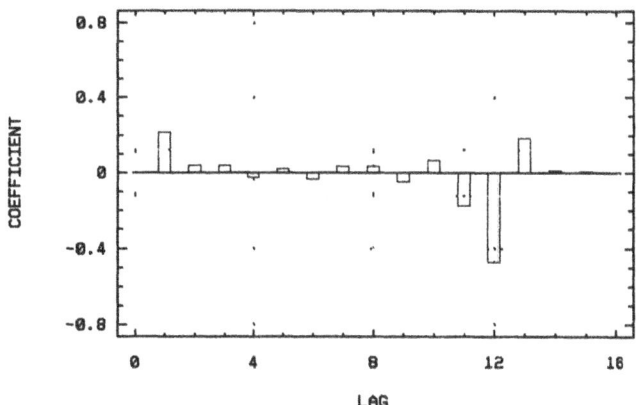

Figure 1. Partial autocorrelation function of monthly minimum average temperature series with seasonal difference of order 1

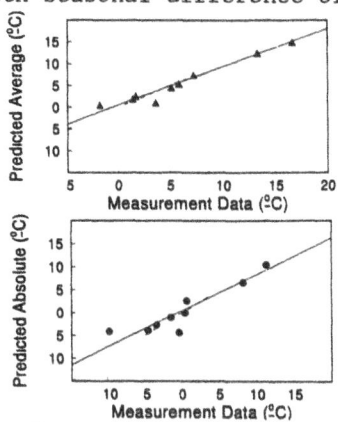

Figure 2. Predicted values versus measurement data for 1985; closed triangle, minimum average temperatures (t_{avg}); closed circle, absolute minimum temperatures (t_{abs}).

3.2. INDEX OF FREEZING

When adjusted to a normal distribution the minimum temperatures that best fit in the index from 1961 to 1985 were obtained with a constant of -1/1.6. Therefore, assuming that the minimum temperatures and their monthly average have a difference of 1.6 times the standard deviation, the numbers of freezing days calculated show an R^2 of 92 % (Table 3).

TABLE 3. R^2 of freezing days calculated with different constant (a) values in a normal distribution function ($s.d. = (t_{abs} - t_{avg})/a$).

a	Number of Observations	R^2 (%)	s.e. estimated (days)
2.8	23	75	8
2.4	23	80	7
2.0	23	89	5
1.6	23	92	4
1.2	23	60	11

So once the minimum temperatures and their average are predicted it is possible by this method to estimate the number of days at or below freezing (fd) per month too. One of the problems that must be solved is the confidence limits for the predicted values (Figures 3 and 4). The lower and upper limits in the forecast were calculated for 95, 90, 85 and 80 % confidence limits (data not shown) giving, approximately, a difference between them of 6, 5, 4.5 and 4 °C. It was observed that only in the last case (80 %) the predicted index could be useful. For example, the model predicts 45 fd, and if we use the lower limit predicted with 80 % of confidence limits for both temperature series, we get a result of 79 fd. But if we will use 90 % of confidence limits, the result will be 182 days (all the days of those 6 months).

3.3. CROP PLANNING

Taking all the values in both series the ARIMA model was applied to estimate for the next six months the percentage of days where the temperature falls below the critical resistance value for tomato and cabbage. The results obtained shows that tomatoes will need 14 days of protection (with the cost that this implies) and cabbages will not need it.

Further research has to be applied to monthly temperatures to see if it is possible to forecast the crop growth and not just the crop damage.

308

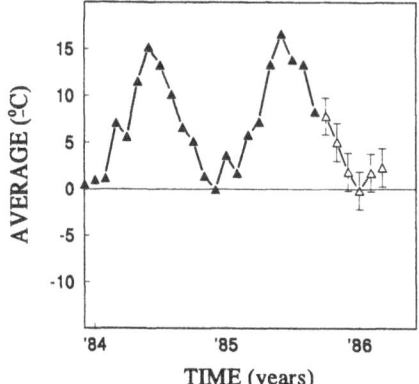

TIME (years)

Figure 3. Predicted minimum average temperatures (t_{avg}) from October '85 to March '86 (open triangles) with a confidence limit of 80 %.

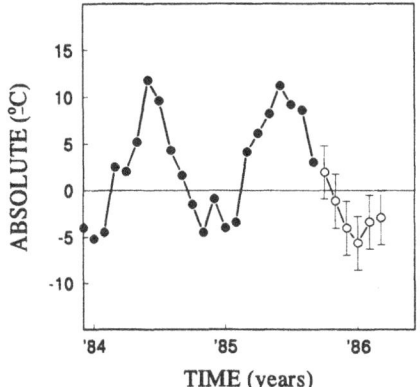

TIME (years)

Figure 4. Predicted absolute average temperatures (t_{abs}) from October '85 to March '86 (open circles) with a confidence limit of 80 %.

4. Acknowledgments

This research was supported, in part, by Comunidad de Madrid Research Project 076/92. The cooperation of the Spanish National Meteorological Institute and the Spanish Ministerio de Agricultura, Pesca y Alimentación (M.A.P.A.) is gratefully acknowledged. Miss Maite Aznar assisted in the preparation of the database on crop critical temperatures.

5. References

1. Box, G.E.P. and Jenkins, G.M. (1976).'Time Series Analysis. Forecasting and Control'. Revised Edition. Holden-Day, California. 575 pp.

2.   Chang, I., Tiao, G.C. and Chen, C. (1988). 'Estimation of time series parameters in the presence of outliers'. Technometrics, 30(2), 193-204.

3.   Jolliet, O. and Bailey, B.J. (1992). 'The effect of climate on tomato transpiration in greenhouses: measurements and models comparison'. Agricultural and Forest Meteorology, 58, 43-62.

4.   Lamperti, J. (1977). 'Stochastic Processes. A survey of the mathematical theory'. Applied Mathematical Sciences, 23, Springer-Verlag, NY, USA.

5.   Peña, D. (1987). 'Estadística. Modelos y Métodos, 2. Modelos lineales y series temporales'. Alianza Ed. Madrid, 691 pp.

6.   Penning de Vries, F.W.T. and Laar, H.H. van (1978). 'Simulation of plant growth and crop production'. Center for Agricultural Publishing and Documentation, Pudoc, Wageningen.

7.   Pérez, S., Gallego-Díaz, J. and Elías, F. (1989). 'Modelos ARIMA para la temperatura media anual en seis ciudades españolas'. Revista Meteorológica A.M.E. 12, 35-48.

8.   Wit, C.T. de, and Bailey, B.J. (1978). 'Simulation of ecological processes'. Center for Agricultural Publishing and Documentation, Pudoc, Wageningen.

9.   Yang, X. and Chen J. (1989). 'A mathematical model for prediction of time course of vegetable supply based on meteorological factors'. Agricultural and Forest Meteorology 49, 35-44.

UNCERTAINTY AND TACTICAL DECISION SUPPORT IN WINTER WHEAT PEST MANAGEMENT

W.A.H. ROSSING[1], R.A. DAAMEN[2], E.M.T. HENDRIX[3] AND M.J.W. JANSEN[4,5]

[1] Wageningen Agricultural University, Dept. Theoretical Production Ecology, P.O. Box 430, 6700 AK Wageningen, The Netherlands
[2] DLO-Research Institute for Plant Protection
[3] WAU, Department of Mathematics
[4] DLO-Agricultural Mathematics Group
[5] DLO-Centre for Agrobiological Research

ABSTRACT. Timing of chemical control of pests and diseases is a decision problem which consists of objectives, strategies, a model of the system, and decision criteria. 'Return on expenditure' and 'insurance' are generally applicable, potentially conflicting objectives of pest control. Uncertainty about the costs of different strategies of chemical control of cereal aphids (especially *Sitobion avenae*) and brown rust (*Puccinia recondita*) in winter wheat is calculated with a deterministic model. Sources of uncertainty, which comprise estimates of initial state and parameters, future weather, and white noise, are modelled as random inputs.

A widely used decision criterion in crop protection is the damage threshold, defined as the level of pest attack where projected costs of immediate control just equal projected costs of no control. No chemical control is the recommended action when the level of pest attack is below the damage threshold. It is shown that ignoring uncertainty leads to wrong recommendations.

It is argued that a consultative decision support approach which emphasizes the consequences of strategies in terms of objectives, is more appropriate than a prescriptive approach which concentrates on applying decision criteria to find the 'optimal' strategy. A consultative framework is proposed in which strategies are analyzed in terms of 'return on expenditure' and 'insurance'.

1. Introduction

Supervised control is the dominant paradigm in tactical decision support in crop protection (Zadoks, 1985). The concept is based on maximization of returns on expenditure for chemical control, but applies also to other methods of pest management where proper timing

P C Struik et al (eds), Plant Production on the Threshold of a New Century, 311–321

is required. The optimal time of pesticide application is considered to be equivalent to the level of pest attack at which the projected costs of chemical control just equal the projected costs of no control. This level is called the damage threshold. Mathematical models are used to calculate costs associated with decision alternatives. The current state of the system, characterized e.g. by pest density or crop development stage, is input for the models and is established by monitoring. The recommended decision is presented to the farmer in a decision support system.

Typically in current decision support systems, recommendations are prescriptive, contain no information on the uncertainty associated with decision alternatives, and are adjusted to be 'on the safe side', i.e. biased to chemical control. Since farmers appear to use information from various sources before arriving at a decision (Tait, 1987), unbiased information on the uncertainty of decision alternatives appears more useful than recommendations in which a farmer's presumed risk-attitude is implicitly accounted for (Rossing et al., 1993a). Therefore, a probabilistic as well as consultative approach to decision support is called for, rather than a deterministic, prescriptive approach.

In this paper the importance of uncertainty for supervised control of aphids and brown rust in winter wheat in the Netherlands is investigated. Two questions are addressed. Firstly, to what extent do damage thresholds change when uncertainty is taken into account. Secondly, how can information on uncertainty about costs associated with decision alternatives be made operational for consultative tactical decision support in crop protection.

2. Research approach

2.1. DECISION MODEL

A deterministic simulation model is used to predict costs of spray strategies at given initial temperature sum and initial levels of pest attack in a winter wheat field in the Netherlands. A spray strategy consists of a series of decisions on chemical control of aphids, brown rust or both, with fixed, one week time intervals. Costs of a strategy comprise the monetary equivalent of yield loss due to pest attack plus the costs of eventual chemical control.

The decision model consists of mathematical relations describing crop development, population dynamics and damage per unit of pest density. Relations are based on data collected during multi-year, multi-location experiments. The decision model represents an upgraded version of parts of the EPIPRE advisory system which has been operational in the Netherlands for over a decade (Daamen, 1991).

Crop development is calculated as a function of temperature sum above a developmental threshold of 6°C, accumulated from crop development stage pseudo-stem elongation (DC 30, Zadoks et al., 1974).

Population dynamics is calculated using observed incidences of aphids and brown rust incidences, i.e. observed percentage of sample units containing the respective pest. Incidence is transformed into density, which is assumed to increase exponentially with time. The relative growth rate of the aphid population decreases with advancing crop development stage. For brown rust the relative growth rate is constant. Aphicide application decreases population density by 85 % and arrests population increase during 12 days. In contrast, brown rust specific fungicides do not affect current population density and arrest population increase during 18 days.

Damage per pest-unit decreases linearly with advancing crop development stage for aphids and is constant throughout the season for brown rust. A maximum level of damage is assumed for both pests.

Uncertainty is represented as random inputs into the model. Four categories of uncertainty are distinguished (Figure 1). Uncertainty about initial incidences is modelled as binomial distributions with parameters depending on sample size and incidence estimates. Parameters in the mathematical relations were estimated using field data and regression. Estimated variance-covariance matrices provide measures of parameter uncertainty. Residual variance was ascribed to measurement effects and was disregarded for prediction. Some data sets were sufficiently detailed to allow estimation of the measurement variance. In those cases the surplus residual variance was ascribed to real, natural variability and was included in the model as mutually independent, identically distributed normal random inputs. This source of uncertainty will be called white noise. Uncertainty about future average daily temperature was described by 36 years of daily minimum and maximum temperature measured in Wageningen between 1954 and 1990.

Uncertainty about parameters and estimates of the initial state represents *controllable uncertainty*, since uncertainty may be decreased by collecting additional data. The categories future average daily temperature and white noise represent sources of *uncontrollable uncertainty*, as long as the structure of the model is unchanged.

Using stratified random sampling from the statistical input distributions in combination with Monte Carlo simulation estimates were obtained of the probability distribution of the major model output, costs of a spray strategy. Details are given in Rossing *et al.* (1993b).

2.2. TOOLS FOR DECISION SUPPORT: PROFITABILITY AND RISK

A decision problem can be decomposed into objectives, representing the goals of the decision maker, strategies, the means available for attaining the goals, a system model, representing the relation between strategies and their outcome, and decision criteria by which a decision maker chooses between alternative strategies. In a pre-scriptive approach to decision support identification of generally applicable objectives is followed by application of decision criteria to arrive at the 'best' strategy which is subsequently recommended to the decision maker. This approach ignores the subjective, variable nature of decision criteria in tactical crop protection as was pointed out by e.g. Tait (1987) who showed that attitudes to uncertainty vary

314

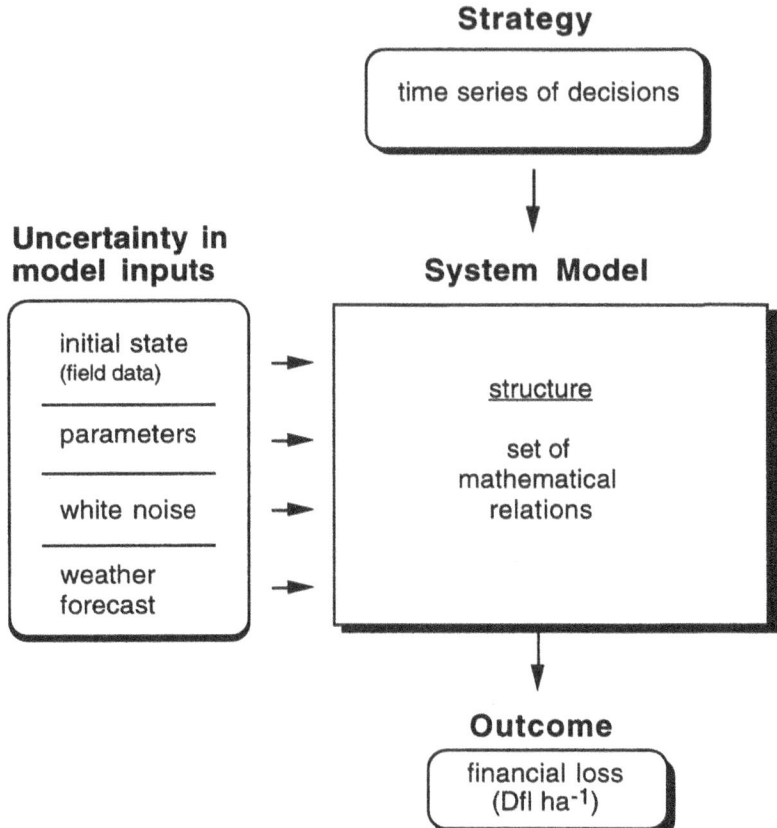

Strategy

time series of decisions

Uncertainty in model inputs

initial state
(field data)

parameters

white noise

weather
forecast

System Model

structure

set of
mathematical
relations

Outcome

financial loss
(Dfl ha⁻¹)

Figure 1. Outline of the decision model.

between farmers and, for individuals, vary between years. In contrast, a consultative decision support approach focuses on presentation to a decision maker of consequences of strategies in terms of generally applicable objectives. Application of subjective decision criteria, i.e. actual choice of the 'best' strategy, is explicitly left to the farmer.

Norton and Mumford (1983) postulated that two generally applicable aspects can be distinguished in objectives of pest control, 'return on expenditure' and 'insurance'. When emphasis is on 'return on expenditure' concern is primarily about positive returns to pesticide input. In this paper this aspect of pest control objectives is made operational by calculating the *profitability of strategy* A *compared to strategy* B, defined as the probability that strategy A results in lower costs than strategy B. When concern is predominantly about avoiding excessive costs, the 'insurance' aspect of pest control is stressed. Here, insurance-related aspects of pest control objectives are made operational by calculating the *risk of a strategy*, defined as the value of costs which is surpassed with an arbitrarily chosen probability of at most 10 %.

Each time a decision is to be made three (groups of) strategies can be pursued: no chemical control at any time, immediate chemical control, or postponing chemical control to some later point in time. Since 'profitability' involves comparison of two strategies, assessment of profitability of one strategy compared to another becomes cumbersome when the number of alternative strategies is large. We postulate that often the set of relevant strategies can be reduced to a manageable number by expert knowledge or, as in the present case, because of the nature of the system dynamics, *viz.* the dominant effect of exponential increase of pest density. For aphids and brown rust in winter wheat only no control at any time and immediate chemical control need be considered. Preliminary analysis showed that postponing chemical control causes profitability to decline and risk to increase, compared to immediate chemical control. Thus, whatever the subjective decision criteria, a rational decision maker (*sensu* Tait, 1987) who aims at maximizing profitability and minimizing risk, will never decide to carry out chemical control at a predetermined time in the future. The decision problem reduces to deciding whether at a given initial state of the system no chemical control results in an acceptable combination of profitability compared to immediate chemical control, and risk. If this is not the case, immediate chemical control is the preferred strategy. This decision process is repeated each time a decision is to be made, i.e. decisions are made with 'rolling planning horizon'.

3. Results and discussion

3.1. UNCERTAINTY ABOUT COSTS OF RELEVANT STRATEGIES

The decision model with random inputs was used to estimate probability distributions of costs of no chemical control and immediate chemical control for a single initial state of the system (Figure 2). Costs associated with no chemical control range from almost 0 Dfl ha^{-1} to 1200 Dfl ha^{-1}. For immediate chemical control costs range between almost 185 Dfl ha^{-1}, the fixed costs of a control operation, and about 500 Dfl ha^{-1}. The large difference in uncertainty about costs associated with these two strategies underlines the usefulness of information on the degree of uncertainty.

For both strategies expected costs are identical, approximately 200 Dfl ha^{-1}. Such initial state of the system where *expected* costs of no control equal *expected* costs of immediate control, will be called a stochastic damage threshold. A stochastic damage threshold represents a decision criterion to select the best among strategies. Since only expected costs of strategies are considered, this criterion implies that the decision maker is risk-neutral, i.e. neither prefers nor avoids high or low costs.

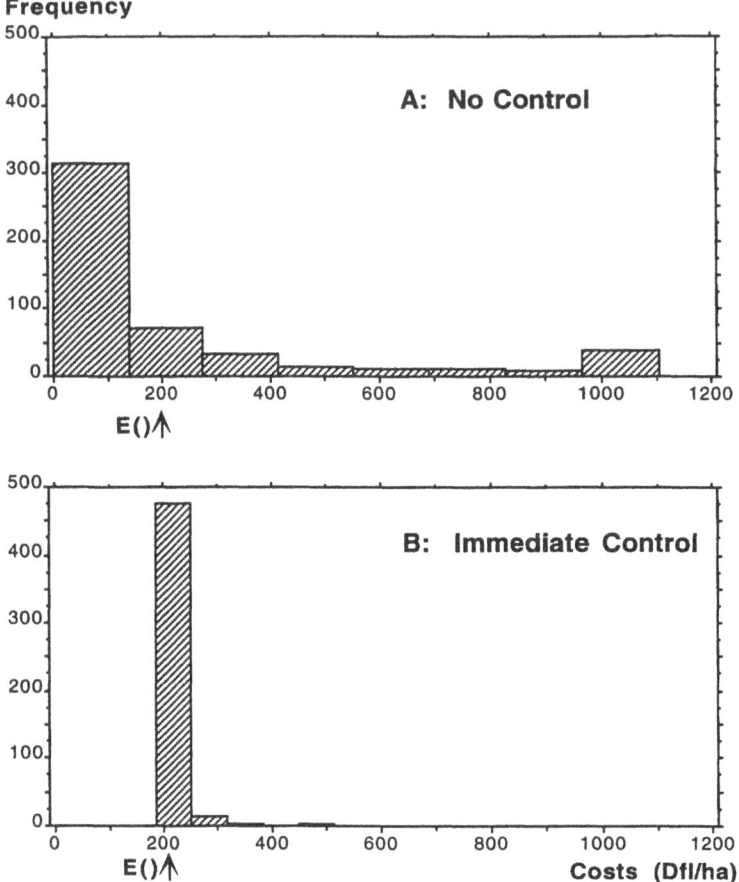

Figure 2. Frequency distributions of costs of no chemical control (A) and immediate chemical control of aphids and brown rust jointly (B) in 500 Monte Carlo runs. Initial state of the system: temperature sum 200 °days, equivalent with DC 58±4 (se), aphid incidence 5 % of 100 tillers, brown rust incidence 2 % of 160 leaves. The arrows indicates the expected value (E()) of costs.

3.2. UNCERTAINTY AND PRESCRIPTIVE DECISION SUPPORT: DAMAGE THRESHOLDS

The decision model with random inputs was used to calculate stochastic damage thresholds for cereal aphids and brown rust at a range of initial crop development stages. In comparison with deterministic damage thresholds which were calculated using mean values of model inputs, the stochastic damage thresholds were lower at all initial crop development stages (Figure 3). The discrepancy is caused by non-linearity, more specifically, by convexity of the decision model.

Compared to the current state of knowledge, perfect knowledge would result in spraying at higher pest incidences, resulting in on average less pesticide use. Thus, uncertainty has its price. Elsewhere, we analyzed the major causes of uncertainty and the way to most efficiently reduce it (Rossing et al., 1993c).

Calculated stochastic damage thresholds were used as references to evaluate the assumptions in the damage thresholds used by the prescriptive decision support system EPIPRE concerning risk-attitude of farmers. With respect to cereal aphids EPIPRE appears to assume that risk-aversion increases with advancing crop development stage (Figure 3a). With respect to brown rust EPIPRE recommendations consider farmers to be largely risk-neutral (Figure 3b).

3.3. UNCERTAINTY AND CONSULTATIVE DECISION SUPPORT: PROFITABILITY AND RISK

Profitability of no chemical control compared to immediate chemical control and risk associated with each strategy are calculated for aphids (Figure 4) and brown rust (Figure 5). The profitability of immediate chemical control compared to no control increases as initial incidences of aphids or brown rust increase, and decreases with advancing crop development stage. In contrast, risk increases with increasing initial incidences, and decreases with advancing crop development stage for both strategies, irrespective of pest organism. Chemical control reduces risk. Clearly, return on investment and insurance are conflicting objectives in these pathosystems. To demand a high profitability of chemical control implies accepting large risk, while minimizing risk is equivalent to accepting a low return on expenditure for chemical control. The nomograms show the 'exchange rate' between these objectives, and provide a basis for decision making.

Stochastic damage thresholds are included in the nomograms as 'yardsticks', since they represent a well defined attitude to uncertainty. Fore both aphids and brown rust the stochastic damage thresholds are equivalent to a profitability of no chemical control compared to immediate control of approximately 70 %. In other words, although on average immediate chemical control is economically rational at initial pest incidences equal to or higher than the stochastic damage threshold, the majority of pesticide applications at the stochastic damage threshold are ineffective. The low probability of economically successful chemical control is caused by the occurrence of very high costs, although with low probability, when no chemical control is carried out (see Figure 2). To increase the effectivity of pesticide applications at the stochastic damage threshold, uncertainty about the costs associated with no chemical control should be reduced.

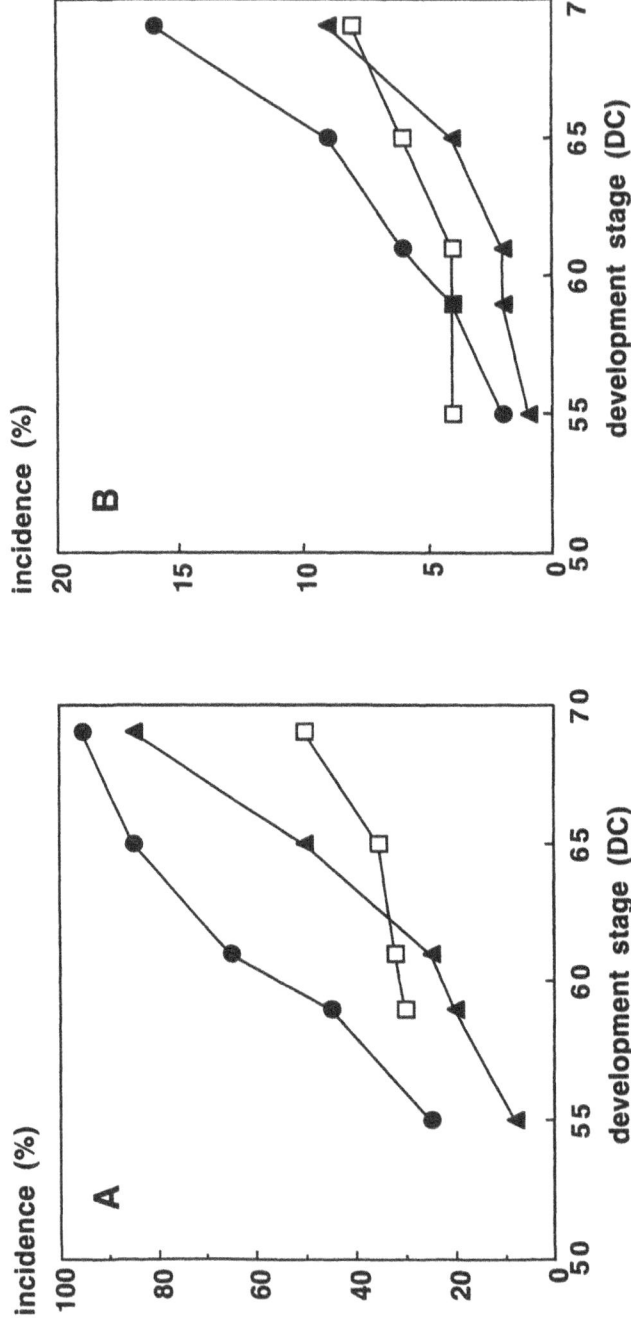

Figure 3. Damage thresholds for aphids (A) and brown rust (B) according to the deterministic version of the decision model (——●——), the stochastic version, based on 500 Monte Carlo runs (——▲——), and according to EPIPRE (——□——).

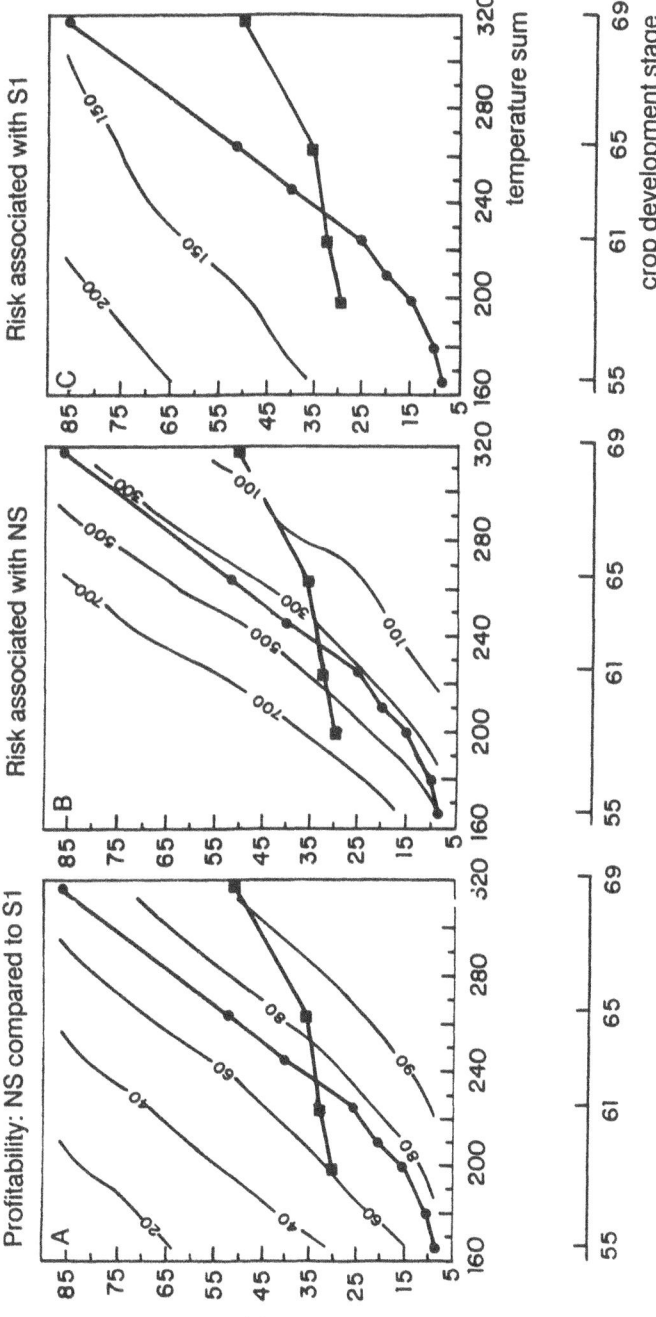

Figure 4. Contour plots of the *profitability* of no chemical control compared to immediate control of aphids (A), and the *risk* associated with no chemical control (B) and immediate chemical control (C), at different initial temperature sums (°day) or equivalent crop development stages (DC), and aphids incidences (%). Brown rust is absent. Profitability (%) and risk (Dfl ha⁻¹) are indicated within the contour lines. Also shown are the stochastic damage thresholds (——●——) and the EPIPRE damage thresholds (——■——).

320

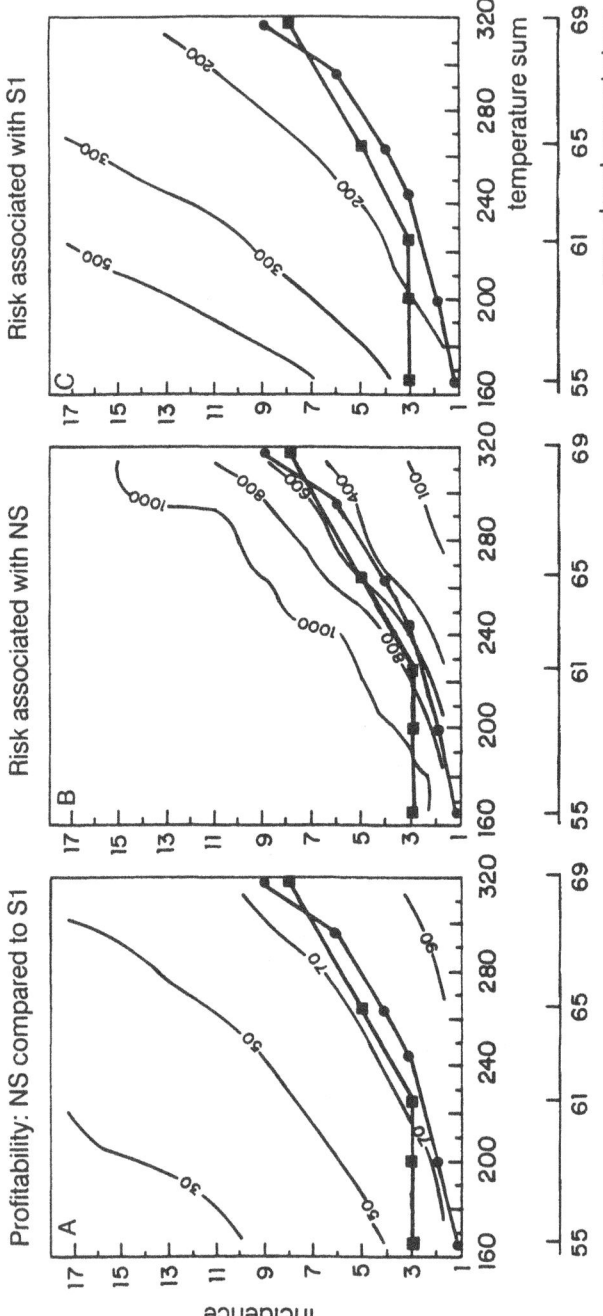

Figure 5. Contour plots of the *profitability* of no chemical control compared to immediate control of brown rust (A), and the *risk* associated with no chemical control (B) and immediate chemical control (C), at different initial temperature sums (°day) or equivalent crop development stages (DC), and aphids incidences (%). Aphids are absent. Profitability (%) and risk (Dfl ha^{-1}) are indicated within the contour lines. Also shown are the stochastic damage thresholds (————) and the EPIPRE damage thresholds (——■——).

3.4. CONCLUSIONS

This paper focused on quantification of uncertainty about costs of pest control strategies and assessment of the consequences for tactical decision support. The results show that neglecting uncertainty will lead to wrong recommendations to farmers in prescriptive decision support systems. Further analysis shows that selection of the 'best' strategy as is done in prescriptive decision support systems is of limited value since generally applicable selection criteria do not exist. In contrast, specification of generally applicable objectives of pest control appears feasible and results in a consultative framework in which strategies are assessed in terms of profitability and risk. Further evaluation of the framework in relation to existing crop management systems is needed.

Reduction of uncertainty may be brought about by strategic crop husbandry decisions such as choice of cultivar or level of nitrogen fertilizer input, and by increasing knowledge of the system dynamics through research on those components which contribute most to uncertainty about costs.

4. References

Daamen, R.A. (1991b) 'Experiences with the cereal pest and disease management system EPIPRE in the Netherlands', Danish Journal of Plant and Soil Science 85(S2161), 77-87.

Norton, G.A. and Mumford, J.D. (1983) 'Decision making in pest control', Advances in Applied Biology 8, 87-119.

Rossing, W.A.H., Daamen, R.A. and Hendrix, E.M.T. (1993a) 'A framework to support decisions on chemical pest control, applied to aphids and brown rust in winter wheat', Crop Prot., in press.

Rossing, W.A.H., Daamen, R.A. and Jansen, M.J.W. (1993b) 'Uncertainty analysis applied to supervised control of aphids and brown rust in winter wheat. 1. Quantification of uncertainty in cost-benefit calculations', Agric. Systems, in press.

Rossing, W.A.H., Daamen, R.A. and Jansen, M.J.W. (1993c) 'Uncertainty analysis applied to supervised control of aphids and brown rust in winter wheat. 2. Relative importance of different components of uncertainty', Agric. Systems, in press.

Tait, E.J. (1987) 'Rationality in pesticide use and the role of forecasting', in K.J. Brent and R.K. Atkin (eds.), Rational pesticide use. Cambridge University Press, Cambridge, pp. 225-238.

Zadoks, J.C. (1985) 'On the conceptual basis of crop loss assessment: The threshold theory', Annual Review of Phytopathology 23, 455-473.

Zadoks, J.C., Chang, T.T. and Konzak, C.F. (1974) 'A decimal code for the growth stages of cereals', Eucarpia Bulletin 7, 42-52.

SPATIAL HETEROGENEITY OF RADIATION INCIDENCE AND INTERCEPTION FOR
KIWIFRUIT VINES, AND IMPLICATIONS FOR FRUIT QUALITY

J.G. BUWALDA[1], E. MAGNANINI[2] AND G.S. SMITH[1]
[1] The Horticulture and Food Research Institute of
New Zealand, Ruakura Research Centre, Private Bag 3123,
Hamilton, New Zealand
[2] Istituto di Coltivazioni, Universita di Bologna,
Via Filippo Re 6, Bologna, Italy

ABSTRACT. Spatial heterogeneity of radiation incidence and interception for canopies of kiwifruit (Actinidia deliciosa) vines was
considered in terms of trellis design, canopy shape, leaf area
distribution, and incident radiation at different depths within the
canopy. Incident radiation was uniform across the surface of a canopy
trained on a horizontal "Pergola" trellis, but was frequently reduced
at distal regions of the canopy surface for a vine trained on a "T-
bar" trellis with faces inclined downwards, due to partial or complete
shading. Leaf area index also tended to decline with distance from the
central cordon, further reducing radiation interception in these
zones. Compared to radiation levels above the canopy, incident
radiation at any position within the canopy varied diurnally according
to the leaf area between the sensor and the radiation source (which in
turn was affected by the shape of the canopy trellis and by canopy
growth during the season) and the fraction of total radiation that was
diffuse. Spatial heterogeneity of radiation incidence and interception
can be linked to heterogeneity of fruit growth and quality.

1. Introduction

Photosynthesis (A, μmol CO_2 m^{-2} s^{-1}) for a plant canopy depends on
spatial and temporal integrals of A for all leaves. As the response of
leaf A to quantum flux density (Q, μmol m^{-2} s^{-1}) is typically non-
linear (Causton and Dale, 1990), spatial heterogeneity of Q on leaf
surfaces leads to spatial heterogeneity of A within the plant canopy
(Buwalda et al., 1993). This heterogeneity may in turn contribute to
heterogeneity of fruit growth and quality (Jackson et al., 1971; Smith
et al., 1992).

In this paper, factors influencing spatial heterogeneity of
incidence and interception of radiation for kiwifruit vines are
discussed. Data presented are all from measurements near Hamilton, New
Zealand (latitude 38.2°S), using vines in rows spaced at 4.5 m (889

323

P. C. Struik et al. (eds.), Plant Production on the Threshold of a New Century, 323–330.
© 1994 Kluwer Academic Publishers.

vines ha^{-1}). The impact of this heterogeneity for fruit development and quality is also considered.

2. Canopy surface shape and leaf area distribution

The arrangement of branches on perennial plants influences the spatial distribution of shoots and leaves. Trellis forms used for kiwifruit vines include a continuous horizontal trellis about 1.8 m above the ground ("Pergola"), and a discontinuous trellis comprising a central horizontal section (ca. 2 m wide at 1.8 m above the ground) and two faces inclined at 30 - 45° below horizontal ("T-bar"). For rows running north-south, these inclined faces usually face east and west respectively. Incident radiation at any time will be similar at all positions on the canopy surface for Pergola vines, but will vary with position on the canopy surface of T-bar vines due to shading and varying angles of incidence of the canopy surface and the radiation source(s).

Incident radiation at three positions on the canopy surface of a T-bar vine was measured throughout a growing season, using quantum sensors (LI 190, LiCOR, Lincoln, USA), on the central horizontal section and at 1 m above the ground on each of the inclined faces, and oriented normal to the trellis surface at each position. For a sunny day, (direct radiation ca. 86 % of total radiation), Q was highest on the eastern face in the early morning, on the central horizontal section in the middle of the day, and on the western face in the late afternoon. For the inclined faces, Q was relatively low for much of the day, due to shading, and the diurnal integral (Q_d) was only ca. 66 % of that for the central section (Figure 1a). For a cloudy day, (direct radiation ca. 42 % of total radiation), Q_d for the inclined faces was ca. 75 % that for the central section (Figure 1b).

The spatial distribution of leaf area index was estimated by counting contacts with leaves as a needle was passed through the canopy at 200 positions on each of four vines (two on a T-bar trellis and two on a Pergola trellis), at an angle normal to the canopy surface at each position. Measurements were made on year day 349 (1992), ca. 79 days after bud burst. The leaf area index at each point was estimated from the counts of leaf intersects, using a cosine adjustment for a typical distribution of leaf inclination angles (Buwalda *et al.* 1992), and the inclination relative to the trellis surface at each measurement point. The total leaf area index averaged 2.5 m^2 m^{-2} for the T-bar vines and 2.4 m^2 m^{-2} for the Pergola vines. For all vines, leaf area index declined significantly with distance from the cordon (Figure 2).

3. Spatial heterogeneity of incident radiation within a developing plant canopy

Spatial heterogeneity of incident radiation within the canopies of two vines, trained on a Pergola and T-bar trellis respectively, was

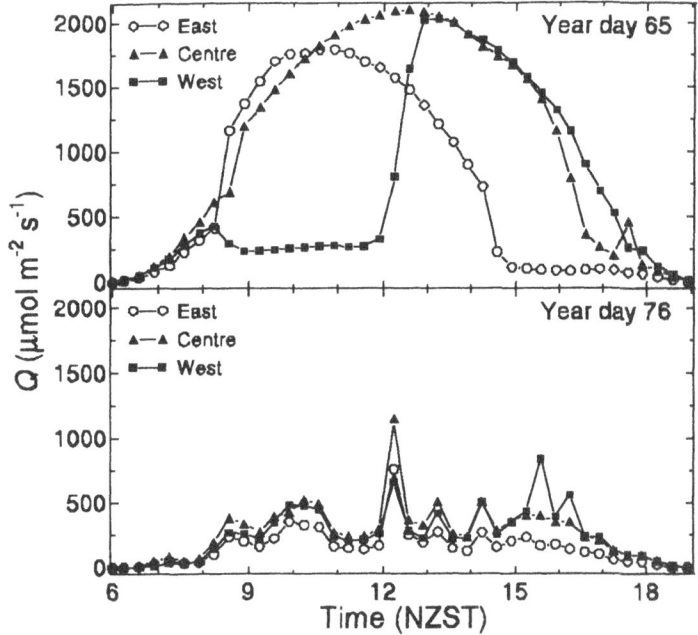

Figure 1. Diurnal trend of Q at three postions on a T-bar canopy surface, for (a) a sunny day and (b) a cloudy day.

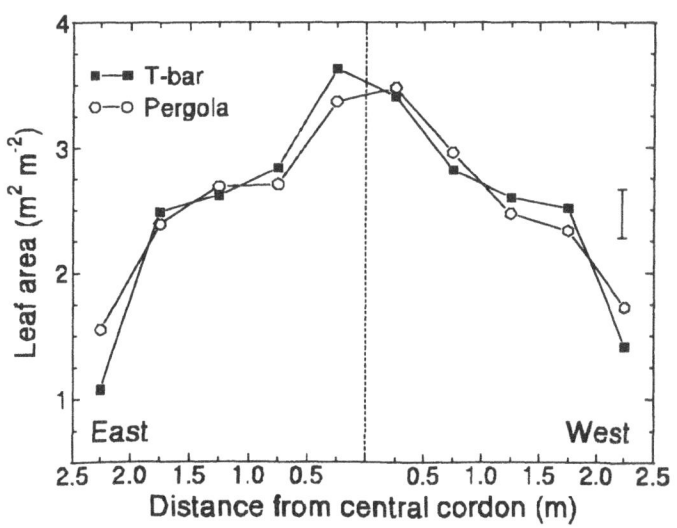

Figure 2. Spatial distribution of leaf area for T-bar and Pergola vines, at 74 days after bud burst. Vertical bar is standard error of mean.

measured during spring 1992, using arrays of quantum sensors. Each sensor comprised a light sensitive selenium cell housed below a diffuser and filters (limiting light to 400 - 700 nm), within a PVC chamber. The voltage output from the sensor was linearly correlated ($R^2 > 0.99$) with Q, as measured with a LI 190 quantum sensor. Each array comprised 24 sensors, with 6 sensors at each of 4 heights (*viz.* 0 m, 0.25 m, 0.50 m, 0.75 m) above the trellis surface. On the T-bar vine, some of the sensors above the inclined faces were lower than the central horizontal section of the trellis, and therefore not directly visible to the radiation source for parts of the day, even when not shaded by leaves. A LI 190 quantum sensor recorded Q above the canopy. Measurements of Q were made between year days 304 and 355.

Shoot growth commenced on year day 270. The leaf area of each vine was estimated on five occasions, by counting total bud number for the vine, and fraction bud burst, leaves per shoot, and average leaf area on a sub-sampled section. The leaf area per unit orchard area allotted to the vines increased from less than 0.5 m^2 m^{-2} at 20 d after bud burst, to 3.1 and 2.2 m^2 m^{-2} for the T-bar and Pergola vines respectively, at 75 d after bud burst. For the T-bar vine, the canopy occupied only ca. 0.66 of the orchard area, so the leaf area per unit projected canopy area was 4.7 m^2 m^{-2} at 75 d after bud burst.

Q_d above the canopy (Figure 3) showed large day-to-day variations, ranging from 17 mol m^{-2} d^{-1} (year day 336) to 62 mol m^{-2} d^{-1} (year day 349). Q_d at each sensor height was expressed as a fraction of Q_d above the canopy ($Q_{rel,d}$), using data for the period from ca. 30 min. after sunrise to 30 min. before sunset only (Figure 3). $Q_{rel,d}$ at all positions generally declined during the season as shading of sensors by the growing leaf canopy increased. At 0 m and 0.25 m above the trellis surface, $Q_{rel,d}$ was only 0.4 - 0.5 when measurements commenced, ca. 34 days after bud burst, and fell to less than 0.1 by year day 330 - 335 for the T-bar vine, but not until after year day 340 for the Pergola vine. At 0.50 m above the canopy, $Q_{rel,d}$ for the Pergola vine was still more than 0.4 by year day 355, but for the T-bar vine was less than 0.2 by year day 355. At 0.75 m above the canopy, $Q_{rel,d}$ was not affected by shading from other leaves until year days 314 and 325 for the T-bar and Pergola vines respectively. The subsequent reduction in $Q_{rel,d}$ was generally less for the Pergola vine than for the T-bar vine, except for the sharp increase in $Q_{rel,d}$ for the T-bar vine after year day 348 associated with inadvertent pruning of some shoots on this vine. On sunny days, Q_{rel} at any sensor position showed considerable diurnal variation, as the number of sensors at each position directly shaded by leaves varied. On cloudy days however, Q_{rel} at any sensor position was typically more stable, reflecting diffuse radiation conditions and hence the more constant position of each sensor relative to the radiation source (the whole sky).

4. Spatial heterogeneity of fruit quality

Spatial heterogeneity of fruit quality for 3 Pergola and 3 T-bar vines was assessed, by recording the spatial coordinates of each fruit

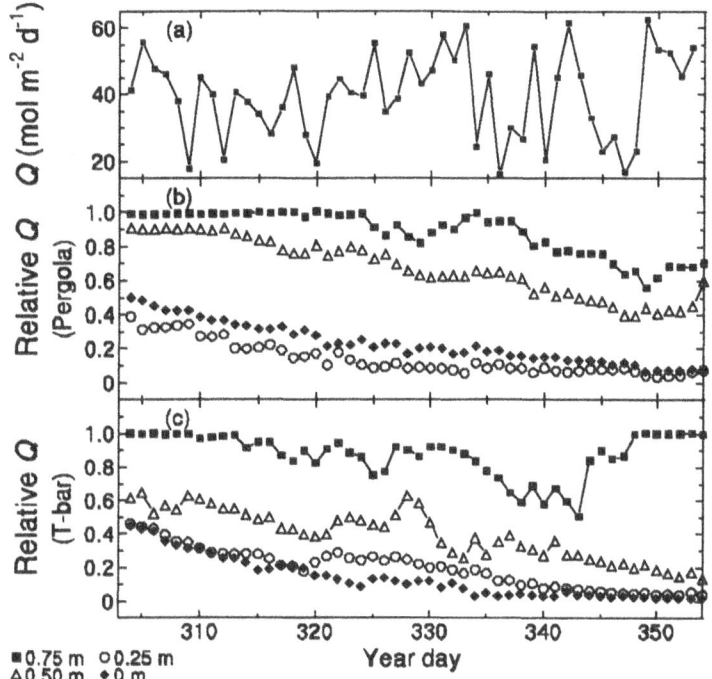

Figure 3. Seasonal trends for Q_d; (a) above the canopy, at different heights above the trellis for (b) Pergola and (c) T-bar vines.

■0.75 m ○0.25 m
△0.50 m ◆0 m

(Smith *et al*. 1992) and measuring the fresh weight (at harvest), and flesh firmness and soluble solids (after 12 weeks storage at 0°C). The total number of fruit on the Pergola vines was 27 % higher than on the T-bar vines, and this difference was associated with lower average fruit weight (7 % less) on the Pergola vines.

For Pergola vines, fruit weight and flesh firmness increased significantly with distance from the cordon, while fruit soluble solids concentration declined (Figure 4). For T-bar vines, trends in fruit attributes differed within the central horizontal section and the inclined sections. Fruit weight showed no significant difference with distance from the cordon in either the central horizontal or the outer inclined sections (Figure 4). Fruit soluble solids concentration increased significantly with distance from the cordon within the central horizontal section, but declined significantly with distance from the cordon in the outer inclined sections. Flesh firmness did not differ significantly with distance from the cordon within the central horizontal section, but declined significantly with distance from the cordon within the inclined sections. For fruit on shoots arising near the distal ends of the laterals, soluble solids concentration and flesh firmness were on average ca. 2° Brix and 0.60 kg less respec-

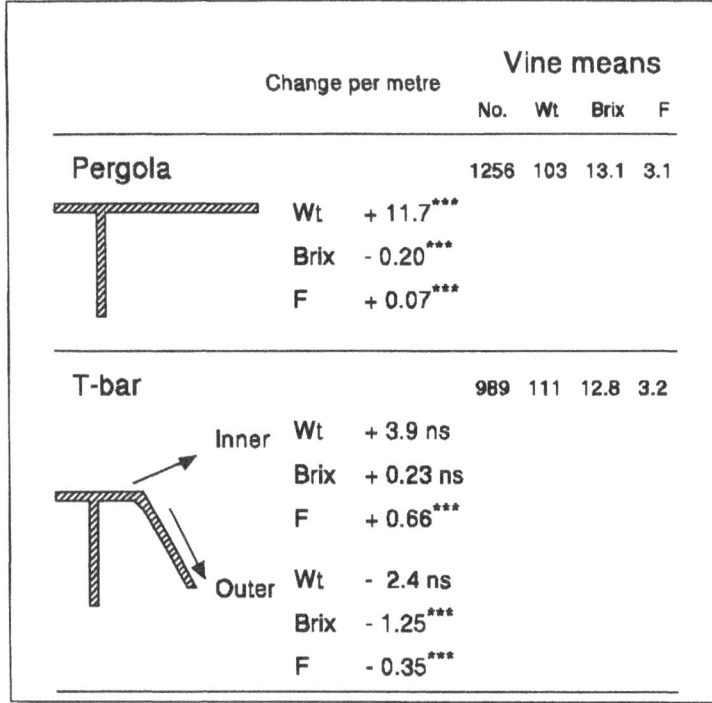

Figure 4. Systematic trends of fruit weight (g), soluble solids (°Brix) and firmness (kg force), for fruit from Pergola and T-bar vines (from Smith et al. 1993).

tively than for fruit on shoots arising within the central horizontal section. This decline in soluble solids concentration was the principal factor associated with the lower mean soluble solids concentration for fruit on T-bar vines compared to fruit on the Pergola vines.

5. Discussion

Variations in fruit quality within a single plant contribute significantly to the overall variation within an entire industry (Smith *et al.*, 1993). Spatial patterns of fruit quality may result from complex interactions of many factors, including environment (e.g. radiation incidence), plant (e.g. time of flowering, rate of canopy growth), and management (e.g. trellis design).

Absolute values for fruit quality parameters cannot be compared directly for groups of vines where factors such as crop load differ. However, trends within each group of vines reveal differences (Figure 4). Flower opening is usually earliest in regions distal to the central cordon for kiwifruit vines on all trellis types, (Lai *et al.*, 1990; Smith *et al.*, 1993), but fruit size at harvest is typically largest in these regions for Pergola vines only (Figure 4). Flesh firmness and soluble solids concentrations in the distal regions of the canopy were also less favourable for the T-bar vines compared to the Pergola vines. For the T-bar vines, zones of the canopy distal to the cordon intercepted less radiation than comparable zones of the Pergola vines. The large difference on sunny days can be easily attributed to direct shading of each inclined face of the T-bar trellis under mostly direct radiation conditions, but even on cloudy days the difference was still substantial, as some of the overhead radiation was still direct and the inclined faces were exposed to less of the whole sky than the horizontal planes. These differences should contribute to higher diurnal integrals of canopy A for Pergola vines (Buwalda *et al.*, 1993). As the horizontal projection of the canopy on the T-bar canopy is less than that for the Pergola canopy, the leaf area on the T-bar vines is essentially concentrated so that the vertical attenuation of radiation is greater than that for a Pergola vine (Figure 4). Consequently, the range of Q on the leaf surfaces at any time is likely to be greater for vines on the T-bar trellis than for vines on the Pergola trellis.

Clumped foliage distributions lead to leaf area and sunfleck heterogeneity (Jarvis and Leverenz, 1983). The systematic decline in leaf area index with distance from the cordon (Figure 2) however can not be attributed to such foliage clumping, but may be related to a general decline in shoot vigour. Nevertheless, radiation interception and hence A should decline with distance from the cordon as the leaf area also declines. This decline should be greater for T-bar vines than for Pergola vines, due to the more marked decline in leaf area and especially to the reduced Q_d associated with the inclined canopy surfaces.

Heterogeneity of fruit quality may be reduced by increasing the uniformity of Q and hence A within the canopy. For kiwifruit vines, this uniformity is clearly greater for horizontal canopies compared to those with inclined faces, and may be improved where the leaf area index shows more uniformity from basal to distal regions relative to the central cordon.

6. References

Buwalda, J.G., Curtis, J.P. and Smith, G.S. (1993) 'Use of computer graphics for simulation of radiation interception and photo-synthesis for canopies of kiwifruit vines with heterogeneous surface shape and leaf area distribution'. Ann. Bot., in press.

Buwalda, J.G., Meekings, J.S. and Smith, G.S. (1992) 'Radiation and photosynthesis in kiwifruit canopies'. Acta Hort. 297, 307-313.

Causton, D.R. and Dale, M.P. (1990) 'The monomolecular and rectangular hyperbola as empirical models of the response of photosynthetic rate to photon flux density, with application to three *Veronica* species'. Ann. Bot. 65, 389-394.

Jackson, J.E., Sharples, R.O. and Palmer, J.W. (1971) 'The influence of shade and within-tree position on apple fruit size, colour and storage quality'. J. Hort. Sci. 46, 277-287.

Jarvis, P.G. and Leverenz, J.W. (1983) 'Productivity of temperate, deciduous and evergreen forest', in O.L. Lange, P.S. Nobel, C.B. Osmond and H. Ziegler (eds.), Physiological plant ecology IV. Ecosystem processes, mineral cycling, productivity and man's influence. Springer-Verlag, Berlin, pp. 233-280.

Lai, R., Woolley, D.J. and Lawes, G.S. (1990) 'The effect of inter-fruit competition, type of fruiting lateral and time of anthesis on the fruit growth of kiwifruit (*Actinidia deliciosa*)'. J. Hort. Sci. 65, 87-96.

Smith, G.S., Curtis, J.P. and Edwards, C.M. (1992) 'A method for analysing plant architecture as it relates to fruit quality using three-dimensional computer graphics'. Ann. Bot. 70, 265-269.

Smith, G.S., Gravett, I.M., Edwards, C.M., Curtis, J.P. and Buwalda, J.G. (1993) 'Spatial analysis of the canopy of kiwifruit vines as it relates to the physical, chemical, and post-harvest attributes of the fruit'. Ann. Bot., in press.

THE USE OF GENETICAL VARIATION IN ROOTS TO IMPROVE PRODUCTIVITY

P.A. VAN DE POL, A.A.M. KLEEMANS AND J.H. OUDENES
Department of Horticulture
Wageningen Agricultural University
Haagsteeg 3
6708 PM Wageningen
The Netherlands

ABSTRACT. The majority of the world's most important fruit and nut crops and many ornamentals are still totally or partly propagated by grafting on seedling rootstocks. Genetical variation between these rootstocks can be used as a source for selection. New propagation techniques like stenting (simultaneously cutting and grafting) and root grafting are presented here being promising tools to select higher yielding root systems for many crops.

Some applications are presented in this paper. New rootstocks for roses with promotive effects on bottom break formation, plant architecture, productivity, flower quality or resistance to nematodes were selected. Root grafting of walnut proved to be possible. This enables rootstock selection among trees grafted on seedlings and showing rootstock induced variation.

In crops where own rooted cuttings are used the plants can be considered as an unintentional compromise between the different functions of shoots and roots. Variation in root quality between cultivars of *Chrysanthemum* indicated that own roots are not always the best for optimal growth and quality development. In breeding programmes of vegetatively propagated crops like *Chrysanthemum*, root quality as such should be optimized.

1. Introduction

Many plant cultivars, selected for their desirable ornamental or fruit qualities, do not have comparably suitable root systems but require grafting onto other roots to give satisfactory plants.

Size control, and sometimes an accompanying change in tree shape, is one of the most significant rootstock effects (Hartmann *et al.*, 1990). That specific rootstocks can be used to influence the size of the trees has been known since ancient times. Theophrastus - and later the Roman horticulturists - made use of dwarfing apple rootstocks that could be easily propagated (Bunyard, 1920). In apples now a complete range of tree size, from very dwarfed to very large, has been obtained

P C Struik et al (eds), Plant Production on the Threshold of a New Century, 331–337
© 1994 *Kluwer Academic Publishers*

with a given scion cultivar grafted to different rootstocks (Preston, 1958a, b; Preston, 1966). A wide assortment of size controlling rootstocks has now been developed for some of the major tree fruit crops (Brase and Way, 1959; Hartmann et al., 1990). Some rootstocks, particularly in citrus, give better size and quality of the fruit of the scion cultivar than others (Samish, 1962).

For many kinds of plants, rootstocks are available that tolerate unfavourable conditions, such as heavy wet soils, or resist soil-borne insects or pathogens better than the plants' own roots (Langfort and Townsend, 1954).

Crops grafted on seedling rootstocks, like many roses or walnuts, have the disadvantage of genetic variation, which may lead to variability in the growth and performance of the scion of the grafted plant (Hartmann et al., 1990). However, genetical variation between root systems can also be used as a source for selection. In the present study techniques are presented to investigate variation in root quality.

In crops where own rooted cuttings are used, like Chrysanthemum every cultivar has unique genetical characters for shoots as well as roots. Because during the breeding procedure in general no attempt was made to optimize shoots separately from roots, these plants can be considered as unintentional compromises between the different functions of shoots and roots. In this work development of Chrysanthemums grafted on own roots or on roots of other cultivars is compared.

2. Material and methods

In the programme to select root material of roses new propagation techniques were developed.

Stenting (Figure 1, left), based on grafting unrooted cuttings, is a quick propagation technique, applicable the year around and therefore ideal for studying scion-rootstock interrelations (Van de Pol et al., 1982, 1986).

Root grafting (Figure 1, right) is based on stenting. Instead of an internode of a stem as a rootstock, a piece of root of around 5 cm length and 3 mm diameter is whip grafted. This technique is developed for screening the root quality of seedlings after grafting, when shoots of the rootstock are not available anymore. Moreover, rootstocks with problems of rooting of their cuttings, like the Rosa canina group, can be successfully screened by root grafting (Van de Pol, 1986). From one root system many pieces of root can be used, so seedlings can be cloned.

For Chrysanthemum a deviating grafting procedure was used. Rootstocks were the Chrysanthemum morifolium cultivars 'Cassa', 'Reagan', 'Refla', 'Majoor Bosshardt' and the botanical variety C. boreale. The scion cultivar was 'Cassa'. Rootstock material, consisting of two internodes and two leaves, was taken from the basal part of side shoots of vegetative mother plants. The rootstocks were dipped in talc powder with 0.1 % indole butyric acid (IBA) and

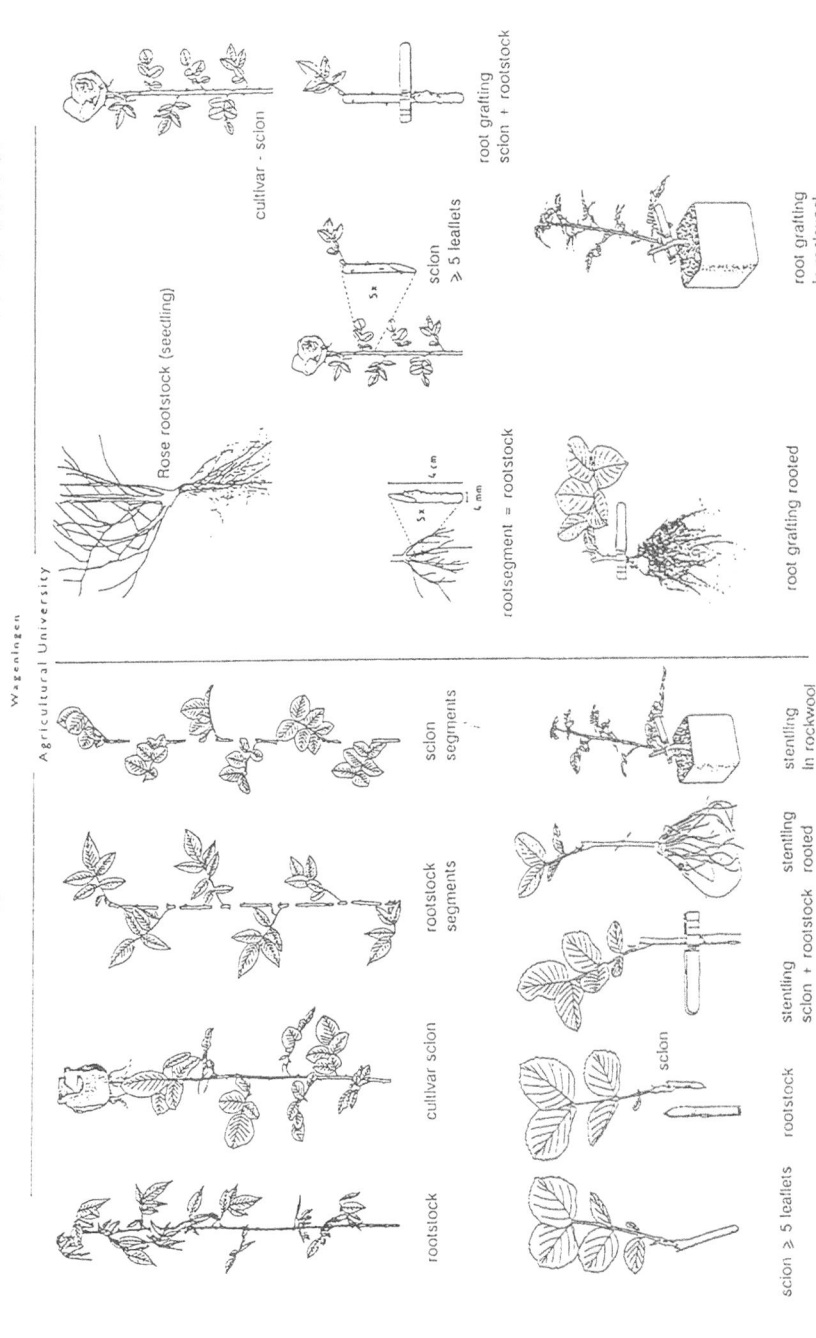

Figure 1. Grafting techniques for roses.

pretreated (Van de Pol, 1993), for development of root primordia, for one week in a propagation bench with electrical bottom heating (21°C) and for rooting substrate a mixture of sand and peat at a ratio of 1:1 V/V at pH 5.5. The scions were vegetative top shoots with three unfolded leaves, which were cleft grafted on the pretreated root-stocks, tied together with tape and placed in the propagation bench. After two weeks the plants were hardened, leaves with axillary buds of the rootstocks were removed and the combinations were transplanted into soil substrate in pots of 14 cm diameter and 1.25 l volume.

The combinations were placed in three blocks, five plants per treatment per block and during five weeks development of the plants was observed. Results are statistically compared by analysis of variance, with mean separation by Student's t-test ($P<0.05$).

3. Results

In the Netherlands stenting proved to be a successful technique for selection and propagation of new rootstocks for glasshouse roses. The new rootstock 'Multic' promoted the appearance of bottom breaks (Kool and Van de Pol, 1991) and improved the plant architecture (Kool and Van de Pol, 1992), resulting in higher yields (Van de Pol et al., 1988). The new rootstock R. multiflora 'Ludiek' selected via stenting proved to be a poor host for the root lesion nematode Pratylenchus vulnus (Schneider et al., 1993).

The technique of root grafting was successfully tried out with various rose rootstocks. In a large scale experiment 550 seedlings of R. canina 'Inermis' were cloned to 16 rootstocks per seedling and root grafted with the cultivar 'Sweet Promise'. During four years the harvested flowers were counted, while the quality during the winter season of 40 clones was measured. The best clones produced two times more than the poorest one. Also quality parameters like stem length and stem diameter showed variation between the clones. Although competition between quantity and quality occurred, some clones showed good production combined with good quality as well. These results indicate that rose production can be increased by selection among the commonly used seedling population (Vonk Noordegraaf, 1990).

In one of our preliminary experiments root grafting of Juglans regia, the walnut, proved to be possible. Root segments of a grafted seedling were successfully grafted and regenerated. This enables root stock selection among trees grafted on seedlings and showing rootstock induced variation.

Variation in root quality between cultivars of Chrysanthemum was tested by comparing the development of cultivars grafted on own roots or roots of other cultivars. Growth parameters after five weeks are presented in Table 1, indicating that own roots of 'Cassa' are not the best for optimal growth, indicated by length and quality, indicated by diameter and weight.

TABLE 1. Development of *Chrysanthemum morifolium* 'Cassa' grafted on 5 root systems and grown in soil (5 weeks after grafting).

Combination	length (cm)	dry shoot weight (g)	diameter (mm)
C*	67.3 a**	3.75 a	4.9 a
Rg	76.7 b	4.97 b	5.1 ab
Rf	74.5 b	4.49 b	5.2 ab
MB	81.9 c	6.11 c	5.6 c
bor	63.2 a	3.43 a	4.6 a

* Abbreviations: C='Cassa', Rg='Reagan', Rf='Refla', MB='Majoor Bosshardt' and bor=*Chrysanthemum boreale*.
** Figures in one column followed by the same latter do not differ at P<0.05.

4. Discussion

Hartmann *et al*. (1990) reported that 32 out of 51 of the world's most important fruit and nut crops and 36 out of 187 species of woody ornamentals, are still totally or partly propagated by grafting on seedling rootstocks. Genetical variation between seedling rootsystems can be used as a source for selection of superior clonal rootstocks. In breeding programmes of vegetatively propagated crops like *Chrysanthemum*, root quality as such, should be optimized.

In *Poinsettia* (Stimart, 1983), tomato (Woolley and Brenner, 1981) and rose (De Vries, 1993) degree of branching of the rootstock plant proved to be graft-transmittable.

De Vries (1993) selected rose seedlings for vigour on the basis of shoot yield during about 8 months. Subsequently their roots were root grafted as clonal stocks for the rose cultivar 'Sweet Promise'. The shoot yield of 'Sweet Promise' on seedling clones over 12 months proved to be significantly correlated with the former shoot yield of the own rooted seedlings. The mean fresh weight of the roots of the seedling clones after 12 months was significantly correlated with the number of harvested 'Sweet Promise' shoots on these clones. These results illustrate the role of genetical variation in root characters.

Explanations for variation in root behaviour can be found in various aspects.

Lambers (1983) reported that genetic differences between roots, even within the same species, are reflected in the consumption of different amounts of carbohydrates in particular with respect to growth and respiration. Jones *et al*. (1983) showed that in apple the graft union can reduce the amount of minerals and cytokinins passing through, and therefore may play a role in the dwarfing effect of a rootstock. Verstappen (1992) tentatively concluded that rose rootstocks, that vary in ability to release axillary buds of the cultivar from inhibition, vary for cytokinin synthesis. Cytokinins from the roots, arriving at the shoot tip would affect shoot growth which, in turn, influences auxin synthesis and its amount that activates the roots. Via this feed-back system, genotypically different stocks are presumed to control the vigour of the grafted cultivar (De Vries, 1993).

5. References

Brase, K.D. and Way, R.D. (1959) 'Rootstocks and methods used for dwarfing fruit trees'. N.Y. State Agr. Exp. Sta. Bul. 783.

Bunyard, E.A. (1920) 'The history of the Paradise stocks'. Jour. Pom. 2, 166-176.

Hartmann, H.T., Kester, D.E. and Davies, F.T. (1990) 'Plant propagation: principles and practices'. 5th ed. Prentice Hall, New Yersey: 308 + 332, 527-566.

Jones, O.P., Waller, B.Y., Hopgood, M.E. and Samuelson, T.J. (1983) 'Rootstock/scion interactions in apple'. East Malling Res. St. Rep., 60 pp.

Kool, M.T.N. and Pol, P.A. van de (1991) 'Invloed onderstam op vroege ontwikkeling roos Madelon'. Vakbl. Bloemisterij 14, 38-41.

Kool, M.T.N. and Pol, P.A. van de (1992) 'Aspects of growth analysed for Rosa hybrida 'Motrea' as affected by six rootstocks'. Gartenbauwissenschaft, 57(3), 120-125.

Lambers, H. (1983) '"The functional equilibrium": nibbling on the edges of a paradigm'. Netherlands Journal of Agricultural Science 31, 305-311.

Langford, M.H. and Townsend, Jr., C.H.T. (1954) 'Control of South American leaf blight of Hevea rubber trees. Plant Dis. Rpt. Suppl., 225 pp.

Pol, P.A. van de, and Breukelaar, A. (1982) 'Stenting of roses; a method for quick propagation by simultaneously cutting and grafting'. Scientia Hortic. 17, 187-196.

Pol, P.A. van de (1986) 'Root grafting and screening super canina rootstocks'. Acta Hort. 189, 81-87.

Pol, P.A. van de (1993) 'Landbouwuniversiteit ontwikkelt variant op stenten van rozen'. Vakbl. Bloemisterij 18, 35.

Pol, P.A. van de, Joosten, M.H.A.J. and Keizer, H. (1986) 'Stenting of roses, starch depletion and accumulation during the early development'. Acta Hort. 189, 51-59.

Pol, P.A. van de, Fuchs, H.W.M. and Peppel, H.F. van de (1988) ''Multic' kan spectaculaire produktieverhoging geven'. Vakbl. Bloemisterij 24, 42-45.

Preston, A.P. (1958a) 'Apple rootstock studies: Thirty-five years' results with Lane's Prince Albert on clonal rootstocks'. Jour. Hort. Sci. 33, 29-38.

Preston, A.P. (1958b) 'Apple rootstock studies: Thirty-five years with Cox's Orange Pippin on clonal rootstocks'. Jour. Hort. Sci. 33, 194-201.

Preston, A.P. (1966) 'Apple rootstock studies: Fifteen years' result with Malling-Merton clones'. Jour. Hort. Sci. 41, 349-360.

Samish, R.M. (1962) 'Physiological approaches to rootstock selection. Advances in horticultural sciences and their applications', Vol. 2. Long Island City, N.Y.: Pergamon Press.

Schneider, J.H.M., s' Jacob, J.J. and Pol, P.A. van de (1993) 'Response of six rose rootstocks to infection by Pratylenchus vulnus', in preparation.

Stimart, D.P. (1983) 'Promotion and inhibition of branching in *Poinsettia* in grafts between self-branching cultivars'. J. Amer. Soc. Hort. Sci. 108, 419-422.

Verstappen, F.W.A. (1992) 'Selecteren onderstammen in ver verschiet. Cytokinineproduktie maat voor groeikracht roos'. Vakbl. Bloemisterij 20, 49.

Vonk Noordegraaf, C. (1990) 'Influence of rootstocks on rose production in glasshouses. Abstr. of Contrib. papers, XXIII Int. Hort. Congr. Firenze (Italy) p. 2325.

Vries, D.P. de (1993) 'The vigour of glasshouse roses. Scion-rootstock relationships. Effects of phenotypic and genotypic variation'. Ph D thesis, Wageningen Agricultural University, pp. 1-69.

Woolley, D.J. and Brenner, M.L. (1981) 'Correlative inhibition of axillary growth'. Plant Physiology 67, p. 3 suppl.

CROP REACTIONS TO ENVIRONMENTAL STRESS FACTORS

A.J. HAVERKORT AND A.H.C.M. SCHAPENDONK
Centre for Agrobiological Research CABO-DLO
P.O. Box 14
6700 AA Wageningen
The Netherlands

ABSTRACT. Agricultural productivity is strongly reduced by biotic and abiotic factors and much effort goes into breeding for stress tolerant varieties and into agronomic measures. Breeding research, to date, concentrated on physiological characteristics which determine tolerance. Unfortunately, simple plant characteristics are seldom reflected in plant growth and even less in crop productivity as the direct link between growth and the characteristic studied is lost in feed-back control mechanisms and in the variance from other environmental and developmental factors. Simple measuring and observation techniques combined with summary models of crop growth quantify the impact of certain stress factors before a costly breeding programme is started. Additionally, more fundamental research will improve our understanding of the feedback mechanisms in plants. The relevance of some recent findings to modelling, breeding and practice is discussed in seven theses.

1. Introduction

1.1. STRESS FACTORS

Crop performance may be expressed in terms of factors that determine yield (e.g. temperature and solar radiation), restrict yield (e.g. lack of water and nutrients) and reduce yield (e.g. pests and diseases). Abiotic and biotic yield restricting and reducing factors are often referred to as stress factors. In the course of the growth cycle a crop may be subjected to several stress factors. For instance, emergence of potato may be hampered by unfavourable soil conditions (drought) or soil- (e.g. *Rhizoctonia solani* and nematodes) or seed-borne diseases (viruses). During canopy development the same stress factors affect crop growth but more hazards (low temperatures or night frosts) are added. The canopy gradually is subjected to air-borne diseases such as *Phytophthora infestans* and aphid-transmitted viruses. Towards the end of the growth cycle depletion of water and minerals become apparent as well as certain diseases such as

P C Struik et al (eds), Plant Production on the Threshold of a New Century, 339–347
© 1994 *Kluwer Academic Publishers*

Verticillium dahliae. The objective of crop management and disease control is to reduce the chance of growth restricting and reducing factors to affect the crop.

1.2. MONITORING STRESS

Crop reactions to environmental stress factors may be divided in short term and long term reactions. Short term (instantaneous) reactions to drought for instance, include the effect on the plant and cell water relations, stomatal conductance, photosynthesis, respiration, nutrient uptake and chemical composition.

Therefore, several techniques such as the use of a pressure chamber, porometry, infrared gas analysis and isotope labelling have been developed. Long term effects of a stress factor are often expressed as growth retardance or an earlier senescence, by wilting, yellowing and leaf shedding. They may become apparent by crop growth analysis, measurement of intercepted radiation and radiation use efficiency, dry matter partitioning and specific leaf area. Other long term effects may be detected through special measurements such as crop reflectance, transpiration efficiency, stable isotope fractionation and mineral uptake.

The objectives of this paper are to review the short term and long term effects of environmental stress factors on crop growth, development and production and to discuss the relevance for increased tolerance in strategies for breeding and cultural practices.

2. Short term effects

2.1. PLANT WATER RELATIONS

The moisture condition of a plant cell is described in terms of the water potential (ψ), an osmotic component (ψ_π) and a pressure component (ψ_p): $\psi=\psi_p+\psi_\pi$, expressed in MPa. When plant or cell moisture content decreases if transpiration exceeds water transport through the roots (in case of a high evaporative demand, a dry soil or infected roots) ψ, ψ_p and ψ_π decrease as well as the relative water content of the cell or the tissue (Turner, 1988). Leaf water potential may be observed with a pressure chamber or psychrometrically (Brown and Van Haveren, 1972) and the osmotic potential with the aid of an osmometer. Vos and Oyarzun (1987) observed typical values of ψ and ψ_p on a clear day of potato in moist soil of -0.5 and -0.8 Mpa respectively. Both drought and infection of roots by potato cyst nematodes led to decreased leaf water potentials from -0.6 MPa to about -1.1 MPa (Haverkort *et al.*, 1991). The specific leaf fresh weight (mg cm^{-2}) mainly depends on the amount of water present in a leaf, reflects the relative water content and is monitored with the aid of β-gauging techniques (Jones, 1973).

2.2. GAS EXCHANGE

Heat and water stress can reduce the photosynthetic rate indirectly by closure of the stomata (early effect) or directly by a reduction of the photosynthetic capacity of the leaves (after a few days). As a side effect of stomatal closure the assimilation/transpiration ratio increases as well as the proportion of stable isotope ^{13}C within the plant (Farquhar et al., 1982). No consensus exists on the primary site of the reduction in photosynthesis (Kaiser, 1987) or whether photoreactions in the thylakoid membranes or biochemical reactions of the Calvin cycle are most affected (Ögren and Öquist, 1985). In vivo fluorescence signals give information on the light use efficiency and on the rate of electron flow in the thylakoids (Schreiber and Bilger, 1985). Combined with gas exchange measurements, analysis of fluorescence signals provides information on rate limitation of processes related to gas phase resistance and to internal photosynthetic processes. The capacity to recover from stress in interaction with the developmental stage of the crop, may be as important as functioning during stress itself. Schapendonk et al. (1989) found in young potato leaves which recovered faster from water stress than old leaves, that mesophyll conductance was greater than in leaves from the control treatment. Conductance in the gas phase was smaller leading to stomatal limitation, reduced internal CO_2 and increased transpiration efficiency showing that recovery of the photosynthetic capacity was faster than that of the stomatal regulation mechanism.

3. Long term effects

3.1. EFFECTS ON MORPHOLOGY

The most convenient way to observe differences in crop development caused by environmental factors is the proportion of the ground covered by the canopy, which is highly correlated with the leaf area index (below 3), with the proportion of absorbed photosynthetically active solar radiation and with the proportion of the incoming infrared radiation reflected by the canopy (Haverkort et al., 1991). This is illustrated in Figure 1 where the proportion of ground cover and infrared reflectance of potato crops affected by drought (unirrigated), potato cyst nematodes (unfumigated) and night frost (7 weeks after planting) is shown. Ground cover and infrared reflectance were reduced more strongly by nematodes than by drought. Night frosts, however, reduced the canopy reflectance properties which did not show in the visual observation of ground cover by green leaves.

Table 1 shows the effects of an early drought period on some morphological characteristics of potato plants grown in containers under a rain shelter. At the end of the dry period the leaf area of the droughted plants was only half of the control because thicker and smaller leaves were formed. The stressed plants remained shorter, had lower tuber yields and lower shoot/root ratios than the control plants. Four weeks after rewatering both treatments had similar leaf areas and weights, although the plants that had been droughted before still had thicker leaves and lower Leaf Area Ratio (LAR)-values. The unstressed plants were senescing at day 70 whereas the plants which were subjected to an early drought stress continued to grow.

Beside such above-ground changes of plant characteristics, which scientists have been able to observe for a long time, the recent construction of the Wageningen Rhizotron (Smit and Groenwold, 1992) also allows a frequent and non-destructive observation of roots of plants subjected to various soil stress factors. Figure 2 shows the number of roots per cm^2 minirhizotron with time (mean values of all minirhizotrons between 0 and 30 cm depth and 30 to 100 cm depth) with and without infection by *Globodera pallida* (40 living juveniles per g soil). Potato cyst nematode infestation led to a higher number in the top-soil, but to less roots in the sub-soil.

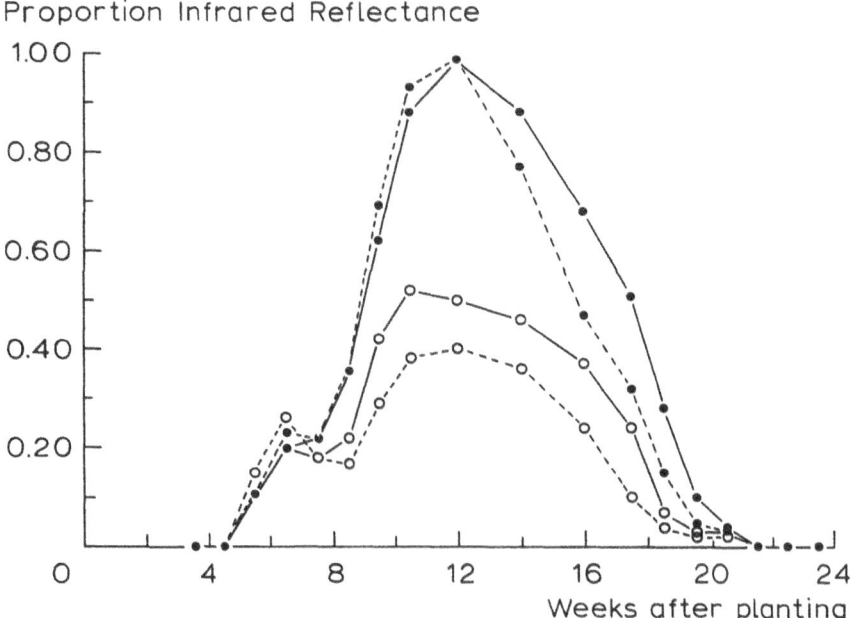

Figure 1. Proportion of infrared reflectance of potato crops subjected to drought and potato cyst nematodes —— irrigated, --- unirrigated, • nematodes controlled, o nematodes not controlled.

TABLE 1. The effect of a drought treatment until 43 days after planting on plant characteristics of potato cv. Mentor. After day 43 water supply was restored (Fasan and Haverkort, 1991).

Characteristic	Harvest (days after planting)			
	43 Treatment		70 Treatment	
	control	dry	control	dry
Leaf, dry (g/plant)	27.4	20.2	22.6	22.3
Leaf area (cm²/plant)	8253	4661	7663	6934
Leaf size (cm²/leaf)	150	108	187	140
SLA (cm²g⁻¹ dry leaf)	301	228	341	297
LAR (cm²g⁻¹ plant dry)	93.1	74.2	33.8	39.0
Stem length (cm)	52	34	62	54
Tuber, dry (g/plant)	42	27	194	140
Shoot/root ratio	17.3	15.1	21.4	20.3

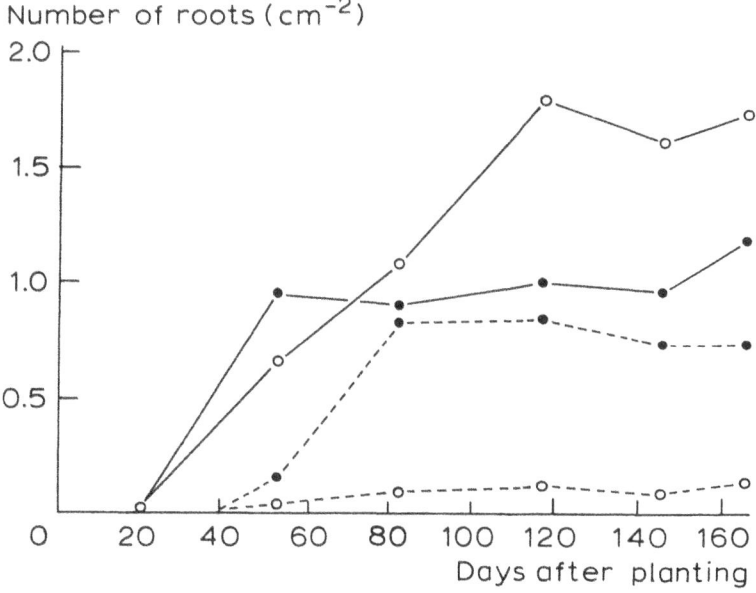

Figure 2. Mean number of roots with time of potato in the Wageningen Rhizotron (—— top soil, --- subsoil, • uninfested, o infested with 40 living eggs of cyst nematodes (*Globodera pallida*) per g soil.

3.2. YIELD FORMATION

The yield of a crop (Y) is determined by the amount of photosynthetically active solar radiation it intercepts (R_i), its conversion efficiency into dry matter (E), the proportion of total dry matter in the harvested parts (H_i) and its dry matter content (D_m). In a formula: $Y=R_i$ E H_i/D_m. The value of each parameter in this simple model of crop growth is determined by a number of underlying processes. R_i depends on the (duration of) leaf expansion rate and light extinction by the canopy, E depends on photosynthesis and respiration rates and H_i depend on dry matter distribution processes related to crop development. In a series of field trials in the North East of the Netherlands in which four cultivars were either (Irr) or not irrigated and either (Fum) or not fumigated periodic harvests were carried out to determine the effect of drought and cyst nematodes on components of yield formation. Table 2 shows these values relative to those of the irrigated and fumigated controls.

Apparently cyst nematodes and drought act similarly on the components of yield formation: yields are mainly reduced because crops intercept less solar radiation (see also Figure 1). Other fresh tuber yield reducing factors are lower efficiencies whereas reduced harvest indices and increased dry matter contents also play a role. The cultivar least tolerant to nematodes (Darwina) also was least tolerant to drought whereas cv. Elles was most tolerant to both stress factors.

TABLE 2. Relative values of yield components of unirrigated and unfumigated plots (mean data 1989 and 1990, values of irrigated and fumigated plots = 100) (Haverkort et al., 1992).

Year	Cultivar	Irr	Fum	Yield =	R_i	x	E	x	H_i	:	D_m
88-90	Darwina	+	-	48	61		90		94		105
	Désirée	+	-	52	71		86		93		102
	Elles	+	-	73	86		95		102		107
	Mentor	+	-	49	57		92		100		101
89-90	Darwina	-	+	55	62		99		94		105
	Désirée	-	+	77	88		99		94		105
	Elles	-	+	80	93		90		95		101
	Mentor	-	+	73	87		97		97		111

Beside long term visible effects of stress factors on crop morphology and measurable effects on intercepted radiation and radiation use efficiency stress factors affect transpiration efficiency, ^{13}C-discrimination and mineral uptake. Haverkort and Fasan (1991) reported transpiration efficiencies of about 7 g kg^{-1} in plants that were not subjected to stress but values over 10 g kg^{-1} of potato plants subjected to drought and/or cyst nematodes. Potato cyst nematodes and drought both decreased ^{13}C-discrimination indicating that nematodes influence plant water relations (Haverkort and Valkenburg,

1992) but whereas drought decreases Ca-uptake, cyst nematode infection led to an increase (Fatemy and Evans, 1986).

4. Discussion

From our own work and from literature the limitations and possibilities of the techniques to monitor stress are illustrated with the aid of the following theses.

1. Lateness of a potato cultivar is the major factor determining tolerance to stress factors such as drought and nematodes (Spitters and Schapendonk, 1990; Haverkort *et al.*, 1992). Late cultivars develop an abundant canopy of which the capacity to intercept solar radiation is not readily affected.

2. When crops are subjected to stress factors, the amount of intercepted radiation is affected more than the relatively conservative light use efficiency (e.g. Haverkort and Bicamumpaka, 1986; Spitters and Schapendonk, 1990) which make the determination of light interception a proper tool to monitor the degree of cumulative stress.

3. Light interception by green foliage is best and most objectively inferred from (in decreasing order of accuracy) reflectance characteristics, proportion ground cover measured with a grid, solarimetry and light extinction calculated from the LAI (Haverkort *et al.*, 1991; Bouman *et al.*, 1992.

4. Measurements of plant water relations, gas exchange, transpiration efficiency and stable isotope fractionation give insight into the degree of stress a plant is (or has been) subjected to but are useless as instruments to determine the degree of tolerance of genotypes.
   Little reaction to drought may indicate a waste of water and a strong reaction may indicate an incapacity of the root system to collect water (Vos and Oyarzun, 1987), moreover, these characteristics also depend on physiological crop age which may be affected by the stress factor (Haverkort and Valkenburg, 1992).

5. Variable chlorophyll fluorescence reflects photosynthetic functioning. Extrapolation of results from optimal conditions to stress conditions easily leads to erroneous conclusions because some stress induced changes in photosynthetic processes at the chloroplast level may not be translated proportionally into changes in CO_2-fixation.

6. Breeding for stress tolerance may lead to a lower productivity under optimum conditions. For instance, cultivars with a low harvest index producing abundant foliage are more tolerant of drought and nematodes but have lower yields under optimum

conditions than genotypes that produce less foliage (Haverkort *et al.*, 1992).

7. Crop reactions to stress may be modelled with simple or complex simulation models but complex models imply (as in reality) such a great number of feedbacks that in order to arrive at the optimal set of parameter values which will give the highest yield under a given set of growing conditions, the use of optimization techniques (multiple goal programming) should be considered (Haverkort and Goudriaan, 1993).

The grower, in practice, has many tools available to steer production, quality and harvest time through the choice of cultivar and cultural practices such as planting time, fertilization, irrigation and crop protection. Cultural practices aim at maximizing the amount of intercepted solar radiation through shortening the time between sowing and emergence and assuring a rapid development and a late senescing canopy. A high conversion efficiency of intercepted radiation is assured when the foliage functions optimally with photosynthesis unaffected by drought, extreme temperatures and diseases. Dry matter partitioning and crop development are mainly determined by cultivar choice in interaction with environmental factors and cultural practices.

5. References

Bouman, B.A.M., Uenk, D. and Haverkort, A.J. (1992) 'The estimation of ground cover by reflectance measurements'. Potato Research 35, 111-125.

Brown, R.W. and Haveren, B.P. van (1972) 'Psychrometry in water relations research'. Logan, Utah: Utah Agricultural Experimental Station.

Farquhar, G.D., O'Leary, M.H. and Berry, J.A., (1982) 'On the relationship between carbon isotope discrimination and intercellular carbondioxide concentration in leaves'. Australian Journal of Plant Physiology 9, 121-137.

Fasan, T. and Haverkort A.J. (1991) 'The influence of cyst nematodes and drought on potato growth. 1. Effects on plant growth under semi-controlled conditions'. Netherlands Journal of Plant Pathology 97, 151-161.

Fatemy, F. and Evans, K. (1986) 'Effects of Globodera rostochiensis and water stress on shoot and root growth and nutrient uptake of potatoes'. Revue Nematologique 9, 181-184.

Haverkort, A.J. and Bicamumpaka, M. (1986) 'Correlation between intercepted radiation and yield of potato crops infested by Phytophthora infestans in central Africa'. Netherlands Journal of Plant Pathology 92, 239-247.

Haverkort, A.J., Boerma, M., Velema, R. and Waart, M. van de (1992) 'The influence of drought and cyst nematodes on potato growth. 4. Effects on crop growth under field conditions of four cultivars differing in tolerance'. Netherlands Journal of Plant Pathology 98, 179-191.

Haverkort, A.J. and Fasan, T. and Waart, M. van de (1991) 'The influence of cyst nematodes and drought on potato growth. 2. Effects on plant water relations under semi-controlled conditions'. Netherlands Journal of Plant Pathology 97, 162-170.

Haverkort, A.J. and Goudriaan, J. (1993). 'Fysiologische aanpassing aan droogte: perspectieven voor verbetering van droogtetolerantie', in H. van Keulen en F.W.T. Penning de Vries (eds.). Agrobiologische Thema's 8, 49-66.

Haverkort, A.J. and Valkenburg, G.W. (1992) 'The influence of cyst nematodes and drought on potato growth. 3. Effects on carbon isotope fractionation'. Netherlands Journal of Plant Pathology 98,12-20.

Haverkort, A.J., Uenk, D., Veroude, H. and Waart, M. van de (1991) 'Relationships between ground cover, intercepted solar radiation, leaf area index and infrared reflectance of potato crops'. Potato Research 34, 113-121.

Jones, H.G. (1973) 'Estimation of plant water status with the beta-gauge'. Agricultural Meteorology 11, 345-355.

Kaiser, W.M. (1987) 'Effects of water deficit on photosynthetic capacity'. Physiologia Plantarum 71, 142-149.

Ögren, E. and Öquist, G. (1985) 'Effects of drought on photosynthesis, chlorophyll fluorescence and photoinhibition in intact willow leaves'. Planta 166, 380-388.

Schapendonk, A.H.C.M., Spitters, C.J.T. and Groot, P.J. (1989) 'Effects of water stress on photosynthesis and chlorophyll fluorescence of five potato cultivars'. Potato Research 32,17-32.

Schreiber, U. and Bilger, W. (1985) 'Rapid assessment of stress effects on plant leaves by chlorophyll fluorescence measurements. NATO ASI series, vol. 615, pp. 1-15.

Smit, A.L. and Groenwold, J. (1992) 'The Wageningen Rhizolab: first results of nutrient uptake studies', in L. Kutschera (ed.), Root ecology and its practical application 2. ISSR Klagenfurt, pp. 769-770.

Spitters, C.J.T. and Schapendonk, A.H.C.M. (1990) 'Evaluation of breeding strategies for drought tolerance in potato by means of crop growth simulation'. Plant and Soil 123, 193-203.

Turner, N.C. (1988) 'Measurement of the plant water status by the pressure chamber technique'. Irrigation Research 9, 289-308.

Vos, J. and Oyarzun, P.J. (1987) 'Photosynthesis and stomatal conductance of potato leaves. - effects of leaf age, irradiance and leaf water potential'. Photosynthesis Research 11, 253-264.

TUBER INDUCTION IN POTATO: THE POSSIBLE ROLE OF HYDROXYLATED JASMONIC ACIDS

HANS HELDER AND DICK VREUGDENHIL
Wageningen Agricultural University
Department of Plant Physiology
Arboretumlaan 4
6703 BD Wageningen, The Netherlands

ABSTRACT. Since the early fifties, a large number of papers have been published on the role of 'classical' plant hormones in tuber induction in potato. However, no (mix of) compound(s) was found that could explain tuber induction.

Attempts to isolate tuber-inducing compounds from leaves of tuber-bearing potato plants are hindered by a strong decrease of biological activity during the isolation process. The occurrence of hydroxylated jasmonic acids, and not jasmonic acid itself, in leaflets of *Solanum demissum* is positively correlated with tuber initiation. Presumably the decrease of the biological activity is caused by a change in the stereoconfiguration of (hydroxylated) jasmonic acids. The biologically most active configuration of (hydroxylated) jasmonic acids is known to be unstable after extraction from leaflets. If hydroxylated jasmonic acids are involved in tuber induction, enzymes that catalyse the hydroxylation of jasmonic acid could play an important role in this process.

1. Introduction

Night length and night temperature are two major factors that are involved in tuber initiation in potato plants. A tuber-bearing *Solanum* species will start to form tubers after a number of cycles in which the duration of the dark period is longer than a species-dependent, minimal night length, provided that the night temperature is adequately low (5°C-20°C). The minimal night length is commonly referred to as the maximal or critical day length. By an unknown mechanism the plant is able to 'remember' the day length conditions it was exposed to in the past. It is conceivable that a given day length is translated into a certain amount of an accumulatable quantity. Actual tuber initiation would start when a certain threshold level is reached.

In potato, information about the duration of the dark period is perceived in the leaves (Gregory, 1956). A pigment in leaves that

349

P. C. Struik et al. (eds.), Plant Production on the Threshold of a New Century, 349–356.

could be involved in the perception of this information is phytochrome. Batutis and Ewing (1982) obtained evidence for the involvement
of phytochrome in the perception of night length information in the
potato plant. Tuberization could be reduced by exposure of the plants
to 5 min. red light in the middle of the dark period, and this effect
could be reversed significantly by 2 min. subsequent far-red light.
Hence, when phytochrome was converted partially into the biologically
active form, tuber formation was reduced.

Gregory (1956) surmised that under short-day conditions a tuber-
inducing factor was synthesized or activated in the leaves and
transported basipetally to the stolon tips. Since 1956 extensive
research was done to identify the(se) tuber-inducing, transportable
compound(s). It was investigated whether the tuber-inducing principle
could be identified as (a combination of) 'classical' plant hormones.
Two approaches can be distinguished: (1) studies on the tuber-inducing
or inhibiting effect of exogenously applied phytohormones and (2)
investigations on changes in the endogenous plant hormone content
during the development of a non-swollen stolon tip into a tuber.

(1) A combination of sucrose and a cytokinin clearly promoted *in
vitro* tuber initiation in *S. tuberosum* explants (Palmer and Smith,
1969). However, *in vivo* application of cytokinins to a
diageotropically growing stolon tip failed to induce tuber formation,
and converted the tip into a negatively geotropic shoot (Kumar and
Wareing, 1972). Both *in vitro* and *in vivo* application of gibberellins
inhibited tuber initiation (Koda and Okazawa, 1983a; Okazawa, 1960).
Applications of other 'classical' phytohormones did not unequivocally
affect tuber initiation.

(2) From studies on the endogenous phytohormone content of stolon
tips, it is clear that a relatively low gibberellin activity in stolon
tips is a prerequisite for tuber initiation in *S. tuberosum* (e.g.
Okazawa, 1960). Levels of other 'classical' phytohormones changed
during tuber development, but these changes occurred only after the
onset of subapical swelling (Koda and Okazawa, 1983b). Hence, it is
unlikely that these hormonal changes are involved in tuber induction.

The role of the 'classical' plant hormones in tuber induction is
discussed *in extenso* in a number of recently published reviews (Ewing,
1987, Ewing and Struik, 1992, Vreugdenhil and Helder, 1992).

2. The possible role of (glycosides of) hydroxylated jasmonic acids as transportable signalling compounds

In 1988, Koda and coworkers succeeded in the isolation of 2.7 mg of a
specific tuber-inducing compound from 100 kg fresh leaves of *S.
tuberosum*, cv. Irish Cobbler (Koda *et al.*, 1988). At the time the
leaves were collected, the plant had formed tubers. The active
principle was determined to be 3-oxo-2-(5'-ß-D-gluco-pyranosyloxy-2'-
Z-pentenyl)-cyclopentane-1-acetic acid, *i.e.* a glucoside of 12-OH-
jasmonic acid (12-OH-JA, Figure 1) (Yoshihara *et al.*, 1989).

Jasmonic acid itself was already described as a representative of
a new group of endogenous plant growth regulators by the group of

Sembdner in Halle, Germany (*e.g.* Sembdner and Gross, 1986). Remarkably, no 11-OH-JA (Figure 1) was detected in leaves of the *S. tuberosum* cultivar Irish Cobbler, since metabolic studies of [2-^{14}C]-(±)-JA and [2-^{14}C]-9,10-dihydro-JA, fed to 6-day-old barley seedlings, revealed that in both cases hydroxylation preferentially took place at C-11 and to a lesser extent at C-12 (Sembdner *et al.*, 1990).

Using cuttings of potato, Struik *et al.* (1987) showed differences in tuber-inducing activity in extracts from leaves of *S. tuberosum* plants grown under short- (SD) and long-day (LD) conditions. If hydroxylated JAs are indeed responsible for, or at least involved in the process of tuber induction, it is to be expected that these compounds are absent or far less abundant in leaves from potato plants grown under LD than under SD conditions, provided the experiment is done with an absolutely SD-dependent *Solanum* species. *S. tuberosum* cultivars such as Désirée and Bintje do not meet this prerequisite. Provided that the night temperature is adequately inductive, these plants form tubers even under 24 h day length conditions. Hence, it was decided not to use *S. tuberosum*. A number of wild tuber-forming *Solanum* species is absolutely SD dependent. Because of its morphological resemblance to *S. tuberosum*, *Solanum demissum* Lindl. was selected among a number of other wild potato species. Under LD conditions (16 h day length) no tuber formation was observed (Helder, unpublished results). After 3-4 weeks exposure to SD conditions tubers were formed. Therefore, we decided to compare the nature and contents of (hydroxylated) JAs in leaflets of *S. demissum* leaflets collected from plants grown under SD and LD conditions.

3. Notes on the stereoconfiguration of (hydroxylated) JAs in relation to their biological activity

The identification of the glucoside of 12-OH-JA as a tuber-inducing principle from potato leaves was complicated by the substantial losses in tuber-inducing activity during the isolation procedure (Yoshihara *et al.*, 1989). This decrease was explained by a possible *cis/trans* isomerization at C-7, which is generally known for α-substituted cyclopentanones (Vick and Zimmermann, 1979; Miersch *et al.*, 1986). They considered the glucoside of 12-OH-(+)-7-*iso*-jasmonic acid (a *cis*-isomer) to be the biologically most active compound. Such an explanation presupposes a relatively high *cis/trans* ratio in the plant. Due to isomerization, occurring under weakly acidic or alkaline conditions, this ratio decreases (Miersch *et al.*, 1987).

Considering two substituents on a cyclopentanone, four stereoisomers are conceivable. For example, commercially available JA, a racemic mixture, consists of two *trans* isomers, (-)-JA and (+)-JA, and two *cis* isomers, (+)-7-*iso*-JA and (-)-7-*iso*-JA (Figure 1).

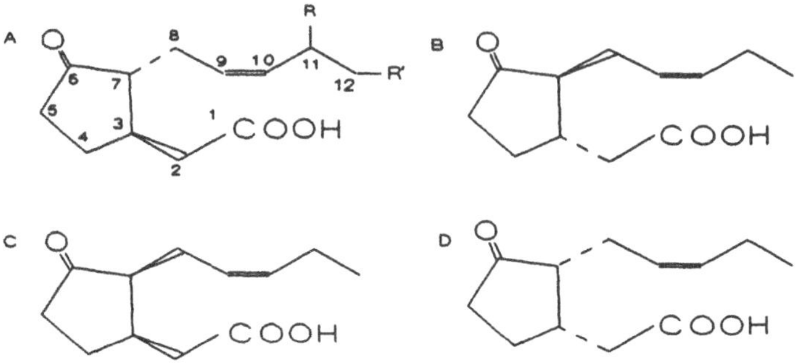

Figure 1. Stereoconfiguration of jasmonic acid (JA), 11-OH-JA (R=OH, R'=H), and 12-OH-JA (R=H, R'=OH). A; (-)-JA, B; (+)-JA, C; (+)-7-*iso*-JA and D; (-)-7-*iso*-JA.

In plants and fungi only one *trans* and one *cis* isomer are detected (in case of, *e.g.*, JA: (-)-JA (A) and (+)-7-*iso*-JA (C)). The *trans* configuration is apparently more stable than the *cis* con-figuration since isomerization under normal conditions results in a *cis/trans* equilibrium of about 1:9. GC-MS analyses of 11-OH-JA and 12-OH-JA from the fungus *Botryodiplodia theobromae* revealed a similar equilibrium ratio between the two pairs of diastereoisomers (1:9) (Miersch *et al.*, 1991). After a very careful isolation procedure of JAs from *Vicia faba* fruits, a (+)-7-*iso*-JA : (-)-JA ratio was obtained of about 1:2 (Miersch *et al.*, 1986). Therefore, it is conceivable that the *cis/trans* ratio of (hydroxylated) JAs *in planta* is higher than 1:9.

Evaluation of the tuber-inducing activity of the two naturally occurring stereoisomers of methyl jasmonate (JA-Me) revealed consider-able difference between (-)-JA-Me (the two substituents on the cyclo-pentanone are in the *trans* orientation) and (+)-7-*iso*-JA-Me (the two substituents on the cyclopentanone are in the *cis* orientation). Koda *et al.* (1992) tested *in vitro* tuber-inducing activities of various isomers of JA-Me including (-)-JA-Me and (+)-7-*iso*-JA-Me on single-node segments of etiolated potato shoots. At concentrations of 10^{-7}, 10^{-6}, and 10^{-5} M in the culture medium the tuber induction rates were 0%, 13% and 47% for (-)-JA-Me, while the rates for (+)-7-*iso*-JA-Me, at the same concentrations, were 27%, 60% and 92%. It can be questioned to what degree (+)-7-*iso*-JA-Me is isomerized to the less active (-)-JA-Me during the three weeks that the testing took. Hence, the difference in the *in vitro* tuber-inducing activity between (-)-JA-Me and (+)-7-*iso*-JA-Me is at least as large as indicated by the data presented by the authors.

The *in vitro* potato tuber-inducing activities of JA, JA-Me, 12-OH-JA and the glucoside of 12-OH-JA are about equal (Koda *et al.*, 1991). It was shown that (+)-7-*iso*-JA-Me had *in vitro* considerably more tuber-inducing activity than (-)-JA-Me. When the *cis/trans* ratio of the glucoside of 12-OH-JA from potato leaves *in planta* was higher than 1:9, like in, *e.g.*, *Vicia faba* fruits (Miersch *et al.*, 1986), it is conceivable that the decrease in potato tuber-inducing activity observed by Yoshihara *et al.* (1989) was indeed due to *cis/trans* isomerization during the isolation procedure.

After extraction of (hydroxylated) JAs from leaflets of *S. demissum* a decrease in tuber-inducing activity was observed (Helder, unpublished results). Since it was for practical reasons impossible to overcome these problems by extraction of a huge quantity of leaves (*e.g.* 100 kg fresh weight) another approach was chosen. The (hydroxylated) JA content of leaflets of *S. demissum* grown under SD conditions was compared with the content of leaflets from plants grown under LD conditions. A disadvantage inherent in this approach is the impossibility to prove that hydroxylated JAs are the only tuber-induction related compounds in leaflets of *S. demissum*.

4. The occurrence of (hydroxylated) jasmonic acid in leaflets from *Solanum demissum*

Leaflets were collected from *Solanum demissum* plants that were exposed either to SD (10 h photosynthetically active radiation (PAR)) or to LD (16 h) conditions. Under LD conditions the plants received 10 h PAR. The photoperiod was extended from 10 to 16 h using low-intensity (< 2 W m^{-2}) incandescent lamps. After 3-4 weeks SD plants formed tubers, whereas no tuber formation was observed in LD plants. At the same day, both LD and SD leaflets were collected. The leaflets were immediately frozen in liquid nitrogen, freeze-dried, and stored at -80°C till further analyses.

Quantification of (-)-jasmonic acid was done by a radio-immunoassay according to Knöfel *et al.* (1990). No difference in endogenous JA levels was detected between LD and SD leaflets. SD leaflets were collected from tuber-bearing plants. The endogenous (-)-JA levels were low at 10±3 and 8±3 ng g FW^{-1}, respectively (Helder *et al.*, 1993).

Hydroxylated JAs were detected by a gas chromatograph equipped with a mass selective detector. Both 11-OH-JA and 12-OH-JA occurred in SD leaflets. It was the first time that 11-OH-JA was detected as a native substance in higher plants. Earlier, it was found to occur in a fungus, *Botryodiplodia theobromae* Pat. (Miersch *et al.*, 1991). The 11-OH-JA concentration was higher than the 12-OH-JA concentration in SD leaflets. Glycosides of 11- and 12-OH-JA were not detected in either SD or in LD leaflets (Helder *et al.*, 1993). From these data it is concluded that the occurrence of hydroxylated JAs, and not JA itself, in the leaflets of *S. demissum* is correlated with tuber formation in *S. demissum*. This implies that enzyme(s) responsible for the metabolisation of JA could play a key role in tuber induction in potato. It is stressed that we only found a correlation between the occurrence of hydroxylated JAs and tuber induction. One of the ways to check whether this relation is causal could be the inactivation of (the) enzyme(s) that catalyse the hydroxylation of JA, to see whether tuber formation is affected.

354

5. Effects of JA on tuberization

The obvious way to test the possible role of 11-OH-JA in tuber induction would be to analyze the effects of applying this compound to potato plants, e.g., in an in vitro system. However, since 11-OH-JA is presently not available in large quantities, we tested in a preliminary experiment the effect of (±)-JA on tuberization in vitro. Similar experiments by Koda et al. (1991) and by Pelacho et al. (1991) have already shown that addition of (±)-JA in vitro eventually results in swelling of stolons. Therefore, we decided to analyze one of the earliest biochemical changes associated with tuber initiation, viz. the induction of the sucrose-splitting enzyme sucrose synthase (Helder et al., unpublished). Figure 2 shows that the level of sucrose synthase in in vitro grown S. achacachense explants is significantly higher in the presence of (±)-JA, as compared to controls without (±)-JA, before visible swelling occurred. This indicates that JAs are not only involved in radial growth of stolons, but also regulate at least one of the early biochemical changes associated with tuber formation.

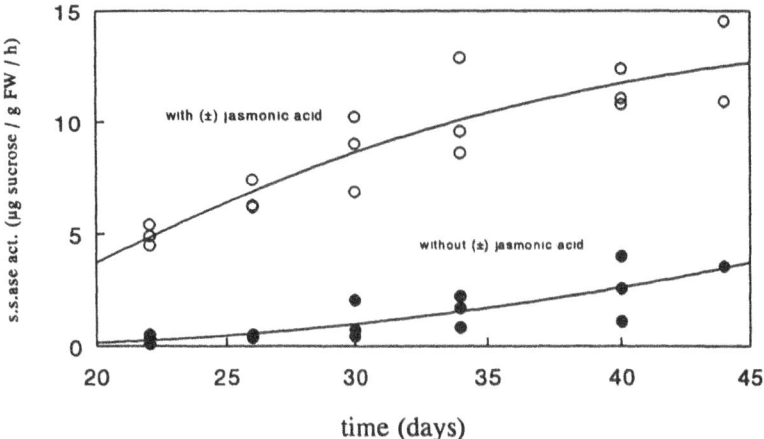

Figure 2. Effect of (±)-JA on tuber formation in vitro in S. achacachense. In the control, 6% sucrose and 1 mg/l BAP was added to MS-medium. The (±)-jasmonic acid concentration was 1 mg/l. Explants were incubated in the dark at 20°C.

6. References

Batutis, E.J. and Ewing, E.E. (1982) 'Far-red reversal of red light effect during long-night induction of potato (Solanum tuberosum L.) tuberization', Plant Physiol. 69, 672-674.

Ewing, E.E. (1987) 'The role of hormones in potato (*Solanum tuberosum* L.) tuberization', in P.J. Davies (ed.), Plant Hormones and their Role in Plant Growth and Development, Martinus Nijhoff Publishers, Dordrecht, pp. 515-538.

Ewing, E.E. and Struik, P.C. (1992) 'Tuber formation in potato: induction, initiation, and growth', Hort. Reviews 14, 89-198.

Gregory, L.E. (1956) 'Some factors for tuberization in the potato', Ann. Bot. 43, 281-288.

Helder, H., Miersch, O., Vreugdenhil, D. and Sembdner, G. (1993) 'Occurrence of hydroxylated jasmonic acids in leaflets of *Solanum demissum* plants grown under long- and short-day conditions', Physiol. Plant., in press.

Knöfel, H.-D., Brückner, C., Kramell, R., Sembdner, G. and Schreiber, K. (1990) 'Radioimmunoassay for the natural plant growth regulator (-)-jasmonic acid', Biochem. Physiol. Pflanzen 186, 387-394.

Koda, Y., Omer, E.A., Yoshihara, T., Shibata, H., Sakamura, S. and Okazawa, Y. (1988) 'Isolation of a specific potato tuber-inducing substance from potato leaves', Plant Cell Physiol. 29, 1047-1051.

Koda, Y., Kikuta, Y., Tazaki, H., Tsujino, Y., Sakamura, S. and Yoshihara, T. (1991) 'Potato tuber-inducing activities of jasmonic acid and related compounds', Phytochemistry 30, 1435-1438.

Koda, Y., Kikuta, Y., Kitahara, T., Nishi, T. and Mori, K. (1992) 'Comparison of various biological activities of stereoisomers of methyl jasmonate', Phytochemistry 31, 1111-1114.

Koda, Y. and Okazawa, Y. (1983a) 'Influences of environmental, hormonal and nutritional factors on potato tuberization *in vitro*', Japan. J. Crop Sci. 52, 582-591.

Koda, Y. and Okazawa, Y. (1983b) 'Characteristic changes in the levels of endogenous plant hormones in relation to the onset of potato tuberization', Japan. J. Crop Sci. 52, 592-597.

Kumar, D. and Wareing, P.F. (1972) 'Factors controlling stolon development in the potato plant', New Phytol. 71, 639-648.

Miersch, O., Meyer, A., Vorkefeld, S. and Sembdner, G. (1986) 'Occurrence of (+)-7-iso-jasmonic acid in *Vicia faba* L. and its biological activity', J. Plant Growth Regul. 5, 91-100.

Miersch, O., Preiss, A., Sembdner, G. and Schreiber, K. (1987) '(+)-7-Iso-jasmonic acid and related compounds from *Botryodiplodia theobromae*', Phytochemistry 26, 1037-1039.

Miersch, O., Schneider, G. and Sembdner, G. (1991) 'Hydroxylated jasmonic acid and related compounds from *Botryodiplodia theobromae*', Phytochemistry 30, 4049-4051.

Okazawa, Y. (1960) 'Studies on the relation between the tuber formation of potato and its natural gibberellin content', Proc. Crop Sci. Soc. Japan 29, 121.

Palmer, C.E. and Smith, O.E. (1969) 'Cytokinins and tuber initiation in potato *Solanum tuberosum* L', Nature 221, 279-280.

Pelacho, A.M. and Mingo, C.A.M. (1991) 'Jasmonic acid induces tuberization of potato stolons cultured *in vitro*', Plant Physiol. 97, 1253-1255.

Sembdner, G. and Gross, D. (1986) 'Plant growth substances of plant and microbial origin', in M. Bopp (ed.), Plant Growth Substances 1985, Springer Verlag, Berlin, pp. 139-147.

Sembdner, G., Meyer, A., Miersch, O., Brückner, C. (1990) 'Metabolism of jasmonic acid', in R.P. Pharis and S.B. Rood (eds.), Plant Growth Substances 1988, Springer Verlag, Berlin, pp. 374-379.

Struik, P.C., Boon, E.J. and Vreugdenhil, D. (1987) 'Effects of extracellular extracts from leaves on the tuberization of cuttings of potato (Solanum tuberosum L.)', Plant Physiol. 84, 214-217.

Vick, B.A. and Zimmermann, D.C. (1979) 'Distribution of a fatty acid cyclase enzyme system in plants', Plant Physiol. 64, 203-205.

Vreugdenhil, D. and Helder, J. (1992) 'Hormonal and metabolic control of tuber formation', in: C.M. Karssen, L.C. van Loon and D. Vreugdenhil (eds.), Progress in Plant Growth Regulation, Kluwer Academic Publishers, Dordrecht, pp. 393-400.

Yoshihara, T., Omer, E.A., Koshino, H., Sakamura, S., Kikuta, Y. and Koda, Y. (1989) 'Structure of a tuber-inducing stimulus from potato leaves (Solanum tuberosum L.)', Agric. Biol. Chem. 53, 2835-2837.

ECONOMIC BACKGROUND TO THE AGRARIAN REFORM IN RUSSIA

A. ANFINOGENTOVA, O. ERMOLOVA AND H. RESHETNIKOVA
Institute of Social-Economic Problems
of the Development in the Agro-Industrial Complex
Russian Academy of Sciences
Moscow Street 94
410600 Saratov
Russia

ABSTRACT. The paper presents the analysis of the main tendencies observed in production and consumption of the agrocomplex's goods and in the change of its structure. It contains a classification of Russian regional agrosystems, made on the basis of the level of development of agriculture and processing industries, consumption and dependence on externalities. The main features of adjustment of food markets are outlined for different stages of market relations.

Food prices dynamics and structure, the subsidy mechanism and the inflation processes in the consumer sector are analysed.

The process of land re-distribution, the new farms' efficiency levels and the contradictions of privatization are characterized.

The authors propose ways in which farmers could be supported and the social sphere of the agro-industrial complex could be reformed.

1. Introduction

The agrarian reform in Russia is realised in the following directions: price liberalization, transformation of property relations and land re-distribution, and reorganization of production sphere. It is a long- term transition period of the agrarian sector integration into the market system. There is a problem of reorganization of the structure of the agro-industrial complex (AIC).

2. Materials and methods

The production of agriculture decreased by 8 % in 1992 as compared with 1991. The crop production and sowing area increased in this period (Table 1) but livestock production decreased by 10 %.

By 1993 the total number of peasant farms was 230 000 with an average land area of 42 hectares per farm. But their share in the total production is low: 2.2 % of grain purchases, 0.5 % of potatoes,

P. C. Struik et al. (eds.), Plant Production on the Threshold of a New Century, 357–359.
© 1994 *Kluwer Academic Publishers.*

0.8 % of vegetables, 0.3 % of meat, 0.3 % of milk.

The growth of consumer prices exceeded the population incomes. The price liberalization has been developed to a great extent without control. The total level of the wholesale prices in the AIC industry has become 60 times as much, while the purchase prices increased 8 times and the consumer prices by 24 times (Table 2).

TABLE 1. Dynamics of gross harvesting for the main agricultural crops in the Russian Federation (1976-1980 = 100 %).

Products	1976-1980	1981-1985	1986	1987	1988	1989	1990	1991	1992
Vegetables	100	107.5	93.0	93.0	97.5	92.5	85.0	98.0	98.0
Potatoes	100	98.0	91.0	87.5	85.0	84.0	82.0	83.0	87.5
Sunflower seeds	100	92.2	84.0	93.1	98.1	102.2	99.0	102.6	115.0
Sugar beets	100	94.1	92.0	92.0	91.3	92.5	91.3	87.5	87.5
Grain	100	92.0	88.0	87.8	87.6	83.0	82.0	80.0	81.7

TABLE 2. Price and population income growth in Russia in 1992 (December 1991 = 1).

Parameters	January	March	July	October	December
Wholesale prices	6.0	10.5	22.0	40.0	60.0
Purchase prices	4.6	4.8	5.0	7.5	8.0
Consumer prices	4.5	6.0	11.0	15.0	24.0
Population income	2.1	3.2	4.1	6.0	12.0

3. Discussion

The priorities of the economic policy are the following: maintenance of the production level; structural reorganization; economic re-integration of Russian regions and the CIS countries on the new political basis; increase of exports to pay off the country's external debts; entering the world economic system.

A most important function of state centralized adjustment of the agrarian sphere should be the creation of a new system of territorial labour division and increase of the level of self-provision with food in the Russian Federation which should result from increase of specialization of the food producing regions.

Regional agrofood complexes are in different starting conditions today, the levels of their development are different. In this situation a quota market and the practice of paying current market prices for agricultural produce in advance could seriously help to form the centralized food resources. The state should be an equal-in-right trade partner on the food market. Its demand is determined by the income part of the budget and realized by means of contracts and future deals.

Privatization should be faced by agricultural producers and processing enterprises in their unity. It is of utmost importance to synchronize the process. Agricultural enterprises should not be turned into stock-holding companies operating independently from the processing ones, and vice versa. Otherwise done, it will lead to disintegration of the production chains.

The potential of state large collective farms has not yet been exhausted so far. And there is no doubt that a mere substitution of the state type of ownership for a private one will not automatically result in an increase of the farm's efficiency.

New Russian farms differ greatly by economic effectiveness, financial state, the level of income and profits and the number of people employed. The food crisis shows that independent farmers are not commercially oriented, and it is self-provision that they are really after.

The steady agrofood systems development requires the introduction of efficient systems of subsidies and state regulation of prices.

It is not agriculture only that the reforms should be focused on, but the whole sphere of vital functions of rural population.

FARMERS' EXPERTISE, A RESOURCE TO IMPROVE SUSTAINABILITY; THE CASE OF PESTICIDE APPLICATION IN SHALLOTS IN BREBES, CENTRAL JAVA

J.S. BUURMA
Agricultural Economics Research Institute LEI-DLO
P.O. Box 29704
2502 LS   The Hague
The Netherlands

ABSTRACT. This poster presents the methodology and results of a case-study on insecticide application practices in shallots (*Allium cepa*) in the district of Brebes, Central Java. A set of key figures is applied to show and explain the differences in insecticide use between plots, seasons and control strategies. The methodology contained group discussions with farmers on the control of onion caterpillars (*Spodoptera exigua*) coupled with a monitoring system for pesticide application. The farmers distinguished physical control and chemical control. For chemical control they made distinction between preventive and curative insecticides. The mix of these elements depended on the pest incidence. As a result the insecticide costs in the dry season were four times higher than in the wet season. In villages with priority for chemical control the insecticide costs were twice as high as in the village with priority for physical control.

1. Introduction

Technical scientists tend to neglect the farmers' expertise. In their view it is a hopeless task to gain access to this unwritten expertise. This poster presents the methodology to reveal such expertise and shows the results for the case of insecticide application in shallots (*Allium cepa*) in Brebes, Centra Java (Buurma and Nurmalinda, 1993).

For this case the following research questions were considered: which pest control methods do farmers apply, which considerations do farmers have for chemical control, what are the amounts and tank mix combinations of insecticides applied in dry season, what is the impact of physical control on insecticide use?

2. Methodology

The methodology contained three elements: (1) exploratory survey on control practices for onion caterpillars; (2) monitoring system for

P. C. Struik et al. (eds.), Plant Production on the Threshold of a New Century, 361–363.
© 1994 *Kluwer Academic Publishers.*

pesticide application; and (3) analysis system for pesticide monitoring data.

The exploratory survey was conducted by a multi-disciplinary survey team and contained group discussions with farmers in four villages differing in planting date and yield levels for dry season shallots. The discussions were structured according to a checklist of information, but with flexibility to explore certain practices or problems in more depth depending on farmers' responses.

The pesticide monitoring system contained a recording of the successive pesticide applications by field extension workers. The analysis system contained a coding system for the pesticides applied, data transfer to personal computer, followed by the calculation of key figures and comparative analysis.

3. Results

The group discussions made clear, that the farmers applied physical control and chemical control for *Spodoptera exigua*. For chemical control they made distinction between preventive and curative insecticides. The composition of the control package depended on the pest incidence. At low incidence they applied physical control and preventive insecticides. When the incidence increased, they added one or two curative insecticides to the preventive basis.

The analysis of the pesticide monitoring data confirmed the discussion results. In the dry season (high pest incidence) the insecticide costs were about four times higher than in the wet season (low pest incidence). In the villages with priority for chemical control the insecticide costs were nearly twice as high as in the villages with priority for physical control. The differences were closely related to curative insecticides.

4. Discussion

The survey and analysis results made clear, that the farmers already applied a kind of threshold spraying. For that reason technical research should be focused on refining the system, e.g. by checking the applied insecticide combinations on synergisms and antagonisms.

A crucial question for future IPM (Integrated Pest Management) research on shallots in Brebes is why physical control is not equally spread over the region. There are indications that the villages with priority for physical control have: (1) a more humid micro-climate, which slows down the population development of *Spodoptera exigua*; and (2) a better soil structure, which decreases the susceptibility of the shallot crop for *Spodoptera exigua*. These factors deserve more research attention to improve the sustainability of the cropping system.

5. References

Buurma, J.S. and Nurmalinda (1993) 'Farmers' practices, a challenge for IPM research', Journal of Plant Protection in the Tropics (submitted).

EXPERIMENTAL INTRODUCTION OF INTEGRATED ARABLE FARMING IN PRACTICE

S.R.M. JANSSENS[1], F.G. WIJNANDS[2] AND P. VAN ASPEREN[3]
[1] *DLO-NL Agricultural Economics Research Institute (LEI-DLO)*
   P.O. Box 29703, 2502 LS Den Haag, The Netherlands.
[2] *Research Station for Arable Farming and Field Production*
   of Vegetables (PAGV), P.O. Box 430, 8200 AK Lelystad,
   The Netherlands.
[3] *DLO-NL Centre for Agrobiological Research, P.O. Box 14,*
   6700 AA Wageningen, The Netherlands.

ABSTRACT. Integrated arable farming systems (IFS) are being developed and introduced in practice through pilot farms. The agronomic, environmental and economic performance of these pilot farms has been evaluated. Although some problems with new control strategies of biotic stresses occurred, the adoption of new farming systems seems to be well on its way. The reduction of pesticides was over 50 %, whereas also the use of nutrients was strongly reduced. First results indicate that economic results of the integrated farming systems were comparable to traditional systems.

1. Introduction

In the Netherlands integrated arable farming systems (IFS) are being developed at three regional experimental farms, with region-specific crop rotations and cropping systems.

To initiate a large scale introduction of IFS in practice a cooperative project (1990-1993) of the agricultural extension service and several research institutes has been set up. In order to obtain sufficient diversity of soil, farm and management conditions, five pilot-groups of about eight selected farms have been started in the major arable production areas of the Netherlands (Wijnands, 1992; Wijnands and Vereijken, 1992).

2. Evaluation

The evaluation of the performance of pilot farms concerns agronomic, environmental and economic aspects. The agronomic and environmental performance of the pilot farms is preliminarily evaluated by the PAGV. Important agronomic criteria are how and to what extent IFS techniques are adopted and how sustainable the IFS practice is. The evaluation of

P C Struik et al (eds), Plant Production on the Threshold of a New Century, 365–368
© 1994 *Kluwer Academic Publishers*

the environmental impact of IFS focuses on the use of pesticides and the emission of N to ground and surface water. For the latter the input levels of fertilizers and pesticides are important indicators.

The economic performance of the pilot farms is evaluated by the LEI-DLO, based on annual bookkeepings. The results of a farm or of the regional groups can be compared to preceding years or to conventional reference groups. The achieved savings in direct fertilizer and pesticides costs create the required financial means to compensate for increased labour demand, to invest in new machinery (e.g. mechanical weed control) and to compensate for possible lower financial results.

3. Preliminary results and discussion

Since the project started in 1990, only preliminary results on a limited number of aspects are available. A full evaluation is only possible after the project has finished.

3.1. AGRICULTURAL STATE OF THE ART

A farm-specific, multi-functional crop rotation was established when necessary, mostly by small adaptations of the current crop rotation. Subsequently, the integrated nutrient management strategy was carefully planned over crops, fields and years. The N input was gradually decreased to offer farmers the possibility of gaining confidence in the followed approach. In integrated crop protection some constraints occurred, such as (re)introduction of mechanical weed control in potato and cereals. Also the adoption of a consistent strategy to reduce fungicide use for the control of potato blight (*Phytophthora infestans*) appeared to be a difficult point. This especially applies to the so much required adoptions in cultivar choice, which mostly means replacing the very disease-sensitive cultivar Bintje. Agronomically the adoption of IFS seems to be well on its way. However, large regional differences occur in extent and speed of the adoption of new techniques. The objectives set for the last year of the project (1993) are: further stabilizing the results reached so far; specifying the IFS strategies regionally; minimizing non-farm specific variations in the techniques and methods used and optimizing the adoption of IFS in general.

3.2. PESTICIDE USE AND COST

In 1992, the total use of pesticides (kg a.i. ha^{-1}) on the pilot farms was on average reduced by 53 %, compared to 1987-1989. For herbicides, fungicides, insecticides and growth regulators these reductions amounted to 58, 50, 56 and 62 % respectively. For herbicides, the reduction varied from 45 % to 66 %. These reductions are based on mechanical control techniques and band spraying and lead to direct cost savings varying from Dfl 80-125 per hectare. The reduction in the use of other pesticides, based on the adoption of the integrated crop protection strategy, varied from 22 % to 65 % and reduced direct costs by Dfl 35-185 per hectare. Nematicide use is excluded from the active ingredient figures because proper evaluation is only possible after data of 1993 have been collected.

3.3. FERTILIZER USE AND COSTS

The P input on the participating farms over 1987-1989 is considered as unnecessarily high related to the fertility status of the soils. As a result the adoption of the integrated nutrient management strategy led to a reduction in P use varying from 15 kg P ha^{-1} to 45 kg P ha^{-1}. The largest reductions in P use appeared in the two sandy regions where organic manure use often passed the stage of agronomical sound practices. Over the years the use of K stayed about the same, again with the exception of the two sandy regions. The mineral fertilizers were substantially replaced by cheaper organic manure. The P use as mineral fertilizer amounted in all regions to only about 5 kg P ha^{-1}. Generally this substitution results in an increased input of N. However, as a result of the moderated N fertilization per crop, the total N input decreased on average by 57 kg N ha^{-1}, varying from 25 kg ha^{-1} to 140 kg ha^{-1}. The direct costs of fertilizer use were reduced by Dfl 30-125 per hectare. It is too early to report on the N emission assessments as the data of 1993 first have to be available.

3.4. ECONOMIC RESULTS

The economic analyses have been started in 1992, since full economic bookkeeping is only available one year after harvest. In 1990 the savings in direct fertilizer and pesticide costs varied from Dfl 60 to Dfl 430 per hectare and in 1991 from Dfl 185 to Dfl 360 per hectare compared to 1987-1989. Due to the regional reference groups the average fertilizer and pesticides costs of the pilot group were lower in 1990 (respectively Dfl 110 and Dfl 175 ha^{-1}). Preliminary results of 1991 showed almost equal differences. Higher seed costs on the pilot farms were mainly caused by using broadly resistant cultivars of ware and seed potatoes. The participating farmers are satisfied with the fact that their harvest (physical quantity and quality) did not fall short of their expectations.

First analyses showed that yearly crop yields and net profits did not decrease compared to the reference groups during the first two years. However, more detailed analyses will have to prove the economic feasibility of IFS.

4. References

Wijnands, F.G. and Vereijken, P. (1992) 'Region-wise development of prototypes of integrated arable farming and outdoor horticulture'. Netherlands Journal of Agricultural Science 40, 225-238.

Wijnands, F.G. (1992) 'Evaluation and introduction of integrated arable farming in practice'. Netherlands Journal of Agricultural Science 40, 239-249.

WAYS TO IMPROVE EFFICIENCY OF PESTICIDE AND ENERGY CONSUMPTION OF ROSES UNDER GLASS

C.J.M. VERNOOY
Agricultural-Economics Research Institute (LEI-DLO)
P.O.Box 29703
2502 LS The Hague
The Netherlands

ABSTRACT. In cooperation with growers and the research stations in Naaldwijk and Aalsmeer, LEI-DLO has searched for "closed production systems" as a way to environmentally friendly farming. An important part of this research is registration of the consumption of water, fertilizers and pesticides on specialized glassholdings.

Registration on glassholdings with roses of the cultivar Frisco during the year 1991/92 showed a large dispersion of yields and of pesticide and energy consumption. On holdings with substrate cultures, roll-containers and artificial lighting, returns and energy-consumption were the highest. However, on these holdings the use of pesticides was about the same or even lower, compared to soil grown roses without artificial light. The consumption of pesticides and energy was more efficient on holdings with intensive growing techniques.

1. Introduction

Production in greenhouses causes pollution of the environment. Especially in the concentrated greenhouse areas high concentration of pesticides were found in the water of ditches and canals. The policy of the Dutch government is focused on reduction of pesticide use by more than 50 percent by the year 2000.

In cooperation with the research stations at Naaldwijk and Aalsmeer, LEI-DLO is investigating production systems which are environmentally friendly without sacrificing crop quality or becoming uneconomic. Registration on holdings of the use of water, fertilizers and pesticides has been an important part of this research.

2. Material and methods

From November 1991 until November 1992 data were collected on 26 glassholdings with roses of the cultivar Frisco. Every four weeks

369

P. C. Struik et al. (eds.), Plant Production on the Threshold of a New Century, 369–371.
© 1994 *Kluwer Academic Publishers.*

growers sent detailed forms to the Institute where they were collected, processed and analysed.

3. Use of pesticides per ha

The average consumption of active ingredients was 108,8 kg per ha. The fungicide Sulphur was the most important component with 51.7 % of total consumption. Twenty percent of the holdings with the highest consumption used 178.3 kg of active agents per ha. The group of holdings with the lowest consumption of pesticides used 49.9 kg per ha. This is a factor 3.6 less. On roses the average use of fungicides (other than sulphur) and sulphur was about 79 percent. Especially on holdings with the highest consumption the share of fungicides and sulphur was higher (Table 1).

TABLE 1. Use of active agents (in kg/ha) for different types of chemicals on roses of the cultivar Frisco, divided in five groups with different levels of consumption in 1991/92.

level of consumption	Active ingredients in kg/ha				
	Insecticides	Fungicides	Sulphur	Other	Total
> 80 %	26.3	40.1	111.9	0.0	178.3
60 - 80 %	23.1	49.1	63.7	7.8	143.8
40 - 60 %	19.0	45.5	34.5	1.7	100.8
20 - 40 %	20.4	15.9	38.7	0.0	74.9
< 20 %	18.5	7.8	22.7	0.0	49.9
Average	21.0	31.8	54.1	1.9	108.8

4. Energy consumption and yields

The yields of the holdings growing Frisco ranged from 289 stems for group E (cultures in soil, without artificial lighting) to 499 and 614 stems per m^2 on holdings with more intensive production systems (substrate cultures, artificial lighting and roll-containers). On these holdings the energy consumption was higher, which was mainly caused by artificial lighting. However, the consumption of active ingredients in kg per ha was not directly related to the yields (Table 2).

TABLE 2. Use and efficiency of pesticides and energy on glassholdings with roses of the cultivar Frisco, divided in five groups with different yields.

Yield group	Number of holdings	Yields		Active ingredients		Energy	
		stems/m^2	Dfl/m^2	kg/ha	mg/stem	m^3/m^2	m^3/stem
A	2	614	200	97.6	15.7	83.2	0.136
B	4	499	153	88.9	18.6	70.1	0.140
C	5	384	121	116.3	30.1	69.3	0.180
D	5	346	103	121.7	36.4	58.1	0.168
E	7	289	75	103.0	38.3	48.8	0.167
Average		379	113	108.8	30.4	64.4	0.170

5. Improving the efficiency

The methods to reduce the use of pesticides on roses under glass are diverse. Growers indicated that reduction of the use of pesticides may be possible by modern types of chemicals, by better climate conditions, by better observation techniques, by curative late spraying and by the right applications at the right time.

Efficiency of pesticide and energy consumption can be improved not only by reduction of the consumption per ha, but also by more intensive and effective growing techniques with higher yields.

BIOLOGICAL CONTROL OF CUCUMBER POWDERY MILDEW BY MYCOPARASITES

M.A. VERHAAR AND T. HIJWEGEN
Department of Phytopathology
Wageningen Agricultural University
P.O. Box 8025
6700 EE Wageningen
The Netherlands

ABSTRACT. In glasshouse experiments the possibilities of the combination of biological control of powdery mildew and the use of a partially resistant cucumber cultivar were tested. The infected leaf area on the partially resistant cucumber cultivar Flamingo could be kept under the economic threshold by the use of biological control treatments with *V. lecanii*.

1. Introduction

Several mycoparasites of cucumber powdery mildew (*Sphaerotheca fuliginea*) have been described (Philipp, 1988; Burge, 1988; Hijwegen, 1988). Usually biological control experiments are done with susceptible varieties. However, it can be assumed that partially resistant varieties will perform better. On partially resistant cucumber cultivars the mildew grows slowly, which gives the myco-parasites a better opportunity to destroy the whole colonies (Verhaar, unpublished).

In two glasshouse experiments the possibilities to control the powdery mildew by a combination of biological control and the use of a partially resistant cucumber cultivar were tested. As biocontrol agents two mycoparasites, with high potential in *in vitro*-tests (Verhaar, unpublished), *Verticillium lecanii* and *Sporothrix rugulosa*, were used.

2. Materials and methods

Experiments were performed with the susceptible cucumber cultivar Corona and the partially resistant cultivar Flamingo. Two weeks after inoculation with powdery mildew the biological control treatments were started. Plants were sprayed weekly with $5*10^6$ spores/ml of *V. lecanii* or *S. rugulosa*.

P C Struik et al (eds), Plant Production on the Threshold of a New Century, 373–374
© 1994 *Kluwer Academic Publishers*

3. Results

Under glasshouse conditions *V. lecanii* controlled the powdery mildew better than *S. rugulosa*. On both cucumber cultivars *V. lecanii* restricted the mildew infected leaf area to about half of that in the *S. rugulosa* treatment.

There was a remarkable difference between the mildew development on both cultivars. Biological control of mildew on the partially resistant cucumber cultivar Flamingo gave very good results. In two experiments we could keep the mildew infected leaf area below the 20 % economic threshold on this cultivar by the use of weekly *V. lecanii* treatments.

4. Conclusions

It can be concluded, that *V. lecanii* gives good biological control of powdery mildew on a partially resistant cucumber variety.

5. References

Burge, M.N. (1988) 'Fungi in biological control systems'. Manchester University Press, 269 pp.

Philipp, W. -D. (1988) 'Biologische Bekämpfung von Pflanzenkrank-heiten'. Ulmer, Stuttgart, 248 pp.

Hijwegen, T. (1988) 'Effect of seventeen fungicolous fungi on sporulation of cucumber powdery mildew'. Netherlands Journal of Plant Pathology 94, 185-190.

NEW TECHNOLOGIES FOR HORTICULTURAL CROPS

G.E. TERAN SARABIA, A. BENAVIDES MENDOZA, F. HERNANDEZ
CASTILLO AND E. QUERO
Centro de Investigacion en Quimica Aplicada (CIQA)
Apartado 379
Saltillo Coahuila
Mexico, 25100

ABSTRACT. Potato experiments were carried out in Saltillo, Coahuila
(North of Mexico) in the springs of 1991 and 1992. Different mineral
and carbonic nutrition treatments were applied in the drip irrigation
system. All plots received the same amount of irrigation, based on the
water evaporation and the coefficient of development (Kc). The daily
CO_2 supply was correlated to total leaf area and ratio of mineral
nutrients change with the crop develop. The concentration of CO_2 in
the air was constant during day time (350-450 ppm) and leaf photosynthesis was increased. In experimental plots of 1992 additional
sulphur nutrition and chemical regulator (MC1) were supplied to
promote diameter of stem and were of great influence on the yield of
potato, and resulted in a more stable response to CO_2/H_2CO_2 enriched
water.

1. Introduction

Various techniques have been developed to bring water efficiently to
the root zone of cultures, one being the drip irrigation system. This
system has strongly developed. There have been innovations in
engineering of plastic materials, in the dripping systems, in the
composition of the irrigation water for nutrition of plants and in the
control of insects and diseases. These innovations strongly reduced
costs, at the same time improving the operation and functioning of the
irrigation system and enhancing the agricultural frontier. The drip
irrigation system turned out to be the pillar of modern agricultural
technology in Mexico and other countries, since it can be applied
together with other techniques of agroplasticulture such as mulching
or cropping in greenhouses, etc.

P C Struik et al (eds), Plant Production on the Threshold of a New Century, 375–380
© 1994 *Kluwer Academic Publishers.*

2. Results

The agroplasticulture applied together with the drip irrigation system modified the soil and atmosphere conditions and the availability of elements. Thus it improved or enhanced processes involved in production. For example, the photosynthesis process was enhanced by improving the maximum assimilation (ASM) and the total assimilation (AST) of CO_2. In a field culture of potato drip irrigation system increased ASM from 1.95 to 4.2 μmol m^{-2}s^{-1}. In combination with mulching the ASM rates were 2.25 and 7.11 μmol m^{-2}s^{-1}, respectively. Leaf photosynthesis improved only temporarily and especially when the atmosphere became richer in CO_2, since the concentration of CO_2 in the canopy of cultures was reduced more than 50 % when the conditions for a better assimilation of CO_2 occurred (Figure 1). We have designed a system and method to dose CO_2 and apply it directly to the canopy of plants (Quero, E. January 1992: patent register no. 24137(2560), Mexico). This method maintains the concentration of CO_2 in the canopy of the culture between 380-500 ppm, during the time of photosynthesis. The method consists of applying a mixture of CO_2/H_2CO_3 in the drip irrigation system at any moment of the day, improving AST and ASM of CO_2 in the above-mentioned cultures (Figure 2).

On the other hand the conditions of alkalinity, sodicity and solubility of nutrients in the soil can be controlled in order to obtain a larger harvest. Yields are, on average, 50 to 80 % larger, some cultures responding even better. The yield in potato crop var. Atlantic was improved by CO_2, sulphur and a chemical regulator. Control plots produced 25 ton ha^{-1} and the best treatment 101 ton ha^{-1} (sulphur and CO_2 supply) (Figures 3, 4, 5 and 6; Table 1).

These results have been applied to the commercial production of potato, tomato, melon, watermelon, cucumber, strawberry, banana, grape, rose, Chrysanthemum and carnation in which drip irrigation system is used. In these cultures the application has been successful since there was an increase in the quality of the harvest as well as in the resistance of plants to diseases and insects. The use of CO_2 and other gases is promising and it can easily be combined with the techniques of agroplasticulture and mineral fertigation.

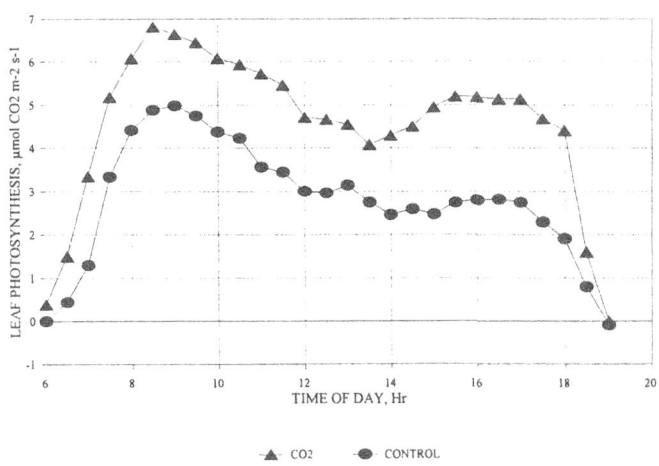

Figure 1 Influence of CO_2/H_2CO_3 in potato (*Solanum tuberosum* L. var. Atlantic).

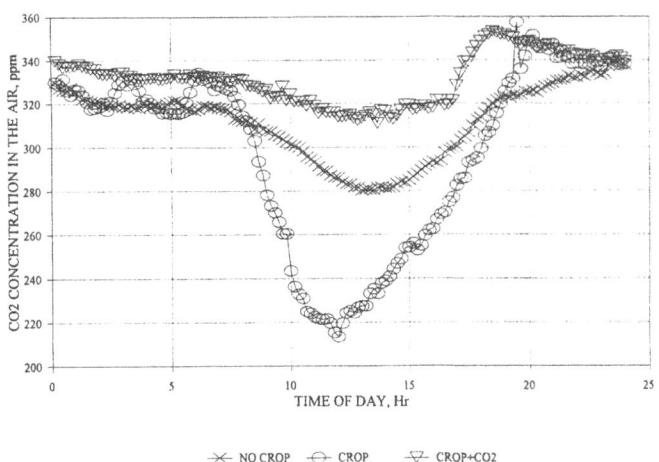

Figure 2. Influence of irrigation with CO_2 enriched water on CO_2 concentration in the canopy air of a potato crop.

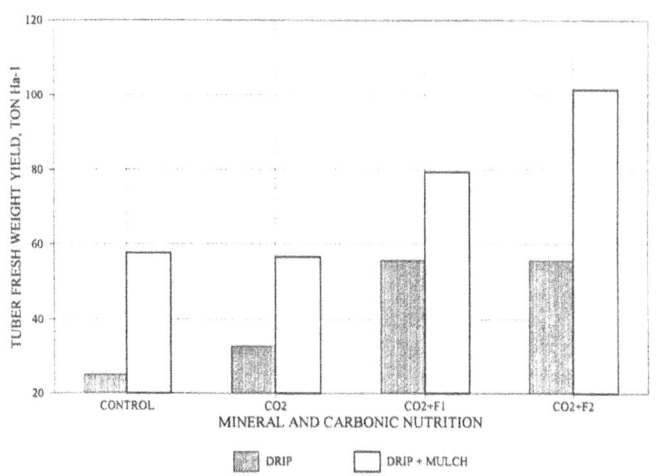

Figure 3. Response of potato to carbonated water (total application of CO_2 = 450 g m^{-2}).

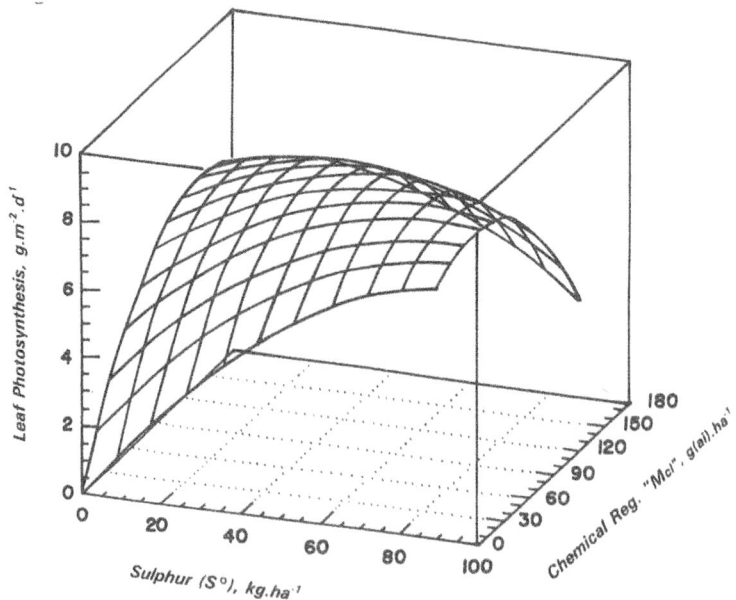

Figure 4. Effects of sulphur and MCl on photosynthesis in potato plant, cultivar Alpha.

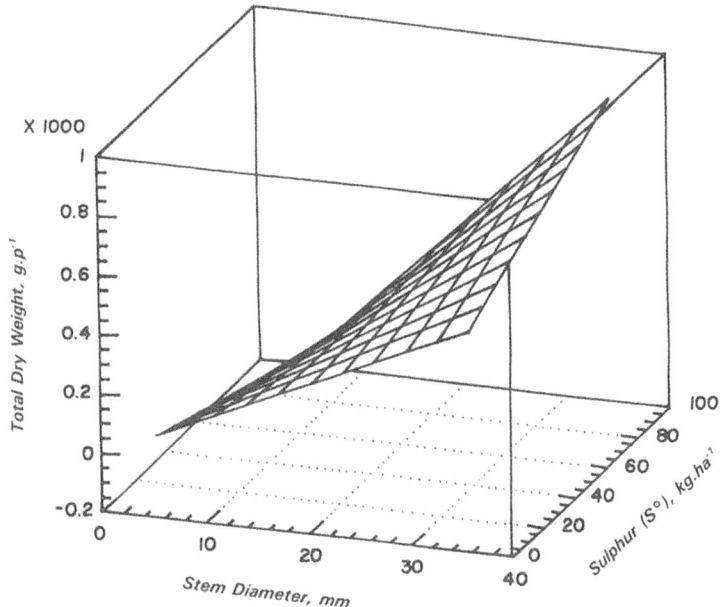

Figure 5   Effects of stem diameter and sulphur on total dry weight in potato plant, cultivar Alpha

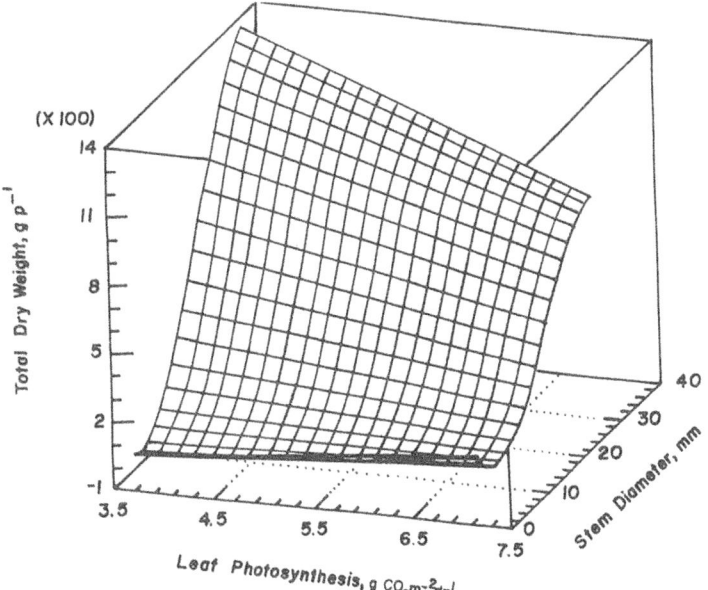

Figure 6   Relation between leaf photosynthesis, stem diameter and total dry weight in potato plant, cultivar Alpha

TABLE 1. Potato plant var. Alpha: total dry weight, g pl^{-1}.

Independent variable	coeff.	std. error	t-value	sig.level
Sulphur, kg S^o ha^{-1}	-0.541	0.296	-1.829	0.080
Plastic no(1)/yes(2)	-13.143	12.254	-1.072	0.294
ASM, g CO_2 m^{-2} d^{-1}	7.626	4.337	1.758	0.092
Stem diameter, mm	0.056	1.106	0.051	0.959
LAI m^2 m^{-2}	-17.344	4.898	-3.540	0.002
Water g pl^{-1}	0.215	0.013	15.745	0.000
(Diameter*ASM*Sulphur)	0.005	0.002	2.272	0.033
(Diameter*LAI*Plastic*ASM)	0.021	0.009	2.220	0.036
(Water*ASM*Plastic)	-0.005	0.001	-5.167	0.000

R-SQ. (ADJ.)=0.9971   SE=19.335025   MAE=11.963456   DurbWat=2.514
Previously: 0.0000   0.00000   0.000000   0.000
32 observations fitted, forecast(s) computed for 0 missing val. of dep. var.

Analysis of variance for the full regression.

Source	Sum of squares	DF	Mean square	F-ratio	P-value
Model	3931706	9	436856	1168.55	.0000
Error	8598.39	23	373.843		
Total	3940305	32			

R-squared = 0.997818   Stnd. error of est. = 19.335
R-squared (Adj. for d.f.) = 0.997   Durbin-Watson statistic = 2.513

Further ANOVA for variables in the order fitted.

Source	Sum of squares	DF	Mean Sq.	F-ratio	P-value
Sulphur	1435191	1	1435191	3839	.0000
Plastic	462415	1	462415	1237	.0000
ASM	19482	1	19482	52	.0000
Diameter stem	1456224	1	1456224	3895	.0000
LAI	379541	1	379541	1015	.0000
Water	160836	1	160836	430	.0000
(Diameter*ASM*Sulphur)	250	1	250	0.67	.4305
(Diameter*LAI*Plastic*ASM)	7784	1	7784	20.82	0.0001
(Water*ASM*Plastic)	9984	1	9984	26.71	0.0000
Model	3931706	9			

PROSPECTS FOR MICROBIAL STABILIZATION IN CLOSED LIQUID HYDROPONIC CULTURES OF TOMATO

B. WAECHTER-KRISTENSEN, U.E. GERTSSON AND P. SUNDIN
Department of Horticulture
P.O. Box 55
S-230 53 ALNARP
Sweden

ABSTRACT. Microbial stabilization of the root zone and nutrient solution in liquid hydroponics can be achieved by the use of plant growth promoting rhizobacteria (PGPR), microbial antagonists and bacteria degrading unwanted organic compounds. Preliminary results from studies using ten different bacterial isolates showed that tomato growth could be stimulated under laboratory and greenhouse conditions. The stimulating effect varied, dependent on tomato cultivar. Two *Pseudomonas* isolates kept their plant growth promoting effect even in presence of *Pythium ultimum* var. *ultimum*.

1. Introduction

In Sweden, environmental laws can soon be expected to impose restrictions on the release of nutrient solutions from greenhouses. Under such legislation, closed hydroponic systems constitute a feasible alternative for the cultivation of tomato [1]. Here, two main problems may arise: the accumulation of organic compounds that may cause yield-reducing damage to the plants, and the multiplication and spread of phytopathogenic [2] or deleterious microorganisms. The latter is of particular significance as Swedish legislation already prohibits the use of chemicals against *Pythium and Fusarium spp.* in vegetable production with continuous harvesting. The specified problems may be avoided if the microflora in the nutrient solution and on the rhizoplane [3] and in the endorhiza [4] can be controlled. The possibility to stabilize closed hydroponic cultures by microbial means constitutes an appealing alternative to be investigated more closely. In microbial control, aimed at improving the stability of a hydroponic culture, plant-growth promoting rhizobacteria (PGPR), microbial antagonists to plant pathogens and bacteria degrading unwanted organic compounds, such as phenolic acids, in the nutrient solution can be utilized [5]. Below, the two former categories are briefly considered.

P C Struik et al (eds), Plant Production on the Threshold of a New Century, 381–383

2. Microbial stabilization of closed liquid hydroponic systems

2.1. THE BACTERIAL BACKGROUND

In liquid hydroponics [6], no rhizosphere in the traditional sense is formed. Instead, the "rhizosphere" is limited to the root surface (rhizoplane), which is intensely colonized by microorganisms, and the endorhiza.

In three hydroponic culture systems with tomato, the viable counts of bacteria from the roots were found to be higher than in the nutrient solution. Also the composition of the microflora in the nutrient solution and on the roots differed. There was a higher number of *Pseudomonas* species (*Pseudomonas fluorescens*, *P. putida*, *P. cepacia*) as well as of *Serratia*, *Xantobacter*, and *Aureobacter spp.* in the root samples than in the nutrient solution. In addition, the *Pseudomonas* species composition differed between the root and nutrient solution. The viable counts of bacteria in the nutrient solution and on the roots showed pronounced variations during the vegetation period. The composition also varied considerably during the vegetation period. The amount of actinomycetes increased in the nutrient solution in the later phase of vegetation.

2.2. PLANT GROWTH PROMOTING BACTERIA (PGPR)

PGPR may act on different plant parameters, such as root growth, stem elongation, internode length, leaf development and flowering. The qualities we are primarily interested in are early emergence, stem elongation, and flowering. In our studies, ten bacterial isolates from nutrient solution were studied under non-sterile conditions. Bacterial inoculation on different tomato varieties showed that the plant responses to different bacterial isolates depend on the cultivar. In a greenhouse experiment, where bacterial isolates in pure cultures or in a mixture of different isolates were incubated in the rockwool cubes before sowing, emergence-promoting qualities as well as effects on plant growth and development could be observed.

2.3. MICROBIAL ANTAGONISTS

In our studies, we have focused our research on antibiosis towards *Pythium ultimum* var. *ultimum*. The ten bacterial isolates tested for plant-growth promotion were also screened for their antagonistic potential towards *Pythium ultimum* var. *ultimum*. In an *in vivo* screening of these isolates, two *Pseudomonas* isolates had a plant growth promoting effect even in the presence of the fungus.

3. Discussion

Microbial stabilization may be a feasible alternative to current control methods when growing tomato in closed hydroponic systems. Further studies have to be conducted in order to gain an increased insight into the mode of action of PGPR and antagonists, as well as into their root colonizing qualities and competitiveness, under greenhouse conditions.

The occurrence of potentially phytotoxic organic compounds, the appropriate threshold levels for these compounds, and their impact on root and endorhiza colonizing bacteria, also deserve further attention. In addition, the potential role of fungi, such as arbuscular mycorrhizal fungi (AM), *Trichoderma spp.* and actinomycetes, as biological control agents has to be taken in account.

For greenhouse cultures other than tomato the concept of microbial stabilization of the root environment should be generally applicable. Naturally, however, the approach has to be modified and accommodated to correspond to the microbial and chemical parameters characteristic to the specific culture.

4. References

[1] Gertsson, U., Hansson, I., Waechter-Kristensen, B., Lundquist, S., Svedelius, G. and Weich, R., Acta Horticulturae (in press).
[2] Wohanka, W. (1991) Gartenbau 38, 9-11.
[3] Clark, F.E. (1949) Advances in Agronomy 1, 241-288.
[4] Kloepper, J.W., Schippers, B. and Bakker, A.H.M. (1992) Phytopathology 82,726-727.
[5] Waechter-Kristensen, B. (1992) SLU Info Rapporter. Allmänt 178, 9-10 (in Swedish).
[6] Jensen, M.H. and Collins, W.I. (1985) Horticultural Reviews 7, 485-553.

ENERGY EFFICIENCY AND CO_2 EMISSION IN THE DUTCH GREENHOUSE INDUSTRY[1)]

N.J.A. VAN DER VELDEN AND B.J. VAN DER SLUIS[1]
Agricultural Economics Research Institute
P.O. Box 29703
2502 LS The Hague
The Netherlands

ABSTRACT. The energy efficiency and CO_2 emission of the Dutch greenhouse industry have to be reduced considerably. This contribution discusses the developments and possibilities in this respect.

1. Introduction

The Dutch greenhouse industry uses a large quantity of fossil fuel ($4\ 10^9\ m^3$ gas in 1992). The target for the sector is an improvement of the energy efficiency by 50 % over the period of 1980-2000. Energy efficiency is defined as the direct energy consumption per unit of physical production. A second target is a reduction of the total CO_2 emission by 3-5 % over the period of 1989/90-2000. The CO_2 emission depends on the total fuel consumption. The fuel consumption and consequently the CO_2 emission can be reduced by the use of extra energy saving measures and alternative energy sources.

2. Energy efficiency

The direct energy consumption consists of mainly natural gas. The quantities of consumption changed in the eighties (Table 1). In the beginning the gas price increased strongly but after 1984 it decreased again. The fuel consumption per m^2 showed an opposite development. After 1988 the gas price was low and therefore the fuel use per m^2 increased autonomously.

The physical production per m^2 increased by 72 % in the period 1980-1992. By these two developments the energy efficiency improved from 100 to 59 % between 1980 and 1985. After 1985 it gets worse to 67 % in 1991 caused by the above-mentioned autonomously increase of fuel intensity. In 1992 it improved again to 65 %. Related to the agreement between growers and government another 23 % has to be

---

[1] financial support by NOVEM

P. C. Struik et al. (eds.), Plant Production on the Threshold of a New Century, 385–387.
© 1994 *Kluwer Academic Publishers.*

realized in the nineties.

TABLE 1. Energy efficiency and CO_2 emission.

Year	Area (ha)	Energy use (MJ/m^2)	Physical product (%)	Energy effic. (%)	CO_2 emiss. (%)
1980	8257	1242	100	100	
1985	8608	920	126	59	
1989/90	9267	1312	160	66	100
1991 *)	9594	1385	166	67	109
1992 *)	9754	1396	172	65	112

*) preliminary data

3. CO_2 emission

CO_2 emission is caused by burning fuel. It amounted to 6.8 tons in 1989/90 and increased to 7.6 tons in 1992, an increase of 12 % which has been caused by an increase of the area under glass of 5 % and an increase of the fuel use per m^2 of glass of 6 %. Through this an extra effort is desired to fulfil the agreement between growers and government, a total decrease of 13-15 % has to be realized in the period 1992-2000. The reduction of CO_2 emission will be more difficult than the improvement of the energy efficiency.

4. Energy saving options

The most important saving measures on the individual holdings are the installation of a condenser, a screen and heat storage. The simultaneous production of heat and power is the most important alternative energy source. This can be realized by using reject heat from power stations or by total energy units installed on individual holding. The penetration of the options is given in Table 2.

TABLE 2. Penetration of energy saving options in 1991*).

Option	Area (%)	Holdings (%)
Condenser	72	57
Screen	64	
Heat storage		10
Total energy	>8	8
Reject heat	>1	1

*) specialized holdings, preliminary data.

All the energy saving options have their own bottlenecks. The condenser has the problem of low temperature energy, the screen the loss of light in the glasshouse and consequently the loss of production and for heat storage the possible energy savings are unknown. The problems of reject heat or total energy are more in the field of governmental and local policy and the economic advantage for the grower.

5. References

Velden, N.J.A. van der and Sluis, B.J. van der (in preparation). 'Energie in de Nederlandse glastuinbouw; Ontwikkelingen in de sector en op de bedrijven', AERI, The Hague.

CROP PROTECTION USING DIVERSIFICATION AND INDUCED RESISTANCE IN LOW-INPUT CEREAL/LEGUME CROPPING SYSTEMS

B.M. COOKE* AND D.T. MITCHELL**
Department of Environmental Resource Management
*Department of Botany***
University College
Dublin 4   Ireland

D. GARETH JONES
Department of Agricultural Sciences
University of Wales
Aberystwyth
Dyfed SY23 3DD   UK

V. SMEDEGAARD
Royal Veterinary & Agricultural University
Department of Plant Biology
(Plant Pathology Section)
DK - 1871 Frederiksberg C
Copenhagen   Denmark

ABSTRACT. A new project to evaluate the effects of a cereal/legume intercropping system on the development of diseases caused by cereal necrotrophs is described.

1. Introduction

In low input cereal systems, input savings will be made on pesticides (primarily fungicides), herbicides and fertilizers. Reduced use of these agrochemicals will result in lower yields, but the resulting lower income will be offset by the reduced input costs. Thus, such a system should help to reduce cereal production in Europe while maintaining farming incomes. Decreased use of pesticides will reduce the opportunity for chemical resistance (insensitivity) to develop in target pathogen populations, and harmful effects on health and the environment will be avoided. The new project described here, financed under the EC CAMAR programme, will investigate the effects of a low-input cereal/legume cropping system (LICLCS) on the development of diseases caused by cereal necrotrophic fungi.

2. Quantitative effects of pathogens

Work is concerned with quantifying the effects of four necrotrophic cereal pathogens (*Drechslera teres*, *Microdochium nivale*, *Septoria tritici* and *Rhynchosporium secalis*) in LICLCS. Initial work, using

P. C. Struik et al. (eds.), Plant Production on the Threshold of a New Century, 389–392.
© 1994 *Kluwer Academic Publishers.*

Drechslera teres, has established basic quantitative relationships between disease and growth and grain yield of the host plant. Results achieved so far clearly indicate the potential of this necrotroph to cause damage to the barley plant. A linear regression model of % disease and % dry matter loss was developed in the form of Y = 0.75 + 0.88x, with r = 0.813; P < 0.001. Further work will investigate how the presence of a clover base might influence such relationships for the four necrotrophs under study by modifying host susceptibility, spore dispersal and disease development patterns. Preliminary data from a wheat/*S. tritici* field experiment indicate that the morphology of the cereal plant is affected by the presence of a clover base; wheat tillers ranged in height from 50-65 cm compared to 40-50 cm without clover. Increased internodal length will necessitate greater upward displacement of splash-borne inoculum of necrotrophs such as *S. tritici* during rainsplash events. The presence of a clover base will increase air humidity at the base of the crop and favour stem base diseases such as *Fusarium* brown foot rot, but restrict or modify the vertical splash dispersal of spores from this pathogen and of others such as *S. tritici* from basal leaf infections.

3. Evaluation of the mycorrhizosphere

3.1. NUTRIENT TRANSFER

Work will involve the evaluation of the role of arbuscular mycorrhizas in the transfer of nutrients from legumes to cereals. Endomycorrhizal properties of soils will be determined in field, glasshouse and controlled environment experiments. Preliminary data suggest that endomycorrhizal spore populations from field sites at Dublin (Ireland), Aberystwyth (UK) and Copenhagen (Denmark) are similar, but variations in arbuscular mycorrhizal spore density occur. Arbuscular endomycorrhizal infection has been found in roots of wheat, clover and lucerne grown in a field trial at Aberystwyth; there is evidence of seasonality in mycorrhizal infection of clover with highest levels of infection occurring during summer. Investigations into the potential arbuscular mycorrhizal infectivity of soils from these sites have begun and preparations of cultures of arbuscular mycorrhizal fungi using *Trifolium repens* as the host plant are in progress. The determination of mycorrhizal spore populations of an Irish farm soil, used for the organic cultivation of cereals, is proceeding.

3.2. DISEASE DEVELOPMENT

Data from a preliminary glasshouse pot experiment showed that leaf infection of barley by *Drechslera teres* was unaffected by arbuscular mycorrhizal infection of the root system, and vice-versa; although root infection levels were generally low (5 %), arbuscules, vesicles and intercellular hyphae were observed. A field experiment to investigate growth, yield and *Septoria tritici* blotch development on winter wheat after the application of either conventional phosphate fertilizer, or the endomycorrhizal inoculant 'Vaminoc' has recently been implemented. The work will be further extended by the addition of a legume fixed nitrogen treatment in the experiment to assess possible *Rhizobium*/endomycorrhizal interactions on disease development.

4. Effects of biologically fixed soil nitrogen on disease development

Work is in progress to evaluate the effects of biologically fixed soil nitrogen on cereal growth, grain yield and components of partial disease resistance to the wheat necrotroph *S. tritici* in cultivar mixtures after cropping with lucerne or white clover inoculated with selected strains of *Rhizobium*. Parallel work has been carried out to demonstrate the effects of applied NH_4NO_3 on *Septoria tritici* blotch development using a detached leaf bioassay. Results showed a greater development of necrosis and pycnidiospores at higher nitrogen doses compared to lower doses (13.8 x 10^6 spores cm^2 leaf at 200 kg N ha^{-1} compared to 2 x 10^6 spores cm^2 leaf at 50 kg N ha^{-1}). Thus low biologically fixed nitrogen inputs are likely to limit the development of the disease, and field experiments have been established to compare such low nitrogen sources with conventionally applied nitrogen fertilizer using cereal / legume intercropping or the incorporation of legumes into crop rotations. Winter wheat has been sown on ground which supported for 1, 2 or 3 years a crop of either white clover (*Trifolium repens*) or lucerne (*Medicago sativa*) inoculated with *Rhizobium trifolii* and *R. meliloti* respectively. Other treatments involve wheat grown with and without applied nitrogen fertilizer (150 kg ha^{-1}), with and without foliar fungicides, and as a wheat/clover intercrop. All treatments will be inoculated with the necrotrophic pathogen *Septoria tritici*. The performance of a 1:1 mixture of two wheat cultivars is also being evaluated under each cultivation regime.

5. Induced disease resistance studies

The importance of induced disease resistance in LICLCS as a non-chemical means of controlling necrotrophic cereal pathogens is being evaluated in growth chamber and field experiments. Experiments to induce resistance to *Bipolaris sorokiniana*, *Rhynchosporium secalis* and *Drechslera teres* in barley, and *Bipolaris sorokiniana*, *Septoria nodorum* and *Septoria tritici* in wheat have been implemented. The most successful treatments on wheat, where inducer organisms were applied at various time intervals before the challengers, were germination fluid used in advance of *B. sorokiniana* and *Pseudomonas fluorescens* used in advance of *S. nodorum*. However, the marked disease reductions observed were not always repeatable in subsequent experiments. On barley, *P. fluorescens* alone or in mixture with *Bacillus subtilis*, *S. nodorum* and germination fluid gave the most successful control of the necrotroph *D. teres*. Future experiments will test microorganisms from clover and lucerne as inducers against cereal necrotrophic pathogens. The mechanisms behind the observed disease reductions in these experiments will be investigated to establish whether true induced resistance or antibiosis from inducer organisms is operating.

PERSPECTIVES OF BREEDING FOR LOW INPUT CONDITIONS

TH. JACOBS
Department of Plant Breeding
Wageningen Agricultural University
P.O. Box 386
6700 AJ Wageningen
The Netherlands

ABSTRACT. An overview is given of the possibilities to reduce inputs and increase input efficiencies by breeding in modern agriculture, and perspectives of breeding for adapted cultivars are discussed.

1. Overview

Farmers operate with various production factors. They can be divided into: capital, land, water, labour and agronomic components like: availability of minerals and agro-chemicals.
     Breeding for low input conditions in the present day context relates to:

Lower input of fungicides, pesticides: breeding for resistance against pests and pathogens contributes considerably to a good solution for these problems, phytosanitation often prevents problems and adds to sustain the durability of resistance. Durable resistance is not often bred for. Integrated Pest Management and assessment based management systems (e.g. EPIPRE) should be used more often. Management of resistance needs to be practised more often, just as cultivar mixtures, mixed cropping, gene deployment, gene rotation and other diversification methods.

Less herbicides: breeding for early ground cover and altered agronomic practices, e.g. closer stands of crops and timely mechanical weeding, will enable reductions.

Less growth regulators: breeding for reduced height lowers the need for application of growth reducing chemicals like CCC in wheat.

Less labour input: There are several methods which require less time: i) zero tillage, ii) natural way of farming, iii) Permaculture. These are different production systems which require adapted cultivars. Some of these methods increase the likelihood of diseases and weeds, more

393

P. C. Struik et al. (eds.), Plant Production on the Threshold of a New Century, 393–395.
© 1994 *Kluwer Academic Publishers.*

resistance is then needed. Growing different crops can reduce the labour requirement as some crops are more labour intensive than others. In the past re-allotment of fields, buildings and farms and a high degree of mechanization have reduced the labour requirements.

Less financial input: A reduction of funds required can be obtained by on-farm production of seed. This excludes the use of hybrids and includes a continuation of the "farmers' privilege".

Smaller area input: This is common in present day agriculture, it involves breeding for high yielding cultivars which are responsive to high external inputs. There is a long history of breeding for higher yields.

Less input of minerals: Modern cultivars yield relatively well under lower input conditions. This seems to be correlated with an altered harvest index (cereals). Optimal application of nutrients (first assess, then split applications) can reduce the loss of nutrients. It is generally believed that the best yielding cultivars under high input conditions outperform the best yielding cultivars under low input conditions (Simmonds, 1979). However Genotype x Environment interactions influence selection under low input conditions (Ceccarelli, 1989; Ceccarelli et al., 1992).

Higher efficiency: The efficiency of inputs can be calculated per unit product, per unit labour or per unit money. In this context only the first point will be discussed. High transformation efficiency of solar energy and more efficient use of nutrients and water will enable high yields under more stressed conditions. There is genetic variation for uptake of macro- and micronutrients by plants as well as for nutrient requirement (Schlegel et al., 1991). A distinction needs to be made between *efficiency and responsiveness* of a cultivar to mineral applications (Blair, 1993). Selection for altered root architecture (size, growth rate, branching, hair number, uptake efficiency) could be part of breeding programmes for low availability of nutrients and water. Several components can be distinguished: root uptake efficiency, transport efficiency (root-to-tuber, root-to-leaf/stem), reallocation efficiency (leaf-to-ear/fruit, stem-to-ear/fruit). Correlations exist between N and P harvest indices and other components like grain yield or edible dry matter yield. Higher yield per unit mineral applied is a good selection criterion.

2. Comments

The negative effects of today's agriculture on the environment, nature values and labour conditions are criticized.

With regard to maximal production of products the present "intensive" way of agricultural production with high inputs of minerals, fungicides, fossil energy, etc. is highly efficient with regard to unit product per unit area and other inputs.

Not necessarily the present day agriculture in the Netherlands is the most efficient in economic terms. There is diversity in the way farmers operate. There is a group of farmers who expand and increase their production continuously, others optimize their labour, a third group maximizes the yield per hectare. Again other farmers invest to safe costs on a long term basis, they do not intend to maximize their production (farming systems analysis, Van der Ploeg, 1993).

3. Perspectives

Breeders can assist in creating cultivars for any type of production system provided the system can be defined in selection criteria, provided that there is variation for the desired characters and within the limits set by the biology of the crops and available breeding technology. The present cultivars quite often do not meet the requirements of low input agriculture. Of the more than 100 potato cultivars on the Dutch cultivars list only 10 seem suitable for integrated low input agriculture. Cultivation of most of these cultivars is not commercially attractive (Vereijken and Van Loon, 1991).

The average time needed between cross and release of a new cultivar of arable crops is more than 10 years. This usually exceeds the time span of policy shifts. Breeders, farmers, consumers and the environment benefit from a production system stable over a long period of time.

4. References

Brail, G. (1993) 'Nutrient efficiency - what do we really mean?', in P.J. Randall *et al.* (eds.) 'Genetic aspects of plant mineral nutrition', Kluwer Academic Publishers, Dordrecht, pp. 205-213.

Ceccarelli, S. (1989) 'Wide adaptation, how wide?' Euphytica 40, 197-205.

Ceccarelli, S., Grando, S. and Hamblin, J. (1992) 'Relationship between barley yield measured in low- and high-yielding environments'. Euphytica 64, 49-58.

Ploeg, J.D. van der, (1993) 'The reconstitution of locality: technology and labour in modern agriculture', in T. Marsden, P. Lowe and S. Whatmore (eds.) 'Labour and locality; uneven development and the rural labour process'. Critical perspectives on rural change series IV. David Fulton, London, pp. 19-43 (ISBN 1-85346-182-2).

Schlegel, R., Werner, T. and Jakob, F. (1991) 'Mineral nutrition and genetical control in cereals'. Vort. Pflanzenzüchtung 20, 85-94.

Simmonds, N.W., (1979) 'Principles of crop improvement'. Longman, New York, 360 pp.

Vereijken, P. and Loon, C.D. van, (1991) 'A strategy for integrated low-input potato production'. Potato Research 34, 57-66.

INTERACTION OF LATE SEASON FOLIAR SPRAY OF UREA AND FUNGICIDE MIXTURE IN WHEAT PRODUCTION

J. PELTONEN
University of Helsinki
Department of Plant Production, Viikki
P.O. Box 27
FIN-00014 Helsinki
Finland

ABSTRACT. To avoid foliar injury or phytotoxic effect of foliar spray of urea, mixing of urea with propiconazole in warm growing conditions should be avoided. In turn, during cool and rainy growing seasons, a mixture of these chemicals is recommended in order to improve rapid absorption of both chemicals.

1. Introduction

Nitrogen (N) fertilization with urea and fungicide treatment at flag leaf emergence (GS 47) are common strategies to improve both grain yield and quality of wheat (*Triticum aestivum* L.). Mixing of urea and fungicide, together, would save time in farm operations. In some cases, the effect of foliar spraying of urea either alone (Pushman and Bingham, 1976) or in combination with a fungicide (Gooding *et al.*, 1991) may, however, give unexpected low response in grain yield and quality when compared to separate application of foliar urea and fungicide. Therefore, it is desirable to determine the possible effect of this mixture on grain yield.

2. Materials and methods

The effects of late season spraying of urea (15 kg N ha^{-1} as an aqueous solution of 110 g L^{-1}) alone or mixed with propiconazole (125 g ha^{-1}) on the grain yield of spring wheat were studied. Field trials (Exp. 1) were carried out on in 1986-1987. The standard N fertilization (110 kg N ha^{-1} as NH$_4$NO$_3$) was selected to correspond to practical usage in Finnish wheat production. Cultivars *Heta*, *Ruso*, and breeding line *Hja 22161*, all susceptible to foliar diseases caused by *Erysiphe graminis* f. sp. *tritici* and *Septoria nodorum* Berk, were used. Greenhouse experiments (Exp. 2) were carried out to measure the leaf photosynthesis efficiency,

397

P. C. Struik et al. (eds.), Plant Production on the Threshold of a New Century, 397–399.
© 1994 *Kluwer Academic Publishers.*

urease enzyme activity, ammonia content and foliar burning damages, after 4, 24, 48 and 96 h of spraying.

3. Results and discussion

Spraying of propiconazole at GS 47 reduced the infected canopy area (*Erysiphe graminis*, *Septoria nodorum*) significantly in both seasons. During 1986, May to June was warmer than the mean of 1961-1990. Therefore, spring wheat reached its maturity in the beginning of August. The foliar application of propiconazole and urea alone at GS 47 increased the mean grain yield significantly in the growing season 1986. In turn, propiconazole-urea mixture resulted in unexpected lower grain yield as compared to the results from single spraying of propiconazole and foliar urea. The growing season in 1987 was colder and more rainy than the mean of 1961-1990, and wheat did not ripen until the beginning of October. The average grain yield increased when propiconazole-urea mixture was applied, but was not affected significantly when these chemicals were applied separately.

TABLE 1. Effect of propiconazole (P), foliar urea (U), and propiconazole-urea (P+U) mixture on grain yield (kg ha^{-1}) in field trials (1986-1987).

Season	Treatment			
	Control	P	U	P+U
1986	3858 b	4109 a	4419 a	3956 b
1987	4659 b	4646 b	4760 ab	5125 a

The means followed by the same letter in the same row showed no significant difference in Duncan's multiple range test at P=0.05.

In the control environment, devoid of foliar diseases, propiconazole decreased significantly the photosynthetic rate of the cultivars. The spraying of propiconazole-urea mixture also caused a significant decrease in leaf photosynthesis in *Ruso* and *Hja 22161*. Leaf photosynthesis of *Heta* was not affected by the spraying mixture. In turn, foliar urea spraying increased leaf photosynthesis in *Heta* and *Hja 22161*, whereas the rate of leaf photosynthesis first decreased in *Ruso*. Cultivar *Ruso* was more susceptible to burning damages than *Heta* and *Hja 22161* after urea or propiconazole-urea application. High-protein cultivar Heta utilized foliar urea application more efficiently than low-protein cultivars *Ruso* and *Hja 22161*. This is indicated by the higher level of urease activity and ammonia content of leaves, as a consequence of increased grain protein concentration. The highest mean leaf urease activity was obtained after application of propiconazole-urea mixture.

To avoid foliar injury or phytotoxic effect of foliar spray of urea, mixing of urea with propiconazole in warm growing conditions should be avoided. In turn, during cool and rainy growing season, a mixture of these two chemicals is recommended in order to improve rapid absorption of both propiconazole and foliar urea. The positive effects on cultivars of these chemicals were attributed to high leaf photosynthesis rate after propiconazole and foliar urea treatment and high urease enzyme activity after foliar urea application.

4. References

Gooding, M.J., Kettlewell, P.S. and Hocking, T.J. (1991) 'Effects of urea alone or with fungicide on the yield and breadmaking quality of wheat when sprayed at flag leaf and ear emergence'. J. Agric. Sci. (Cambridge) 117, 149-155.

Pushman, F.M. and Bingham, J. (1976) 'The effects of a granular nitrogen fertilizer and foliar spray of urea on the yield and bread-making quality of a ten winter wheats'. J. Agric. Sci. (Cambridge) 87, 281-292.

SUPPRESSION OF *RHIZOCTONIA SOLANI* ON POTATO BY MYCOPHAGOUS SOIL FAUNA

M. LOOTSMA
Department of Agronomy
Wageningen Agricultural University
Haarweg 333
6709 RZ Wageningen

ABSTRACT. Possibilities were studied to control *Rhizoctonia* stem infection of potato by stimulating the mycophagous soil fauna. A bioassay showed that the collembole *Folsomia fimetaria* and the mycophagous nematode *Aphelenchus avenae* are capable of suppressing stem infection at relatively low soil temperature. Field research pointed out that stem infection can be reduced significantly by organic manure and that the number of collembola in early spring can be manipulated.

1. Introduction

Rhizoctonia solani is one of the most common soil pathogens in Dutch potato crops. Stem canker of potato can occur by soil-borne infection and in case of a heavy infection it can reduce yield and quality of potatoes. Looking for non-chemical control methods of *Rhizoctonia* stem canker on potato the possibility is studied to increase the natural suppressiveness of soils against it.

A bioassay was used to study the effect of low temperature, comparable with soil temperature in early spring, on the ability of the collembole *Folsomia fimetaria* and the mycophagous nematode *Aphelenchus avenae* to suppress *Rhizoctonia* stem infection.

A field experiment was carried out to investigate, whether organic manures, applied before growing a potato crop, can stimulate the mycophagous fauna and thus reduce *Rhizoctonia* stem infection.

2. Materials and methods

For testing the activity of mycophagous soil fauna the collembole *F. fimetaria* and the nematode *A. avenae* were added at densities of respectively 250 and 20,000 individuals per litre soil to 15 litre of a soil mixture, which was inoculated with sclerotia of *R. solani*. *Rhizoctonia* suppression was measured by growing 20 potato stems through the soil and by screening the stem infection after emergence of the potatoes.

P. C. Struik et al. (eds.), Plant Production on the Threshold of a New Century, 401–403.
© 1994 *Kluwer Academic Publishers.*

In a field trial *Rhizoctonia* inoculation of the soil was realized by planting seed potatoes with black scurf in the first year. In September the potatoes were harvested and different manure treatments were started. *Rhizoctonia* suppression was measured by growing potatoes again in the second year and screening the stem infection 10 weeks after emergence.

3. Results and discussion

At both 10 °C and 15 °C *Rhizoctonia* was suppressed by the combined activities of *F. fimetaria* and *A. avenae* (Table 1). The effects of the single species were not consistent over the two temperatures.

TABLE 1. Effect of temperature on suppressive effect of *F. fimetaria* and *A. avenae*.

	Rhizoctonia index (0-100)	
	10 °C	15 °C
Control	61 a *)	72 a
A. avenae	60 a	55 b
F. fimetaria	49 b	44 c
A. avenae + F. fimetaria	18 c	40 c

*) Means followed by a different letter are significantly different at P<0.05.

In the field trial an early application of cattle manure, white mustard and the combination of the two increased the number of collembola in early spring significantly (Table 2). All treatments, except early manure reduced *Rhizoctonia* infection. Growing of oats and the combination of manure and white mustard proved most effective. The number of collembola in early spring was not related to *Rhizoctonia* suppression.

4. Further research

Further research will be carried out to study the population dynamics of collembola and other mycophagous soil organisms as influenced by organic manuring. Also more attention will be paid to the relative importance of mycophagous soil fauna in suppression of *Rhizoctonia*, like it is found in the field.

Bioassay studies will be used to get more insight in how the suppressive activity of soil fauna is influenced by soil moisture, soil pH, food competition and soil disturbance. Also the mechanisms of suppression will be studied.

TABLE 2. Effect of green manure crops and cattle manure, applied 6 months (early) or 1 month (late) before planting potatoes, on the number of collembola (1 week before planting date) in the soil and on *Rhizoctonia* stem infection (10 weeks after planting date).

	Collembola (number per litre soil)	*Rhizoctonia* (% heavily infected plants)
Control	288 a	26 c
White mustard	548 b	15 ab
Rape	282 a	13 ab
Oats	367 ab	10 a
Early manure	548 b	20 bc
Late manure	297 a	18 abc
Early manure + White mustard	715 c	10 a
Late manure + White mustard	485 ab	10 a

BIOLOGICAL CONTROL OF *FUSARIUM* WILTS

M.L. GULLINO, Q. MIGHELI, M. MEZZALAMA, C. ALOI, A. MINUTO, AND A. GARIBALDI
Dipartimento di Valorizzazione e Protezione delle Risorse agroforestali - Patolologia vegetale
Via Giuria 15
10126 Torino
Italy

ABSTRACT. The biocontrol activity of antagonistic *Fusarium spp.* against several *formae speciales* of *Fusarium oxysporum*, their survival in the soil and methods for strain characterization are described.

1. Introduction

In Italian *Fusarium*-suppressive soils, suppressiveness has been related to the presence of high amounts of antagonistic, non pathogenic *Fusarium spp.* Antagonistic *Fusarium spp.* are highly rhizosphere competent: their ability to compete with pathogenic *Fusarium oxysporum* could be due to their ability to precede the pathogen at the infection sites (Garibaldi *et al.*, 1990).

The biocontrol activity of saprophytic *Fusarium spp.*, their survival ability and some of their molecular characteristics are briefly described.

2. Materials and methods

Biocontrol activity has been repeatedly tested on carnation, cyclamen and basil, grown in steamed or natural soil, artificially infested with *F. dianthi*, *F. cyclaminis*, and *F. basilicum*, respectively. Survival in the soil was studied in the presence and in the absence of carnation plants and of *F. dianthi*. Different preparations of antagonistic *Fusarium spp.* were tested in order to evaluate the best suitable method for antagonist introduction in the soil-plant system. Benomyl and prochloraz resistant strains, as well as orange and dark red pigmented mutants, were obtained from antagonistic *Fusarium spp.* by physical and chemical mutagenesis. These markers have been used in studies of population dynamics.

Karyotyping was accomplished by the contour-clamped homogeneous electric field (CHEF) gel electrophoresis.

P C Struik et al (eds), Plant Production on the Threshold of a New Century, 405–406

3. Results and discussion

Antagonistic, saprophytic *Fusarium spp.* considerably and consistently reduced *Fusarium* wilt incidence on carnation, cyclamen and basil. In the case of carnation, dipping roots in a conidial suspension ($5x10^7$ cfu/ml) before transplanting generally gave the best results (Garibaldi *et al.*, 1992).

In the case of cyclamen, antagonists applied as soil dusting ($3x10^5$ cfu/g of soil) and dipping root of plantlets ($5x10^7$ cfu/ml) strongly reduced wilt incidence. Both in the case of carnation and cyclamen, the use of benomyl resistant mutants of antagonistic *Fusarium* permits to combine a treatment with a benzimidazole with the use of the antagonist. On basil, soil dusting (10^5 cfu/g of soil), combined with seed dressing with $3x10^{10}$ cfu/kg of seeds offered the best control. The use of antagonistic *Fusarium spp.* represents one of the few control options against *F. basilicum*.

In natural and disinfected soils antagonists, introduced at 10^5 cfu/g of soil were recovered at the rate of 10^3 cfu/g four months after infestation. The presence of *F. dianthi* (at $2x10^3$ cfu/g of soil) did not affect the antagonist population. Different preparations of antagonistic *Fusarium spp.* (conidial suspension, chlamydospores in talc powder, sodium alginate pellets) did not show differences in survival ability in the presence of carnation plants: six months after application, 10^3 cfu/g of soil were recovered, regardless of the preparation used. However, the population of the antagonists increased more rapidly after transplanting when applied as alginate pellets.

The electrophoretic karyotyping of antagonistic *Fusarium spp.* revealed a high degree of polymorphism both in the number and size of fungal chromosomes (Migheli *et al.*, 1993). This method can be used in combination with fungicide resistance and colour markers, to fingerprint antagonistic *Fusarium spp.* strains for ecological and patenting purposes.

4. References

Garibaldi, A., Aloi, C., Parodi, C. and Gullino, M.L. (1992) 'Biological control of *Fusarium* wilt of carnation', in E.C. Tjamos *et al.* (eds.), Biological control of plant diseases, Plenum Press, New York, pp. 105-108.

Garibaldi, A., Guglielmone, L. and Gullino, M.L. (1990) 'Rhizosphere competence of antagonistic *Fusaria* isolated from suppressive soils'. Symbiosis 9, 401-404.

Migheli, Q., Berio, T. and Gullino M.L. (1993) 'Electrophoretic karyotyping of *Fusarium spp.*' Exp. Myc., in press.

THE SCREENING OF MAIZE RESISTANCE TO APHIDS AS A CONTRIBUTION TO INTEGRATED PEST MANAGEMENT

TH. HANCE[1], O. DELANNOY[1] AND G. FOUCART[2]
Catholic University of Louvain
[1] *Unité d'Écologie et de Biogéographie*
   Place Croix du Sud, 5
[2] *Centre d'Information et de Promotion*
   de la Culture Fourragère
   Place Croix du Sud, 3
   1348, Louvain-la-Neuve
   Belgique

ABSTRACT. The sensitivity of maize strains to *Rhopalosiphum padi* was analysed by the determination of the demographic properties of the aphid on each strain and the study of natural infestation in field. Strong differences in aphid population development were observed according to the strains. These results show the potential of plant resistance for the implementation of Integrated Pest Management.

1. Introduction

The use of resistant plant strains is one of the most ecological, low energy budget and inexpensive solutions to control insect pests. However, it is still not applied on a large scale, probably because of the difficulty of screening great numbers of plant strains and of the lack of knowledge concerning the stability of this resistance. In this context, the aim of our work was to develop a reliable and rapid method of plant resistance screening.

2. Materials and methods

In order to compare the different maize strains, in controlled conditions in the laboratory, 50 first larvae of *Rhopalosiphum padi* were introduced in small cages (two individuals per cage) clipped on whole maize plants. Their survival and fecundity were recorded daily until they were 14 days old. Indeed, at this age they had already contributed to the establishment of 99 % of the growth potential of the population (Delaney, 1992). A test of slope heterogeneity was used to analyse the survival curves. Cumulative fecundity was compared by applying the method described by Van Impe and Hance (1993). A screening of the

P. C. Struik et al. (eds.), Plant Production on the Threshold of a New Century, 407–409.
© 1994 *Kluwer Academic Publishers.*

susceptibility of 27 half precocious and late maize strains and of 18 precocious strains was also done on the field by means of a completed random block design (4 reps/ strain). The level of infestation was estimated with a damage index (Delannoy, 1992). Data were analysed with an Anova and a Newman-Keuls test.

3. Results

3.1. LABORATORY EXPERIMENTS

A test of slope heterogeneity shows that the aphid survival was influenced by strains (F = 146.8, dl = 13.56, P < 0.0001). This could be the consequence of an antibiosis factor. Significant differences were also observed for the cumulative fecundity, meaning that after one generation some strains have to support three times more aphids than other strains (Figure 1).

3.2. FIELD EXPERIMENT

For the precocious strains as for the half precocious or late strains, differences in aphid abundance index were observed (P < 0.0001). A classification of the strains with regard to aphid susceptibility was proposed.

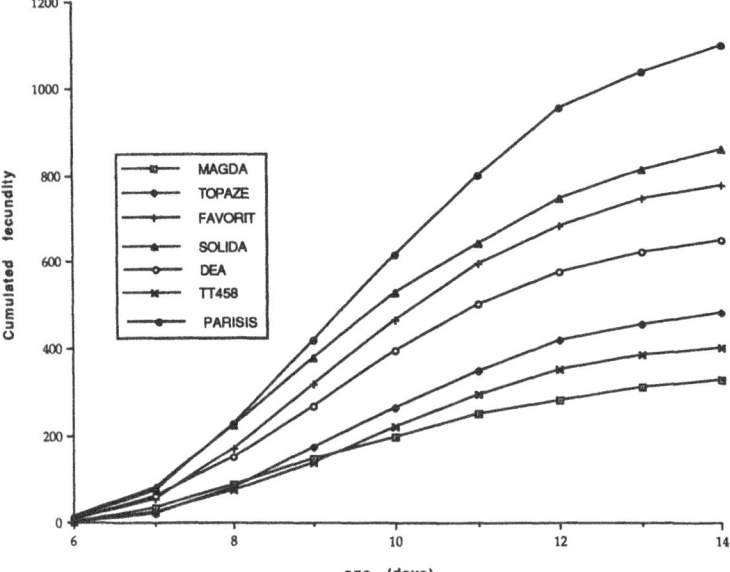

Figure 1. Cumulated fecundity of *R. padi* on 7 maize strains according to the age of the aphid females.

4. Conclusion

Many studies have already been done on the susceptibility of cereal strains to aphid (Auclair, 1989). However, few applications have been carried out in Europe. The development of a rapid method of screening the susceptibility should be a useful tool for improving this possibility. Aphids seem, however, capable to overcome plant resistance. In the future, we plan to use an RAPD-PCR method for the study of DNA polymorphism to characterize aphid strains. This should give valuable information when assessing the risk of plant resistance overcoming.

5. References

Auclair, J.L. (1989) 'Host plant resistance', in A.K. Minks and P. Harrewijn (eds.) Aphids, their biology, natural enemies and control, 2c, Elsevier, Amsterdam, pp. 225-265.

Delannoy, O. (1992) 'Étude de la sensibilité variétale du maïs au puceron *Rhopalosiphum padi*'. Thesis, Université Catholique de Louvain, Louvain-la-Neuve, Belgique.

Impe, G. van and Hance, Th. (1993) 'Une technique d'évaluation de la sensibilité variétale au tétranyque tisserand *Tetranychus urticae* (Acari: Tetranychidae), Application au haricot, au concombre, à la tomate et au fraisier', Agronomie, 13, in press.

CHARACTERISATION OF RNA-MEDIATED RESISTANCE TO TOMATO SPOTTED WILT VIRUS
IN TRANSGENIC TOBACCO PLANTS

M. PRINS
Department of Virology
Wageningen Agricultural University
Binnenhaven 11
6709 PD Wageningen
The Netherlands

ABSTRACT. Tobacco plants were transformed with different TSWV N gene constructs to obtain a generally applicable source of resistance to TSWV.

1. Introduction

Among plant viruses tomato spotted wilt virus (TSWV) is unique in its particle morphology and genome structure, and moreover, it is the only plant virus transmitted by thrips species. Virus particles are enveloped, spherically shaped and are studded with surface projections. The genome consists of three species of single stranded linear RNA (L, M, and S). The RNA segments are tightly associated with nucleoproteins to form stable nucleocapsids.

The S RNA is 2916 nucleotides (nt) long and contains the N protein (28.8K) gene and a non-structural protein (NSs, 52.4K) gene in an ambisense gene arrangement. The M RNA comprises 4812 nt and also exhibits an ambisense character. It encodes the glycoprotein precursor (127.4K) gene and another non-structural protein (NSm, 33.6K) gene. The L RNA is 8897 nt in length and is completely of negative polarity. It codes for the putative viral RNA polymerase (331.5K).

The worldwide distribution of TSWV, together with the current dramatic expansion of one of its major vectors, *Frankliniella occidentalis*, makes this virus one of the most harmful plant viruses.

More than 400 plant species are known to serve as host for TSWV and considerable yield losses have been reported in the cultivation of many important crops. Mainly due to the resistance of thrips species to insecticides, TSWV infection cannot be adequately controlled. For only a few plant species, resistance genes have been identified, which are of potential use in breeding programmes.

To obtain a more generally applicable source of resistance to TSWV, a molecular approach was undertaken. For this purpose, tobacco plants were transformed with several TSWV N gene constructs.

P. C. Struik et al. (eds.), *Plant Production on the Threshold of a New Century*, 411–412.
© 1994 *Kluwer Academic Publishers.*

2. Results

Tobacco var. SR1 was transformed with three different TSWV N gene constructs. Upon mechanical inoculation a number of plants transformed with constructs pTSWV N-A, expressing the TSWV N gene with its viral leader sequence, and pTSWV N-B, expressing the TSWV N gene preceded by a translation enhancing leader, showed high levels of N protein and high resistance.

To determine whether the observed resistance was based on the presence of N protein or on N gene RNA transcripts, the pTSWV N-B construct was modified by distortion of the ATG start codon and the translational reading frame by site-directed mutagenesis.

Levels of N gene transcripts in pTSWV N-C transformed plants were comparable with those of plants containing translationally active N genes.

Upon mechanical inoculation, four out of twenty-three lines of S1 progeny showed reduced susceptibility to TSWV. Both the ratio of transformed plants as the level of resistance were similar to results found in N-A and N-B transformation.

Furthermore, in both protein and non-protein expressing transgenic plants, the levels of resistance were increased in homozygous transgenic tobacco lines (S2), reaching 100 %.

Three homozygous lines were subsequently used to analyse their response to viruliferous thrips. In addition to protection to mechanical inoculation, all transgenic plants (N-A,N-B, and N-C) were resistant to inoculation with viruliferous thrips.

3. Conclusions

High levels of resistance to TSWV were obtained by expressing its N protein in tobacco. The obtained resistance was, at least for a major part, RNA-mediated. In homozygous (S2) plants the resistance levels were increased (reaching 100 %). Transgenic plants were also TSWV-resistant upon inoculation using viruliferous thrips.

This approach is currently applied to other important crops such as tomato, lettuce, melon, etc.

4. Reference

Peter de Haan, Jan J.L. Gielen, Marcel Prins, Ineke G. Wijkamp, Alie van Schepen, Dick Peters, Mart Q.J.M. van Grinsven and Rob Goldbach (1992) 'Characterisation of RNA-mediated resistance to tomato spotted wilt virus in transgenic tobacco plants'. Bio/Technology 10, 1133-1137.

INSECTICIDAL ACTIVITY OF *AUTOGRAPHA CALIFORNICA* NUCLEAR POLYHEDROSES VIRUS EXPRESSING DIFFERENT *BACILLUS THURINGIENSIS* CRYSTAL PROTEIN CONSTRUCTS

J.W.M. MARTENS, M. KNOESTER, F. WEYTS, A.J.A. GROFFEN,
H.J. BOSCH, B. VISSER AND J.M. VLAK
Department of Virology
Wageningen Agricultural University
Binnenhaven 11
6709 PD Wageningen
The Netherlands

ABSTRACT. Baculoviruses are insect pathogens that cause fatal disease in many Lepidopteran larvae. The major drawback in the use of these viruses as biological insect control agents is their slow speed of action. Improvement of this property is sought by introducing an insecticidal gene in the genome of the virus. The product of this new gene should interfere at an early stage with larval metabolism. Several crystal protein gene constructs of *Bacillus thuringiensis* (cryIA(b)) including a minimal toxin fragment, were introduced into the p10-locus of *Autographa californica* nuclear polyhedrosis virus (AcNPV). This gene is dispensable for the production of non-occluded virus and contains a strong promoter. The recombinant viruses expressed the crystal protein constructs at high level. The viral crystal proteins produced via AcNPV in insect cells, were highly toxic. Polyhedra produced by these p10 recombinants are being tested to measure enhancement of insecticidal action against a susceptible host (*Heliothis virescens*) as compared to polyhedra from wild-type virus.

P C Struik et al (eds), Plant Production on the Threshold of a New Century, 413

ENGINEERING OF BIOSAFE RECOMBINANT BACULOVIRUSES

RUUD M.W. MANS AND JUST M. VLAK
Department of Virology
Wageningen Agricultural University
Binnenhaven 11
6709 PD Wageningen
The Netherlands

ABSTRACT. Recombinant *Autographa californica* nuclear polyhedrosis viruses (AcNPV) were constructed to reduce their persistence in the field. A two-step cotransfection procedure was used to generate specific deletions in the genome of AcNPV. Used as an insecticide, such viruses are likely to cause less danger and damage to the environment when heterologous genes are introduced.

1. Introduction

Insect pests can be controlled through release of pathogenic baculoviruses into infested ecosystems. The use of these naturally occurring viral insecticides has the advantages of being environmentally safe, ecologically sound and harmless to non-target animals, including beneficial insects and man. However, their slow speed of action has restricted a widespread commercial use. This limitation may be eliminated by introducing genes for toxins, hormones or enzymes into the viral genome which are deleterious to the insect metabolism. Inherent to this approach, however, is the risk that the genetically engineered baculovirus will acquire undesirable properties like extension of its host range to include non-target insects. The aim of this study was to reduce the potential risk of the field release of genetically modified baculoviruses by reducing the persistence and spread of the virus in the field by deletion mutagenesis.

2. Research approach

Several genes of *Autographa californica* nuclear polyhedrosis virus (AcNPV) have been described whose inactivation reduces the persistence of the virus and are not essential for viral replication. The phosphorylated polyhedral envelope protein pp34 is involved in the morphogenesis of the polyhedral envelope [4]. Deletion of the pp34 gene generates virus particles with a higher sensitivity to alkali [6]. More

415

readily degradable viruses can, therefore, be anticipated in the field. The ecdysteroid UDP-glucosyl transferase gene (*egt*) inhibits moulting of the infected host [2]. Viruses without the *egt* gene yield less progeny and need less time to kill their host [3], two beneficial character-istics of a baculoviral insecticide. One of the functions of protein p10 is to release polyhedra from infected cells by disintegration of the nucleus [5,1]. Viruses with a deletion of the p10 gene can, therefore, be expected to have a reduced spread in the field. AcMNPV viruses lacking one or more of these genes were isolated and analysed.

Due to the fact that viruses containing deletions in these genes cannot be detected phenotypically, the gene of interest is inactivated first by insertion or substitution of a marker gene. This marker gene, the bacterial β-galactosidase gene (lacZ), allows convenient screening of recombinant viruses by blue colouring of their plagues using the chromogenic indicator X-gal. Subsequent deletion of the lacZ gene and, if still necessary, the coding sequence of the gene leads to the desired deletion mutant that, consequently, gives rise to colourless plagues. Deletion of the gene thus requires a two-step procedure.

3. Preliminary results

Preliminary experiments determining the LD_{50} and ST_{50} values of the double and the triple deletion mutants showed that the deletion mutants lost some of their infectivity reflected by higher LD_{50} values as compared to the *wild-type* AcMNPV (strain E2). The first ST_{50} values showed a more rapid mortality in viruses that had the *egt* gene deleted. P10 deletion mutants seemed to have lost the ability to disintegrate the nucleus of the cell, shown by polyhedra retained in the nucleus by phase-contrast microscopy.

This research is carried out in the framework of BRIDGE project Biot-CT91-0291 (Biosafety of genetically modified baculoviruses for insect control) in collaboration with the NERC Institute of Virology and Environmental Microbiology, Oxford, UK and the Federal Biological Research Center for Agriculture and Forestry (Institute for Biological Pest Control), Darmstadt, FRG.

4. References

[1] Oers, M.M. van , Flipsen, J.T.M., Reusken, C.B.E.M., Sliwinsky, E.L., Goldbach, R.W. and Vlak, J.M. (1993) 'Functional domains of the p10 protein of *Autographa californica* nuclear polyhedrosis virus', J. Gen. Virol. 74, 563-574.

[2] O'Reilly, D.R. and Miller, L.K. (1989) 'A baculovirus blocks insect molting by producing ecdysteroid UDP-glucosyl transferase', Science 245, 1110-1112.

[3] O'Reilly, D.R. and Miller, L.K. (1991) 'Improvement of a baculovirus pesticide by deletion of the *egt* gene', Bio/Technology 9, 1086-1089.

[4]   Whitt, M.A. and Manning, J.S. (1988) 'A phosphorylated 34-kDa protein and a subpopulation of polyhedrin are thiol linked to the carbohydrate layer surrounding a baculovirus occlusion body', Virology 163, 33-42.

[5]   Williams, G.V., Rohel, D.Z., Kuzio, J. and Faulkner, P. (1989) 'A cytopathological investigation of *Autographa californica* nuclear polyhedrosis virus p10 gene function using insertion/deletion mutants', J. Gen. Virol. 70, 187-202.

[6]   Zuideman, D., Klinge-Roode, E.C., Lent, J.W.M. van and Vlak, J.M. (1989) 'Construction and analysis of an *Autographa californica* nuclear polyhedrosis virus mutant lacking the polyhedral envelope', Virology 173, 98-108.

THE INFLUENCE OF TREES IN PARKLANDS ON SORGHUM YIELDS

JAN-JOOST KESSLER
Department of Forestry
Wageningen Agricultural University
P.O. Box 342
6700 AH Wageningen
The Netherlands

ABSTRACT. Sorghum grain yields under karité and néré trees are reduced by an average of 50 % and 70 % respectively, in comparison with yields in the open field. Soil fertility, limiting primary production in the Sahel region, is at least as favourable under the tree canopies as in the open field. Reduced light intensity, to a minimum of 20 % under the néré canopy, is probably largely responsible for low sorghum production under the tree canopies. Benefits from the tree products are more valuable than losses in cereal yields, explaining why trees are maintained on the agricultural fields. Pruning of tree branches, selection of (shade-) crops and tree selection could reduce crop yield losses but cannot be expected to both increase tree- and crop production.

1. Introduction

Scattered trees are characteristic of a large part of the African agricultural landscape. The (farmed) parklands may be defined as the regular presence of well-grown trees on cultivated and recently fallowed land. The trees are associated with the agricultural environment because of their specific products. The study was carried out in the Sudan zone in Burkina Faso, where average annual rainfall is 900 mm. Here, dominant trees in the parklands are karité (*Vitellaria paradoxa*) and néré (*Parkia biglobosa*), whereas sorghum is the main cereal crop. Karité nuts are made into butter and néré fruits are made into a highly nutritious and preferred spice. There are indications that the number of trees in the parklands is declining. Whereas trees in agroforestry systems are commonly suggested to have advantages only, farmers claim that the trees suppress cereal yields, which may be one reason to reduce their densities in parklands. This study investigated the influence of trees on sorghum yields, the ecological factors involved, and the socio-economic consequences.

P. C. Struik et al. (eds.), *Plant Production on the Threshold of a New Century,* 419–421.
© 1994 *Kluwer Academic Publishers.*

2. Methodology

Of 18 selected karité and néré trees canopy parameters were measured, and crop production and site characteristics were investigated in plots along transects in the four cardinal directions from the trunk, and with a total length twice the distance from the trunk to the limit of the canopy. In each direction two plots were located under the canopy, one at the edge of the canopy, and two in the open field. Per plot, as site characteristics were measured soil fertility parameters (nutrient and organic matter concentrations and pH), as well as soil humidity and light intensity. Light intensity per plot resulted from measurements throughout the year at different times of the day. Agricultural practices and tree management practices were closely monitored.

3. Results

Most soil properties determining soil fertility were better under the tree canopies than in the open field. Soil humidity under the tree canopies was also better, as a result of the stabilising influence of trees on the micro-climate (reduction of evapotranspiration). However, light intensities under the tree canopies were much lower than in the open field.

There was a significant correlation between sorghum yields and light intensity per plot suggesting that light availability was the major factor responsible for lower sorghum yields under the tree canopies. Important factors determining the extent of shading by a tree were the height of the lowest branches, the shape of the canopy, the leaf density and the phenological pattern. The néré canopy provided most shade but under a pruned néré tree sorghum yields were suppressed much less (Figure 1).

In a parkland with average tree densities of 10 large trees per hectare the reduction of sorghum yields due to shading would be about 6 %. However, the benefits of the tree-products in terms of monetary incomes are about twice as high as the sorghum losses, indicating the rationality of the existing system from the farmer's point of view. Moreover, the trees have other advantages, such as wood production and risk reduction in dry years.

4.   Discussion and further research

Improved soil fertility under trees is not limited to the well-known *Faidherbia albida*. Improved yields under this tree, as a result of the unique phenological pattern of leaf shedding in the summer, could therefore also be obtained under other trees if they would be adequately pruned. However, trees also utilise nutrients and water and competitive effects between trees and crops cannot be excluded, in spite of more favourable conditions under the tree canopies. Long-term and large-scale studies are required to obtain more insight into the processes that lead to soil fertility enhancement under tree canopies. The available

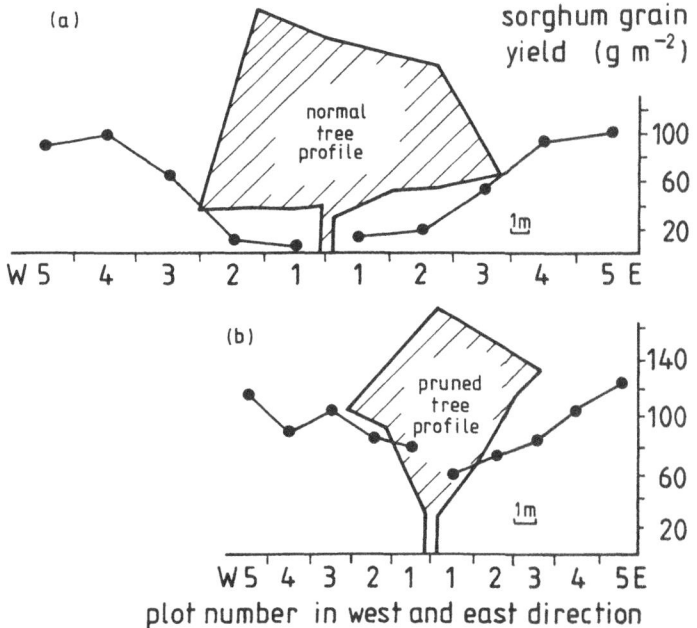

Figure 1. Sorghum yields in relation to the distance from (a) a normal and (b) a pruned néré tree.

evidence suggests that trees do not enrich the agro-ecosystem, but concentrate resources for their own benefit mainly. The expectations of soil improving effects of trees (e.g. *Faidherbia albida*) have so far not been justified by field results. Research should focus on the tree - crop interface of present agrosilvicultural systems and to the possibilities of improved management of the trees. It should be realised that farmers are generally very well aware of the interactions between specific trees and crops, even in relation to soil types and rainfall. Instead of repeating similar studies for various tree-crop combinations, more use could be made of available farmers' knowledge.

MOLECULAR MECHANISMS OF PHOTOINHIBITION, AN ABIOTIC STRESS FACTOR LIMITING PRIMARY PLANT PRODUCTION

VICTOR B. CURWIEL, GERT SCHANSKER, OSCAR J. DE VOS
AND JACK J.S. VAN RENSEN
Department of Plant Physiology
Wageningen Agricultural University
Arboretumlaan 4
6703 BD Wageningen
The Netherlands

ABSTRACT. Molecular mechanisms of photoinhibition are clarified on the basis of experiments with leaves and isolated chloroplasts of peas and triazine-resistant and susceptible *Chenopodium* plants.

1. Introduction

Photoinhibition (PI) is a process which occurs when a plant is exposed to light irradiances higher than those which it can convert or dissipate without harm. PI manifests itself as a reduction in the rate of photosynthesis: it can lead to a decrease in quantum yield of up to 50 %, thus limiting primary plant production.

The damage by PI is usually ascribed to reactions associated with photosystem II (PSII) of photosynthesis. Especially, the D1 protein of PSII seems to be affected. We investigated molecular mechanisms of PI in leaves and isolated chloroplasts of peas and triazine-resistant and susceptible plants.

2. Results

Triazine-resistant and susceptible *Chenopodium album* plants were grown at low and high irradiance. At the lower irradiance the dry matter production of the resistant and the susceptible plants were almost similar. At the higher irradiance the resistant biotype had a significantly lower production.

In vitro electron transport dependent on photosystem I (PSI) alone was much less affected by PI than that dependent on both photosystems PSI and PSII. There was a smaller difference in susceptibility to PI between the photophosphorylation activity dependent on PSI alone and the one dependent on both PSII and PSI. Since photophosphorylation activity always decreased faster upon PI than the rate of electron transport, we

423

P. C. Struik et al. (eds.), Plant Production on the Threshold of a New Century, 423–425.

conclude that photoinhibition causes a gradual uncoupling of electron transport with phosphorylation. Since the extent of the light-induced proton gradient across the thylakoid membrane decreased upon PI, it is suggested that photoinhibition causes a proton leakiness of the membrane. We have found no significant differences in response to PI of the various reactions measured in chloroplasts isolated from triazine-resistant and susceptible plants. We have also not observed any significant differences in response to PI of the photophosphorylation reactions in chloroplasts of plants grown under low irradiance, compared with those grown under high irradiance. However, the electron transport reactions in chloroplasts from plants grown under low irradiance appeared to be somewhat less sensitive to PI than those grown under high irradiance.

Fluorescence studies *in vivo* showed that the photochemical yield and PSII electron transport rate were lower in the resistant plants. Also in most cases the photochemical and non-photochemical quenching of fluorescence were lower in the resistant (R) type, indicating a smaller ability of the R-type to dissipate an abundance of light energy leading to damage of the photosynthetic apparatus. It could be demonstrated in intact leaves that the lower productivity of the R-type is caused by a higher sensitivity to PI. The differential effect of photoinhibition *in vivo* and *in vitro* may be caused by the fact that *in vitro* energy conversion by photosynthesis, energy dissipation by protective mechanisms (e.g. photorespiration, Mehler reaction) and recovery processes do not function or function less efficiently. The higher sensitivity of R-types *in vivo* must be due to one or more of these processes, which are all protective against photoinhibition.

Silicomolybdate (SiMo) is an electron acceptor that has many characteristics, the ignorance of which makes an interpretation of the results quite troublesome. In photoinhibition experiments the photosystem II (PSII) activity can be best monitored if 1 μM 2,5-dibromo-3-methyl-6-isopropyl-p-benzoquinone (DBMIB) is added after the photoinhibitory treatment and SiMo is added in the light. Diuron (DCMU) may complicate interpretation of the results as it is also a competitive inhibitor of SiMo binding at pH 7.6. The binding niche of SiMo is located at the stroma side between the parallel helices of the D1 and D2 proteins of PSII close to the non-heme iron.

The whole chain activity was much more affected by the photo-inhibitory treatment than the PSII activity itself. Uncoupling of electron flow by addition of ammonium chloride accelerated the rate of photoinhibition. Photoinhibitory treatment decreased not only the Hill reaction activity at photon saturation, but also decreased the quantum yield and increased the photon flux density yielding half maximum rate of electron flow (K_m). Decrease of quantum yield indicates that the photochemistry of PSII was affected; increase of K_m indicates a conformational change of the SiMo binding site. In experiments on PSII activity monitored with SiMo, DCMU had no protective effect on the damage of the electron transport chain between water and Q_A.

References

Curwiel, V.B. and Rensen, J.J.S. van (1993) 'Influence of photo-inhibition on electron transport and photophosphorylation of isolated chloroplasts'. Physiol. Plant. (in press).

Curwiel, V.B., Schansker, G., Vos, O.J. de, and Rensen, J.J.S. van (1993) 'Comparison of photosynthetic activities in triazine-resistant and susceptible biotypes of *Chenopodium album*'. Z. Naturforsch. 48c, 278-282.

Schansker, G. and Rensen, J.J.S. van (1992) 'Evaluation of the use of silicomolybdate as an electron acceptor for photosystem 2 in photo-inhibition research'. Photosynthetica 27, 145-157.

Schansker, G. and Rensen, J.J.S. van (1993) 'Characterization of the complex interaction between the electron acceptor silicomolybdate and photosystem II'. Photosynth. Res. (in press).

COMPARISON OF PHOTOSYSTEM II ELECTRON TRANSPORT AND OXYGEN EVOLUTION IN TWO MARINE ALGAE

C. GEEL AND J.F.H. SNEL
Department of Plant Physiology
Wageningen Agricultural University
Arboretumlaan 4
6703 BD Wageningen
The Netherlands

ABSTRACT. We compared electron transport determined by the saturating pulse method with the oxygen evolution of samples of two marine algae, *Tetraselmis spp.* and *Rhodomonas spp.*, grown in continuous-batch cultures. The relation between oxygen evolution and electron transport is linear over a range of light intensities from 0 to 500 μmol m^{-2} s^{-1}. The fluorescence measurement is quite easy and could be made suitable for application *in situ*, which is a large advantage to the measurement of oxygen evolution and ^{14}C incorporation.

1. Introduction

Chlorophyll fluorescence can give information about the physiological state of photosystem II. The actual efficiency of photosystem II (Φ_P) is determined in situ by measuring the fluorescence yield in the light adapted state (F) and during saturating illumination (F_M'):
$\Phi_P = (F_M' - F)/F_M'$ (Genty *et al.*, 1989). The rate of electron transport (J_e) is defined by the product of the absorption cross-section of photosystem II, Φ_P and the photon flux density:
$J_e = \sigma_{PSII} * \Phi_P * PFD$. In the short measurements here, we assume that the absorption cross-section of photosystem II is identical in all measurements.

We compared oxygen evolution to electron transport in *Tetraselmis spp.* and *Rhodomonas spp.* with a different organised photosystem. Both algae contain core chlorophyll complexes next to the reaction centre in the membrane. Besides *Tetraselmis* contains light harvesting complexes based on chlorophyll, absorbing blue and red light and located in the membrane. *Rhodomonas* contains phycobilisomes, light harvesting complexes which absorb green light and are located on the outside of the membrane. We were interested whether this would influence the relation between oxygen evolution and efficiency of photosystem II.

427

P C Struik et al (eds), Plant Production on the Threshold of a New Century, 427–429
© 1994 *Kluwer Academic Publishers*

2. Materials and methods

The algae were grown in continuous-batch cultures at 20 °C under a PFD of 200 μmol m^{-2} s^{-1}. Fluorescence and oxygen evolution were measured simultaneously in a DW2/2 cuvette (Hansatech, Norfolk, UK) with the LS2 actinic light source and a PAM fluorometer (Walz, Effeltrich, FRG). Samples were concentrated by centrifugation and dark adapted for 30 minutes. The cells were adapted to each PFD for 5 minutes before the steady state was reached. For each PFD a new sample was taken.

3. Results

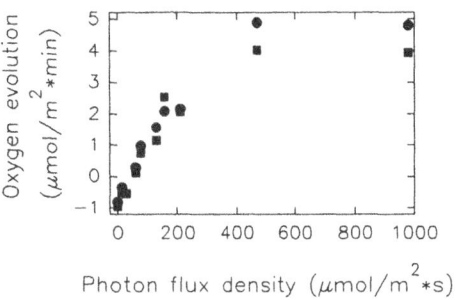

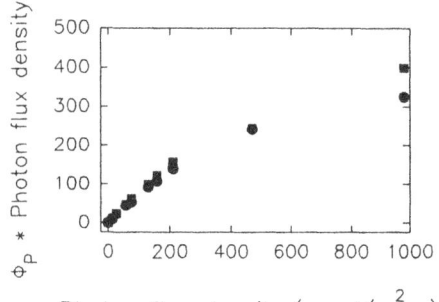

Figure 1. Oxygen evolution of *Tetraselmis* (•) and *Rhodomonas* (■) as a function of the photon flux density.

Figure 2. Estimated electron transport as a function of the photon flux density. Symbols as in Figure 1.

The light response curve of oxygen evolution (Figure 1) had the shape of a quasi-Blackman curve (Leverenz, 1987). From 0 to 200 μmol m^{-2} s^{-1} oxygen evolution changed linearly with PFD. From the slope the quantum yield for oxygen evolution can be calculated. Above 500 μmol m^{-2} s^{-1} oxygen evolution had reached a maximum indicating light saturation of oxygen evolution.

The rate of electron transport (Figure 2) also had the shape of a Blackman curve but the maximum for Φ_P*PFD was not reached at the highest PFD.

The rate of electron transport and the oxygen evolution show a parallel behaviour at low photon flux densities resulting in a linear relationship (Figure 3). At high PFD's a discrepancy occurred. Electron transport was not fully saturated as oxygen evolution was. This could be caused both by photorespiration and Mehler reaction (Edwards *et al.*, 1983). Both reactions wasted energy in oxygen consumption at high photon flux densities.

These two species behaved in the same way under our conditions. Although regulation of photosynthesis is organised differently in these two species, the relation between electron transport and oxygen evolution was similar.

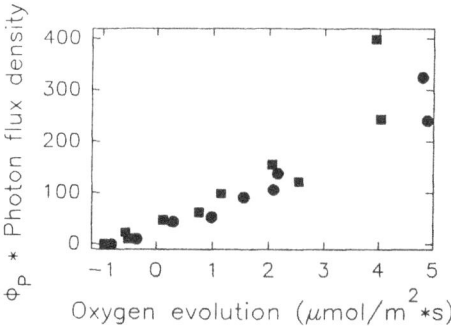

Figure 3. Estimated electron transport as a function of the oxygen evolution. Symbols as in Figure 1.

4. References

Edwards, G. and Walker D.A. (1983) 'C3, C4: Mechanisms, and cellular and environmental regulation of photosynthesis'. Blackwell Scientific Publications, Oxford.

Leverenz, J.W. (1987) Physiol. Plantarum 71, 20-29.

Genty, B., Briantais, J.-M. and Baker, N.R. (1989) Biochim. Biophys. Acta 990, 87-92.

SIMULATION OF TREE GROWTH WITH THE PINOGRAM AND TREEARCH PROGRAMMES

R.P. LEERSNIJDER
Department of Forestry
Agricultural University Wageningen
Gen. Foulkesweg 64
P.O. Box 342
6700 AH Wageningen

ABSTRACT In silviculture growth potentials and conditions necessary for establishment and growth need to be known in order to forecast behaviour of trees under different circumstances. In forests we find an infinite number of situations. Each situation has its own effect on the growth potential of trees. In order to forecast growth and development of trees, models need to be created for different purposes and at different hierarchical levels. This paper discusses a model on stand level and a model on tree level. The first (the PINOGRAM model) simulates growth, competition and appearance of individual Scots pines in monoculture influenced by different planting and thinning regimes (Leersnijder, 1992). The second simulates the branch pattern of Scots pine and poplar at different ages on different sites.

1. The PINOGRAM model

1.1. BASIC CONCEPTS

When trees are released in one or more directions, competition will decrease in these directions and the area used by a tree, the *growth area*, will increase. In fact growth area is the major factor governing competition. Growth area and height of a tree define the *growth volume* (the three dimensional space in which a tree lives). Growth volume and age (or history) define the potential tree appearance.

1.2. DATA PROCESSING AND RESULTS

Growth and dynamics of 158 individual Scots pine trees are studied in 13 stands on the Veluwe, the Netherlands. Ages varied from 8 to 111 years. Height, dbh, growth area, crown length and crown projection of each tree were measured and of some trees also the stem volume and mean branch diameter in the lower part of the crown. By means of non-linear multiple regression growth equations for diameter, crown length and crown width

431

P C Struik et al (eds), Plant Production on the Threshold of a New Century, 431–434
© 1994 *Kluwer Academic Publishers*

are derived, dependent on age, height and growth area. The growth formulas derived are well suited to construct a growth model. The model was used to create the PINOGRAM programme, which is an interactive simulation programme written in "C-language". With this model growth of individual trees can be simulated by giving planting distance, site-index value (= S-value) per tree, age and thinning regime. The model shows stems and abstract forms of crowns in a stand of 20 * 50 meter in a three dimensional way and so, it gives a good impression of the possible forest structures. Figure 1 shows a side-view and a three dimensional design of a 40 years old stand.

2. The TREEARCH model

2.1. BASIC CONCEPTS

The TREEARCH programme was developed to describe and show the number, length, diameter, and spatial distribution of shoots of trees of different species given their height, age and height to the living crown. No external factors are involved. This model is a stochastic model. The base unit is the annual shoot. The model was built using mathematical equations based on correlations between different variables. Dependent variables are estimated within a 95 % confidence range. The model calculates the results using a random deviation of the estimated average within this confidence range.

2.2. DATA PROCESSING AND RESULTS

Two 20 years old poplars (*Populus canadensis* "Robusta") were measured in the "Horsterwold" (S. Flevoland) and one twenty years old Scots pine (*Pinus sylvestris* L.) in the "Speulder en Sprielder Bos" (province Gelderland). Data of poplar were collected on 2 stems, on 3 branches of order 1, 3 branches of order 2, 3 branches of order 3 and 2 branches of order 4. Data of Scots pine were collected on 1 stem, on 6 branches of order 1, 5 branches of order two and 1 branch of order 3. The branches were collected from the bottom, the centre and the top of the crown. Also tree height and height to crown were measured. Per order the next data were collected:
- the age of the branch order,
- the type of the branch order (long shoot, short shoot),
- the state of the branch (dead or alive),
- the position of the branch order along its parent order,
- the angle of the branch with its parent order,
- the rotation of the branch around its parent order,
- the length of the branch order,
- the diameter of the branch order,
- the number of child branches per parent order.

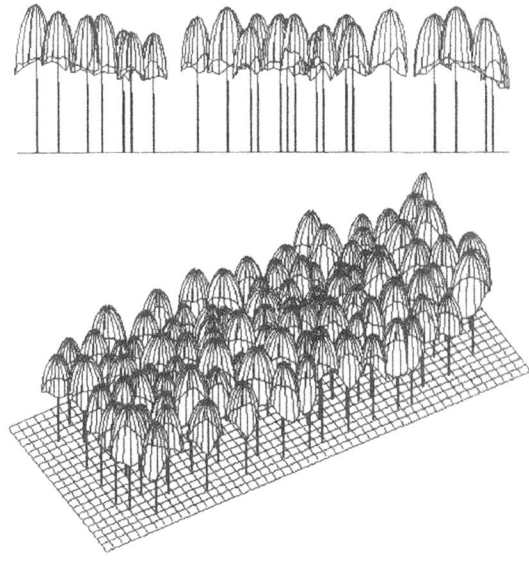

Figure 1. Side-view and three dimensional design of a simulation of a 40 years old Scots pine stand with the PINOGRAM programme.

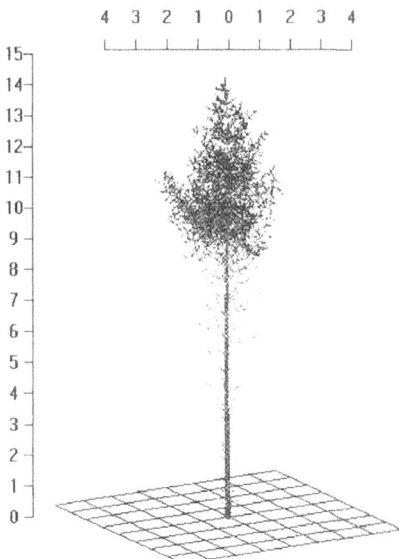

Figure 2. Three dimensional design of a simulation of a 28 years old Scots pine with the TREEARCH programme.

Age, order, type, state, position, angle, rotation, length, diameter and number of children are in fact the variables that the TREEARCH model needs to build a tree. The TREEARCH programme calculates these variables for each shoot. The equations were found by non linear regressions between the variables. Since only a very small number of trees were measured the equations and parameters found are only rough estimations and their use is in principle not permitted to trees of other ages and other areas. Figure 2 shows a simulated 28 years old Scots pine.

3. Future research

Both the models PINOGRAM and TREEARCH will be improved when data are included collected on more trees of different ages, species and provenances on different well-defined sites (i.e. different soils, climates, thinning regimes, pollution, etc.).

At the moment a model is developed to show the growth of some tree species of the tropical rain forest in French Guyana. This model shows great resemblance with the TREEARCH model, but includes growth and mortality of shoots, dependent on light conditions.

Next year it is intended to start creating a model for mixed and uneven aged stands, which is quite similar to the PINOGRAM model.

4. Reference

Leersnijder, R.P. (1992) 'PINOGRAM; a Pine Growth Area Model'. Ecological Modelling 61, 1-152.

MONITORING OF GAS EXCHANGE AS RELATED TO SEASONAL CHANGES IN
MICROCLIMATE OF A KIWIFRUIT VINE

F. SUCCI, E. MAGNANINI
Dept. of Hort. and Forestry
Via Filippo Re 6
40126 Bologna
Italy

G. COSTA
Dept. of Plant Production
and Crop Management
Via Fagagna 208
33100 Udine
Italy

ABSTRACT. The aim of this study was the monitoring of some climatic
parameters and gas exchanges of a kiwifruit vine (*Actinidia deliciosa*)
to examine the seasonal changes in microclimate-plant interactions. The
data were collected by an apparatus which is composed of a polyethylene
chamber housing the whole canopy, some sensors for air temperature,
global solar radiation, water vapour content and a gas analyser (IRGA)
for CO_2 concentration measurements. The results represent a comprehensive
examination of the effects of microclimate on gas exchanges. Due to the
fact that microclimatic variations are characteristic of a short period
in relation to the considered growing season, and are randomly
distributed, they do not affect the CO_2 accumulated quantity. The average
daily CO_2 fixed by the vine is not constant but it gradually increases
from May to October (the considered growing season).

1. Introduction

Gas exchanges in kiwifruit are affected by light intensity, air temperature, air CO_2 concentration, air relative humidity, and canopy
architecture.
    The aim of this study was to determine the behaviour of gas
exchanges on a whole canopy basis as related to microclimate during the
entire growing season.

2. Materials and methods

The experiment was carried out on a kiwifruit vine included in a Hayward
orchard with 10 years old vines. The vines were T-bar trained, South-
East North-West oriented with the following distances: 4.0 x 4.5 m. The
total leaf area was 31 m^2 and the LAI was 2.5. There was hail protection
(nets, 15 % shading). The average yield was 30 t/ha. An open chamber

435

P. C. Struik et al. (eds.), Plant Production on the Threshold of a New Century, 435–436.
© 1994 *Kluwer Academic Publishers.*

made of thin polyethylene film housing the whole canopy vine was used to monitor gas exchange and microclimate parameters.

3. Results

While the total radiation tends to gradually decrease from May to October, the available energy is constantly increasing. Also the temperature follows the radiation trend.

The transpiration and the amount of transpired water trends followed in general the total radiation and the available energy trends.

The CO_2 average amount fixed by the vine during the day was not constant but it varied in relation to the main microclimatic parameter variations. Since the microclimatic variations are typical of a short period and randomly distributed, they do not strongly affect the CO_2 accumulation rate, which gradually increased from May to October. The vine accumulated 26 kg ($\pm$ 10 %) CO_2 during a 150 days period.

4. Discussion

The results defined the modification of the values of photosynthesis and transpiration during the growing season and represent the starting point for further modelling studies devoted to find the best canopy architecture for maximizing light interception and improving fruit quality and productivity.

5. Acknowledgements

The research work was financially supported by M.U.R.S.T (Ministry of University and Scientific Research), Special Grant 40 %.

6. References

Succi, F., Magnanini, E. and Costa, G. (1990) 'Interrelations of light, transpiration and CO_2 flux exchange in kiwifruit vine', Proceedings 3rd International meeting on "Regulation of photosynthesis in fruit crops", pp. 61-69.

PLANT WATER STATUS STUDIES AND SIMPLE FIELD SCREENING METHODS FOR SELECTING DROUGHT TOLERANT GENOTYPES IN WHEAT (*TRITICUM AESTIVUM* L.)

LASZLO CSEUZ AND LASZLO ERDEI*
Cereal Research Institute
P.O. Box 391
H-6701 Szeged, Hungary
*   Institute of Biophysics*
  Biological Research Centre
  P.O. Box 521
  H-6701 Szeged, Hungary

ABSTRACT. Water retention ability of excised wheat leaves and post-anthesis stress induced by chemical desiccation were assessed in the nursery to study a wide range of breeding material's cuticular resistance against water loss and translocation ability of stored stem reserves. Using the measured water potential and relative water content data, pressure-volume curves were drawn and osmotic potential and water saturation deficit at zero turgor were calculated. Genotypes performing well in the field tests, had lower water potential (osmotic potential) values and reached the zero turgor point at a higher water saturation deficit Correlation between the ranking of entries in the different tests, however, was low.

1. Materials and methods

Water retention ability was tested by the determination of the fresh weight of 20 excised leaves of 78 genotypes harvested from the field early in the morning. After keeping them in a controlled environment room for 24 hours their weight was measured again and after a total desiccation the dry weight of the leaves was determined. From these data the total loss of initial water content could be defined. Chemical desiccation tests were done to evaluate the translocation ability of the stem reserves in 114 genotypes. Desiccant spraying was done 14 days after anthesis of each entry. Kernel weight reduction due to the post-anthesis stress was assessed by comparing treated and control plots for each entry.

The water relation characteristics of 16 genotypes of different drought tolerance grown in the field were measured by a Scholander pressure bomb on flag leaves.

P C Struik et al (eds), Plant Production on the Threshold of a New Century, 437–438

2. Results and discussion

As in earlier investigations (Blum, 1988; Clarke, 1983) significant differences were found among the tested genotypes in both field tests. The water loss of excised leaves ranged between 35 and 71 % among the 78 genotypes tested, (LSD=13.9).

Depression in thousand kernel mass due to the desiccant spraying was between 15 and 56 % of the untreated control plots (LSD=12.6).

From the pressure-volume technique total water potential, osmotic potential at zero and full turgor, water saturation deficit at zero turgor and turgid weight/dry weight ratio (TW/DW) were calculated. Among the water relation characteristics listed, water saturation deficit and osmotic potential at zero turgor showed the largest differences among the genotypes tested. Also, these traits had the highest correlation with the results of both field tests and earlier field records. No correlation was found between the TW/DW ratio data and other characteristics.

Genotypes that showed good water retention ability (NE 83/T, Tiszataj m, Mv 8) or resistance to post anthesis drought stress in field tests, (OK 84343, Tiszataj m, Pitic 62) had generally lower water potential (osmotic potential) values and reached the zero turgor point at a higher water saturation deficit.

The pressure-volume technique gave less variable results than the field screening methods, so in spite of its relative slowness, it could be a convenient complementary screening tool in wheat breeding.

3. References

Blum, A. (1988) 'Plant breeding for stress environments', CRC Press, Inc. Boca Raton, Florida.

Clarke, J. (1982) 'Excised-leaf water retention capacity as an indicator of drought resistance of Triticum genotypes', Can. J. Plant Sci. 62, 571-576.

A STRATEGY FOR THE BREEDING OF ELITE FAMILIES OF OIL PALM (*ELAEIS GUINEENSIS* JACQ.)

I. BOS AND C.J. BREURE[1]
Department of Plant Breeding
Wageningen Agricultural University
P.O. Box 386
6700 AJ Wageningen
The Netherlands
[1] *Home address: Bowlespark 12*
   6701 DP Wageningen, the Netherlands

ABSTRACT. A strategy for effective breeding of elite oil palm families is outlined. The strategy consists of three steps: (i) apply mass selection among *dura* and among *pisifera* palms, (ii) apply additional selection on the basis of progeny tests, (iii) identify, among the families obtained by intercrossing of the selected palms, elite families and apply clonal propagation of their parents for large scale production of elite families.

1. Introduction

A three-step selection strategy to improve exploitation of the potential increase of oil-and-kernel yield per hectare (Y) in oil palm (*Elaeis guineensis* Jacq.) is outlined:
(i)   Apply, on the basis of phenotypic values, mass selection in a source population consisting of *dura* palms as well as in a source population consisting of *pisifera* palms.
(ii)   Apply additional selection among these palms on the basis of a progeny test, i.e. on the basis of the breeding values.
(iii)   Cross the palms selected in step (ii) in all *dura* x *pisifera* combinations and test the families obtained to identify the elite families. The parents of the elite families are clonally propagated for large scale production of the elite families.
It will be shown how this strategy can be optimized on the basis of statistical considerations and specific quantitative genetic information.

2. STEP (i): selection on the basis of the phenotypic value

Prospective female parents are selected in a source population consisting of *dura* palms on the basis of their phenotypic value for

P. C. Struik et al. (eds.), Plant Production on the Threshold of a New Century, 439–441.

traits like: magnesium status (absence of symptoms of magnesium deficiency on the leaves), components of oil-and-kernel extraction rate or yield (possibly over many years). Simultaneously, male *pisifera* parents are selected. This selection may be done in crosses between well performing *tenera* palms and concerns number of spikelets. Superior palms with high values for height increment should not be selected because their superiority may be due to a superior ability to compete for light.

3. STEP (ii): selection on the basis of breeding value

The *pisifera* palms are crossed onto *dura* palms. Because of limited manpower unbalanced incomplete factorial mating designs are usually applied. For a certain intended number of crosses, involving a given number of *dura* females and a given number of *pisifera* males, incomplete partially balanced mating designs, with the smallest range and the lowest average for the standard error of the difference between the offspring of two *dura* parents or two *pisifera* parents, are recommended. (The residual standard deviation depends furthermore on: the considered trait, the experimental site, the plot size and the number of replications).

The progeny test, which may consist of six 4-palm plots for each family, yields mean data per family. Each mean value can be partitioned into two general combining ability (=GCA) values, of the *dura* and the *pisifera* parent, and a specific combining ability (=SCA) value. For many traits the relative contribution of GCA to the total genetic variance is much higher than the proportion due to SCA, see Breure and Bos (1992). For these traits the expected performance of families, that were not actually produced, can reliably be predicted from the (parental) GCA-values.

If μ represents the population mean, the total relative progress (in %), possible when exploiting all genetic variation, is approximately equal to $200(\sigma_{GCAdura} + \sigma_{GCApisifera} + \sigma_{SCA})/\mu$; see Breure and Bos (1992). On the basis of contributions to this total one may decide whether the emphasis of the selection should be on the female parents, on the male parents or on pairs of parents. The narrow sense heritability indicates how efficiently this potential progress can be achieved.

The selection for a high value for Y can be supported by additional selection for a high harvest index (=HI). Relevant characters for indirect selection for Y or HI can be identified by means of multiple linear regression of Y or HI on GCA-values for a number of predictors (like: growth parameters, magnesium status). Thus it may come out that an increased leaf area ratio, a better magnesium status and a reduced frond production (or a reduced vegetative dry matter production) are to be pursued if it is aimed to increase Y and/or HI of *dura* x *pisifera* families.

Parents are selected on the basis of their progenies, i.e. their GCA-values ($\approx$ breeding values).

4. STEP (iii): production of elite families

All possible *dura* x *pisifera* crosses among the palms selected in step (ii) are made. The families obtained are tested to identify the (parents of) the best families. A mixture of the progenies tested in step (ii) can be included in the progeny test as a check.

The recently introduced practical possibility of clonal reproduction allows large scale production of the proven elite families. Ideally, the clones corresponding to the parents of elite families should be flowering by the time their progeny performance is evaluated. Cloning of all palms involved in the step (iii) progeny test should, therefore, already start by the time the progeny test is planted in the nursery.

The elite families may be used as a source population for selection of *tenera* palms, which are asexually propagated to produce commercial clones.

5. Reference

Breure, C.J. and Bos, I. (1992) 'Development of elite families in oil palm (*Elaeis guineensis* Jacq.)'. Euphytica 64, 99-112.

POTENTIAL OF TRITORDEUM AS A NEW CEREAL CROP

J. BALLESTEROS, D. RUBIALES AND A. MARTÍN
Instituto de Agricultura Sostenible
Apdo. Correos 3048
14080 Córdoba
Spain

ABSTRACT. Tritordeum is an amphiploid between *Hordeum* and *Triticum*, that shows potential as a new cereal crop. There has been a quick response to selection giving rise to yields close to those of commercial wheats after only 12 years of breeding from material with a narrow genetic basis. More effort still is needed to improve some agronomic characteristics. New primary tritordeums are being produced and that variability is introduced into the advanced tritordeum lines.

1. Introduction

Interspecific and intergeneric hybridization is a common tool for widening the genetic basis of crops, and occasionally the result of these crosses could be the starting point of new crops. In the tribe *Triticeae* both results have been obtained. Alien variation from wild species have been used successfully for breeding, mainly wheat, but also a new crop, triticale, has been produced after crossing wheat with rye. Intergeneric hybrids between wheat and barley are abundant, although fertile amphiploids have only been obtained when using wild *Hordeum* species. The amphiploid *Hordeum chilense* x *Triticum turgidum* (AABBH^{ch}H^{ch}) shows characteristics of a potential crop (Martin and Sánchez-Monge Laguna, 1982). We have called this amphiploid tritordeum (x*Tritordeum* Ascherson et Graebner).

2. Results

From the beginning, primary hexaploid tritordeums showed unexpected good fertility and chromosomal stability (Martin and Cubero, 1981) for an intergeneric amphiploid and a plant morphology similar to that of wheat. Interesting was the high protein content which was not due to grain shrivelling as in primary triticales. These characters prompted us to study the potential of this new species.

The prospective of a plant as a future crop is difficult to evaluate, as a comparison of amphiploids which are currently cultivated

P. C. Struik et al. (eds.), Plant Production on the Threshold of a New Century, 443–445.

with a primary amphiploid, that should be considered as raw material for breeding, is unfair. For this reason, to evaluate the potential of tritordeum, we designed a trial in which two primary amphiploids (synthetic bread wheat and tritordeum) were evaluated together with their durum wheat parents. Both primary tritordeum and synthetic bread wheat were later, less fertile, had smaller kernel and lower biomass, yield and harvest index than their respective durum wheat parental lines. The protein content of the synthetic bread wheat (15 %) and of the primary tritordeum (22 %) was higher than that of their wheat parents (12 and 14 %, respectively). After correcting for harvest index differences the protein content of synthetic bread wheat (14 %) became lower than that of its wheat parent (17 %), whereas the protein content of the primary tritordeum (19 %) was still higher than of its durum wheat parent (17 %). This corroborated the potential of tritordeum to become a crop after a sufficient breeding effort. In principle, primary tritordeum was at least as good as a primary bread wheat.

The first two hexaploid primary amphiploids obtained (HT22 and HT24) were crossed and a pedigree programme was followed. The best six recombinant lines from that programme showed yield (800-1300 kg/ha) around 20 % of that of commercial bread wheats, with a test weight (67-70 kg/hl) similar to that of a good triticale and a high protein content (21-25 %). The seeds were well filled and plump.

One of these recombinant tritordeum lines (HT8 (6x)) was crossed with a secondary tritordeum (HT9 (6x)) obtained from a cross between hexaploid and octoploid tritordeums (HT22 (6x) x HT18 (8x)). Selection was made following a single-seed-descent scheme. The grain yield of the best six lines derived from this cross (2800-5600 kg/ha) was up to 80 % of that of commercial wheat cultivars (average 7000 kg/ha), while total dry matter production was superior in some cases. The protein content (15-20 %) although still higher than that of commercial wheats (13-15 %) decreased when yield increased.

3. Discussion

The response to selection indicates the high potential of this new species, especially if we consider the narrow genetic basis present at that moment (only two *H. chilense* and three wheat lines). In order to broaden the genetic basis of tritordeum about 160 new primary amphiploids have been produced. They are evaluated for winter growth, earliness, plant height, biomass, fertility, grain weight and resistances to biotic and abiotic stresses. This new variation is continuously introduced to tritordeum advanced lines to improve the agronomic performance. Substantial breeding effort is still needed.

4. Acknowledgements

We are gratefully indebted to C. Martínez and M.C. Ramírez for the technical assistance; to CIDA, Córdoba for allowing the use of their facilities; and to CICYT (project AGF92-0184) for the financial support.

5. References

Martín, A., and Cubero, J.I. (1981) 'The use of *Hordeum chilense* in cereal breeding'. Cereal Res. Comm. 9, 317-323.

Martín, A. and Sánchez-Monge Laguna, E. (1982) 'The cytology and morphology of the amphiploid *Hordeum chilense* x *Triticum turgidum* conv. *durum*'. Euphytica 31, 262-267.

EFFECT OF TEMPERATURE ON SALINITY TOLERANCE OF DESMODIUM AT GERMINATION
STAGE

H.A. ESECHIE
College of Agriculture, Sultan Qaboos University
P.O. Box 34
Al-Khod 123, Muscat
Sultanate of Oman

ABSTRACT. A laboratory experiment was conducted to evaluate the effect
of temperature on salinity tolerance of two desmodium cultivars,
greenleaf desmodium [*Desmodium intortum* (Mill.) Urb] and silverleaf
desmodium [*D. unicatum*] at germination stage. Seeds were germinated in
petri dishes with saline solutions of varying concentrations (electrical
conductivities of 0.01, 6.4, 12.2, 17.4, 22.6, 27.2, 32.1 and 37.2 dS/m)
prepared with NaCl. The effects of temperature (15, 20, 25, 30, 35 and
40°C) on germination was determined over a period of 8 days. Germination
rate index (GRI) was computed from germination counts taken every 2
days. The optimum germination temperature for all the salinity
treatments was 25°C. Germination rates declined with increasing
salinity, the adverse effect of salinity being more severe at higher
temperatures (>30°C). Although greenleaf desmodium was slightly more
salt tolerant than silverleaf desmodium at the germination stage, both
had very low tolerance to salinity.

1. Introduction

The effect of high temperatures in reducing seed germination at various
salinities has been well documented in alfalfa (Esechie, 1993) and
clover (Ahi and Powers, 1938) but there has been little or no report on
desmodium. The objective of this study was to investigate the effect of
temperature on salinity tolerance of two desmodium cultivars, silverleaf
desmodium (*D. unicatum*) and greenleaf desmodium (*D. intortum*) at the
germination stage.

2. Materials and methods

Saline solutions of different concentrations (60, 120, 180, 240, 300,
360 and 420 mmol l^{-1}) were prepared with reagent grade NaCl. The
electrical conductivities and the osmotic potentials of these solutions
were determined with a conductivity meter, Model PCM 3 (Jenway,

P. C. Struik et al. (eds.), Plant Production on the Threshold of a New Century, 447–449.

Felstead, Essex) and the advanced tigemetric osmometer, Model 3T-II (Advanced Instruments, Boston, Mass.), respectively.

Three circles of Whatman No.1 filter paper were placed in several 90-mm triple vent petri dishes containing 5 ml distilled water (control) or the various NaCl solutions. Fifty seeds of greenleaf or silverleaf desmodium were sown in each petri dish and dispersed with agitation, after which they were covered and arranged in an incubator in a randomized complete block design. The number of seeds that germinated at different temperatures (15, 20, 25, 30, 35 and 40°C) were counted every 2 days for 8 days. Temperatures, generally maintained at ± 1°C of target levels, were replicated twice in each incubator. Distilled water equal to the mean loss of water from the control was added to each petri dish on day 2 through 6 to maintain salt concentration near target levels throughout the germination period.

Seed germination was determined by radicle protrusion through the seed coat, in accordance with the Association of Official Seed Analysts definition of germination (AOSA, 1970). Seedling vigour was evaluated by a germination rate index (GRI) following the procedure described by Bouten *et al.* (1976).

3. Results and discussion

The optimum germination temperature for the control was 25°C for both cultivars, and agrees with the recommended temperature for measuring germination in legumes (AOSA, 1970). Higher temperatures ($\geq$ 30°C) resulted in decreased germination. At 40°C, for instance, germination percentages were very low, being 36.0 and 13.3 in greenleaf and silverleaf desmodium, respectively. Therefore it seems likely that 40°C is outside the suitable germination temperature range for these crops. Similar results have been reported for alfalfa (Esechie, 1993).

High temperatures also aggravated the deleterious effects of high salinities, especially in silverleaf desmodium. In greenleaf desmodium, however, there were slight increases in germination percentages with increase in temperature ($\geq$ 35°C) at high salinities ($\geq$ 32.1 dS m^{-1}). Earlier work by Epstein (1972) has shown that membrane leakage is frequently enhanced by Na$^+$ and it is likely that this leakage is increased by higher temperatures. Consequently, Na$^+$ and Cl$^-$ would accumulate in toxic amounts in the seeds, especially at higher salinities, thus reducing germinability. Hoffman and Jones (1978) also reported that factors which increase evaporative demands, such as temperature, tend to aggravate the effects of salinity.

At the optimum germination temperature (25°C), the salinity levels that reduced germination by 50 % in greenleaf and silverleaf were 4.8 and 3.0 dS m^{-1}, respectively. Based on the salt tolerance categories established by Mass and Hoffman (1977), these salinity levels fall within the low salinity range. Accordingly, both cultivars would be classified as having low tolerance to salinity, with greenleaf desmodium being slightly more tolerant than silverleaf desmodium.

Based on these results, neither greenleaf nor silverleaf desmodium seems suitable as a forage legume for most of the Arabian Peninsula, especially the Batinah Coast of Oman, where the soil is generally saline and temperatures are very high, particularly during the long summer period.

4. References

Ahi, S.M. and Powers, W.L. (1938) 'Salt tolerance of plants at various temperatures'. Plant Physiol. 13, 767-789.

Association of Official Seed Analysts (1970) 'Rules for testing seeds'. Proceedings of the Association of Official Seed Analysts 60, pp. 1-116.

Bouten, J.H., Dudeck, A.E. and Smith, R.L. (1976) 'Germination in freshly harvested seed of centipedegrass'. Agron. J. 68, 991-992.

Esechie, H.A. (1993) 'Interaction of salinity and temperature in the germination of alfalfa cv. CUF 101'. Agronomie 13, in press.

Hoffman, G.T. and Jones, J.A. (1978) 'Growth and water relation of cereal crops as affected by salinity and relative humidity'. Agron. J. 70, 765-769.

Mass, E.V. and Hoffman, G.J. (1977) 'Crop salt tolerance - current assessment'. J. Irrig. Drainage Div. Am. Soc. Civ. Eng. 103, 115-134.

INFLUENCE OF LIGHT STRESS DURING THE ACCLIMATIZATION OF *IN VITRO* PLANTLETS

J.M. VAN HUYLENBROECK
University of Gent, Faculty of Agricultural
and Applied Biological Sciences
Department of plant production, Laboratory of horticulture
Coupure links 653
9000 Gent
Belgium

ABSTRACT. Exposure of micropropagated *Calathea* 'Maui Queen' plantlets to high light intensities (350-450 μmol.m^{-2}.s^{-1}) during the first weeks of the acclimatization causes photoinhibitory effects and photobleaching. Pigment development was inhibited, the lower part of the leaf became white, electron transfer in the photosystem was blocked and the photosynthetic rate decreased in comparison with plantlets growing at 60-90 μmol.m^{-2}.s^{-1}. On the other hand, when acclimatized plants (rooted during four weeks at low light intensities) were placed under high light intensities, none of these effects were observed.

1. Introduction

The transfer of *in vitro*-cultured plantlets to *in vivo* conditions is one of the most critical stages of the micropropagation process. Nurseries, specialised in the acclimatization, pay a lot of attention to climate control (relative humidity (RH), temperature, light). Nevertheless problems do occur. *Spathiphyllum spp.* and some *Marantaceae* showed sometimes aberrant, partly white leaves. Exposure to sudden high light intensities (especially during winter and early spring, when periods of cloudy, dark weather can be followed by bright periods of full sunlight) can cause damage to the young plantlets.

2. Materials and methods

Regularly micropropagated *Calathea* 'Maui Queen' cultures were maintained at 40 μmol.m^{-2}.s^{-1} under a 16 h photoperiod. Plantlets were transplanted to a soil mixture of 2/3 peat and 1/3 perlite in speedling trays with 104 holes. Directly after transplanting plants were placed under the following controlled climate conditions: temperature 22±3°C, RH 90 %,

451

P C Struik et al (eds), Plant Production on the Threshold of a New Century, 451–453

daylength 16 h and photosynthetic photon flux density (PPFD) 60-90 or 350-450 μmol.m^{-2}.s^{-1}.

Pigment content was followed, whereas chlorophyll-a fluorescence and photosynthesis measurements were performed to evaluate the photo-inhibition effect. Pigments were determined as described by Lichtenthaler and Wellburn (1983). Leaves were extracted in diethylether and absorbance was measured with a Kontron Uvikon 930 spectrophotometer. The intensity of chlorophyll fluorescence from cut and dark adapted leaves was measured with a fluorometer model SF-30 (Richard Branker Research Ltd.). Leaves were illuminated with light of 9 W.m^{-2} (1 W=5.59 μmol.s^{-1}) during 100 s. Fluorescence induction curves were stored in a microprocessor unit and the fluorescence parameters F_I, F_P and F_T from the Kautsky curve were calculated (Briantais et al., 1986).

Photosynthesis measurements were done with an LCA-3 (ADC) portable infrared gas analysing system.

3. Results and discussion

Pigment content. Five to ten days after transplanting, plantlets grown under high PPFD showed aberrant leaves. The photobleaching effects became visible as white spots, mostly on the lower part of the leaf. During this period, total amount of Chl increased significantly in low light grown plantlets, due to a significantly increase of Chl a content. On the contrary, pigment content of plants grown at high PPFD didn't change. The total amount of carotenoids, which is an important protection mechanism against high light stress, increased only significantly in low light acclimatized *Calathea* plants. Plantlets from *in vitro*, with low amounts of pigments, were unable to withstand the damaging effect of the high light intensity. Plants rooted under low PPFD (during four weeks) and subsequently placed under high light intensities, showed no photobleaching effects.

Chlorophyll-a fluorescence. The F_T/F_P ratio was significantly higher for plants grown under high PPFD during the acclimatization. Already two days after transplanting the ratio was 0.97 vs. 0.74 for plants grown at low intensities. The electron transport of the *in vitro* plantlets was totally blocked. On the other hand, acclimatized plants showed no inhibitory effect (ratio between 0.67 and 0.74 for both treatments). Probably the natural protective mechanism of *in vitro* plantlets to light stress is not well developed.

Photosynthesis. Some days after transplanting, maximum net photo-synthesis rate was very low (0.5 μmol.m^{-2}.s^{-1}) for both treatments. Two weeks later, plants grown at low light intensity had the highest maximum net photosynthesis rate (3.5 vs. 2.5 μmol.m^{-2}.s^{-1}). Plantlets grown at the lowest PPFD saturated at the lowest intensity (approximately 200 vs. 300 μmol.m^{-2}.s^{-1}).

4. Conclusion

Micropropagated *Calathea* plantlets are very sensitive to high light intensity during the first weeks of the acclimatization. Photoinhibition and photobleaching took place when plants were illuminated 16 h a day at PPFD of 400-450 μmol.m^{-2}.s^{-1}. In practice the exposure to high light intensities is much shorter. Further research is required to see if short periods of light stress can cause the same damage.

5. References

Briantais, J.M., Vernotte, C., Krausse, G. and Weiss, E. (1986) 'Chlorophyll a fluorescence of higher plants: chloroplasts and leaves', in Govindjee *et al.* (eds.), Light emission by plants and bacteria. Academic Press, New York, pp. 539-593.

Lichtenthaler, H.K. and Wellburn A.R. (1983) 'Determination of total carotenoids and chlorophylls a and b of leaf extracts in different solvents'. Biochem. Soc. Trans. 603, 265-275.

INTERACTION BETWEEN FOLIAR AND ROOT NUTRITION OF GREENHOUSE TOMATO

A. KOMOSA
Department of Horticultural Plant Nutrition
Agricultural University
Dabrowskiego 169/171
60-594 Poznán
Poland

ABSTRACT. The efficiency of foliar nutrition of greenhouse tomato cv. Ostona was studied in four-year pot experiments using peat moss as substrate on the basis of three levels of root fertilization: low, standard and high. Root fertilization was given in a single rate before planting, or during the whole time of plant growing - systematically, on the basis of three-week intervals substrate analyses. For foliar nutrition, polycompound solutions were used which contained macro- and micronutrients. Plants were sprayed seven times at one-week intervals. It was found, that with the two ways of root fertilization, the efficiency of foliar nutrition decreased with the increase of plant nutrition status, however, tomato fruit yield showed an opposite effect. In a single rate of root fertilization, the highest yield was with the high level of root fertilization, but in systematic application of nutrients with the low one. The highest tomato fruit yield which was obtained by the combination of low systematic level of root fertilization with foliar nutrition was 12.9-28.8 % higher than on standard root fertilization recommended for horticultural practice at present. The best synergistic effect between foliar and root fertilization was found when the rates in foliar applications were: 22.4-42.7 % for N, 15.3-31.2 % for P and 32.4-37.8 % for K in relation to rates of these nutrients applied in the low level of root fertilization.

1. Introduction

Many authors have pointed out, that efficiency of foliar nutrition depends upon the status of plant nutrition usually being the result of root fertilization. Jelenic *et al*. (1986) determined the interaction between foliar nutrition and activity of root system, but Eibner (1986) and Faber *et al*. (1988) emphasized the promoting effect of foliar nutrition on the nutrient uptake by roots. The main purpose of this research was to determine the interaction between foliar and root nutrition of greenhouse tomato.

P. C. Struik et al. (eds.), Plant Production on the Threshold of a New Century, 455–457.
© 1994 *Kluwer Academic Publishers.*

2. Materials and methods

The research was carried out in four-year pot experiments using peat moss as substrate. Root fertilization was varied from low (75 mg N, 50 mg P, 125 mg K/dm³ of substrate) through standard (150 mg N, 100 mg P, 250 mg K/dm³) to high level (300 mg N, 200 mg P and 500 mg K/dm³). These rates were given in one single root fertilization before planting - so a differentiated status of plant nutrition was maintained only at the beginning of plant growth, or during the whole time of plant growing by systematic fertilization on the basis of three-week intervals substrate analyses.

For foliar nutrition four polycompound solutions were used: S_1 consisted of KNO_3 (1.4 %), $CO(NH_2)_2$ (0.15 %), $NH_4H_2PO_4$ (0.2 %), K_2SO_4 (1.4 %), S_2 - KNO_3 (1.4 %), $CO(NH_2)_2$ (0.15 %), $NH_4H_2PO_4$ (0.2 %), $Ca(NO_3)_2.4H_2O$ (1.4 %), S_3 - KNO_3 (1.4 %), $Ca(NO_3)_2.4H_2O$ (1.4 %), $Ca(H_2PO_4)_2.H_2O$ (0.4 %) and S_4 - Florogama "S" (6.0 %) - the fertilizer containing all macro- and microelements. Foliar nutrition was applied seven times at one-week intervals. Plants were grown for three clusters.

3. Results

Two ways of root fertilization application influenced yields in opposite directions. Increasing root fertilization from the low to the high levels applied in a single rate caused an increase of yield in contrast to systematic fertilization where the yield decreased. In spite of different results in yield quantity, the efficiency of foliar nutrition was the highest with the low level of root nutrition given in a single rate as well as in the systematic rates.

For horticultural practice, the results indicating a relation between foliar and systematic root nutrition have special importance. Increase of yield which can be obtained by the combination of low level root fertilization with foliar nutrition, can reach 12.9 to 28.8 % in relation to the standard root fertilization. The best synergistic effect between foliar and root fertilization was found when the rates in foliar applications were: 22.4-42.7 % of N, 15.3-31.2 % of P and 32.4-37.8 % of K applied with low levels of these nutrients in root fertilization.

4. Further research

The combining of foliar and root nutrition in adequate quantity proportions is very important for horticultural practice. There is a need to determine for horticultural plants the optimum nutrient rates applied in root and foliar nutrition providing the highest intensity of synergistic interaction. This effect allows to obtain higher yield than in the standard root fertilization recommended for practice at present.

5. References

Eibner, R. (1986) 'Foliar fertilization - importance and prospects in crop production'. Proc. First Int. Symp. Foliar Fert., pp. 3-14.

Faber, A., Kesik, K., and Winiarski, A. (1988) 'Ocena skuteczności krajowych nawozów dolistnych'. Mat. Sem. Nauk., IUNG-Pulawy, pp. 170-179.

Jelenic, Dj.B., Licina, V., and Gajic, B. (1986) 'Improving the nutritional status of plants by Mg, B and Zn fertilization with the plant growth bioregulators'. Proc. First Int. Symp., Foliar Fert., pp. 453-483.

ROLE OF POLLINATION CONDITIONS IN THE REPRODUCTIVE BIOLOGY OF *VICIA FABA*

F. MONDRAGAO-RODRIGUES*, M.J. SUSO**, M.T. MORENO*
AND J.I. CUBERO***
*   Centro de Investigación y Desarrollo Agrario,
     Apartado 4240, 14080 Córdoba, Spain.
**   Instituto de Agricultura Sostenible,
     Apartado 4240, 14080 Córdoba, Spain.
*** Escuela Técnica Superior de Ingenieros Agrónomos y Montes,
     Apartado 3048, 14080 Córdoba, Spain.

Yield stability together with flower and young pod abortion are considered a major problem of the faba bean crop. The present work is an attempt to help in breeding for yield reliability through the study of the influence of pollination conditions.

Twelve genotypes highly divergent for characters related to the reproductive biology of *Vicia faba* were studied by both univariate and multivariate analysis. These genotypes were studied with and without pollinator insects presence. Thirty characters were individually studied by ANOVA following a RCB design. Results showed that most of the variation in reproductive traits is due to both genotypes and inter-action between genotype and pollinator conditions, rather than to the presence or absence of pollinator insects, which was only statistically significant for very few characters.

Eleven characters were chosen out of the initial 30 as the most representative ones to describe the reproductive biology *Vicia faba*. The principal component analysis grouped the eleven variables into two components explaining 83 % of the total variation. Component 1 explained 62 % of the total variation and was associated with young pods, and pods per flower, pods and nodes bearing pods on the main stem. Component 2 accounted for 21 % of the variation and was associated with seeds per pod and number of flowers on the main stem. Autofertility characters are much more important than pollination conditions in determining the total variation.

The response to the pollination conditions was analysed by a discriminant analysis performed by using the most sensible genotypes to these conditions. This analysis explained the manner in which the pollination conditions separated self vs. open pollinated plants as well as the contribution of each trait to that separation.

It is interesting to describe the behaviour of the three most sensible genotypes to pollinations conditions. Genotype 6 is charac-terized by differences in young pod setting. Genotype 903 is, on the contrary, characterized by differences in seeds per ovule but not in

459

P C Struik et al (eds), Plant Production on the Threshold of a New Century, 459–460
© 1994 *Kluwer Academic Publishers*

young   pod flower. Finally genotype 34 shows differences both in young pod per flower and seeds per ovule. The only general response in the highly sensitive genotypes to pollination conditions is the increase of total seed weight and a lowering of the first floral node bearing pods in open-pollinated plants. This increase of the total seed weight can be achieved by different ways depending upon the genotype, i.e., by increasing young pod per flower, or seed per ovule or both.

Francisco Mondragao-Rodrigues was supported by a J.N.I.C.T. fellowship.

NAKED OATS AND THEIR FUTURE CHALLENGES TO PLANT BREEDING IN NORTHERN
GROWING CONDITIONS

P. PELTONEN-SAINIO
Department of Plant Production
P.O. Box 27
FIN-00014 University of Helsinki
Finland

ABSTRACT. Grain yield of naked oat (*Avena sativa* var. *nuda* L.) is lower
than that of conventional hulled cultivars. This study determined the
yield component differences between naked and conventional oat and
assessed the potential of naked oat as a grain crop for northern
environments. The field study was conducted at the Viikki Experimental
Farm, University of Helsinki, Finland in 1991 and 1992. Grain yield and
16 morpho-physiological traits of ten naked and six hulled lines were
compared. Due to the high hull content of conventional lines, they had
only 10 % higher groat yields than the naked lines. The naked lines had
fewer panicles m^{-2} due to their lower emergence. Thus, 10 % increase in
seeding rate of naked lines may result in equal groat yields of naked
and conventional lines. The naked lines had higher vegetative phytomass
and lower harvest index than the conventional lines. The naked lines had
fewer spikelets per panicle and more grains per spikelet when compared
with the hulled lines. To achieve high grain yield and increase the
competitive ability of naked oat, improvement in partitioning
assimilates into grains is needed.

1. Introduction

Naked oat, the caryopsis of which threshes free from both lemma and
palea, has excellent nutritional quality (Givens and Brunnen, 1987), but
is lower yielding than conventional cultivars. Increasing grain yield by
up to 25 %, either through breeding or through improved crop husbandry,
is probably easier than improving the quality of other cereals
(Valentine, 1987). The present study was carried out to compare the
yielding ability and morpho-physiological traits of naked and
conventional oat lines, and to identify characteristics limiting grain
yield of naked oat.

P C Struik et al (eds), Plant Production on the Threshold of a New Century, 461–463
© 1994 *Kluwer Academic Publishers.*

2. Materials and methods

Studies, including three naked spring oat cultivars, seven Finnish naked breeding lines, and six conventional oat cultivars, were conducted at the Viikki Experimental Farm, University of Helsinki, Finland in 1991 and 1992 in a randomized complete block design with three replications. Plot size was 10 m² and the seed rate was 500 viable seeds m⁻². The plots were fertilized with 80 kg N ha⁻¹ as NH_4NO_3, and weeds were controlled with MCPA.

The following morpho-physiological traits were measured in each plot: grain yield (kg ha⁻¹), hull content (%), groat yield (kg ha⁻¹), number of sprouts m⁻², days to heading, days to yellow ripeness, length of grain-filling period, number of panicles m⁻², phytomass and vegetative phytomass (g plant⁻¹), panicle weight (g), number of spikelets and grains per panicle, number of grains per spikelet, spikelet and grain weight (mg), and harvest index (%). Significant differences between lines in morpho-physiological traits were tested with ANOVA.

3. Results and discussion

An equitable comparison between hulled and naked lines can be made by dehulling grain and comparing the actual groat yields (Table 1), because one quarter of the grain yield produced by the conventional line is hull. The difference between naked and conventional lines in groat yield was only 10 % and the highest yielding naked oat, Rhiannon, out-yielded the bred landrace. Further increases are likely to be achieved in naked oat, if intensive breeding programmes are carried out and management of naked oat improves.

TABLE 1. Significance of difference between hulled and naked lines in yield components.

Trait	Mean		Significance
	Hulled	Naked	
Grain yield (kg ha⁻¹)	3270	2210	***
Hull content (%)	25.6	5.0	***
Groat yield (kg ha⁻¹)	2460	2100	**
Sprouts (m⁻²)	642	558	***
Panicles (m⁻²)	591	523	***
Spikelets/panicle	16	13	***
Grains/spikelet	1.9	2.5	***
Harvest index (%)	53	43	***

The yield advance of conventional lines was associated with more panicles m⁻², which resulted from greater emergence and not from improved tillering when compared with naked lines (Table 1). Thus, a 10 % increase in seeding rate may result in equal groat yields of naked and conventional lines. Conventional lines had a higher harvest index than the naked lines. Naked lines produced higher vegetative phytomass at the expense of panicle weight. Further improvement in partitioning assimilates into grain is needed to increase the competitive ability of naked oat.

Naked lines produced significantly fewer spikelets per panicle and more grains per spikelet. Hence, the number of spikelets per panicle should be increased more relative to number of grains per panicle to reduce the number of grains per spikelet. Such changes in spikelet-grain interaction are likely to result in increased allocation of assimilates into grain. In Finland, naked lines had limited ability to produce spikelets with more than two to three grains, whereas, in Canada, the number of grains per spikelet varies between three and five (Burrows, 1986). The significant differences between naked lines in the number of spikelets and grains per panicle ($P \leq 0.001$) suggest that these traits can be manipulated in breeding programmes. Moreover, naked oats are early enough to be cultivated in northern growing conditions. These findings indicate that naked oat is a potentially useful crop for northern environments.

4. References

Burrows, V.D. (1986) 'Tibor oat'. Can. J. Plant Sci. 66, 403-405.

Givens, D.I. and Brunnen, J.M. (1987) 'Nutritive value of naked oats for ruminants'. Anim. Feed Sci. Technol. 18, 83-87.

Valentine, J. (1987) 'Breeding cereals of high nutritional quality with special reference to oats and naked oats'. Aspects Appl. Biol. 15, 541-548.

DROUGHT TOLERANCE IN OATS - PHYSIOLOGICAL METHODS TO ASSESS VARIETAL
DIFFERENCES

P. PELTONEN-SAINIO AND P. MÄKELÄ
Department of Plant Production
P.O. Box 27
FIN-00014 University of Helsinki
Finland

ABSTRACT. Several methods have been developed to assess differences in
genotypic response to drought. This study consisted of 19 oat (*Avena
sativa* L.) lines and three species of wild oats (*Avena fatua*, *A.
sterilis*, and *A. abyssinica*) and focused on comparison of several
physiological screening methods with yield losses. Early drought,
occurring at 4-5 leaf stage, or close to pollination resulted in
significantly reduced vegetative phytomass, panicle weight, and harvest
index. Drought stress significantly increased stomatal resistance and
accumulation of ABA and proline, whereas disparity between air and leaf
temperatures and relative water content (RWC) decreased due to water
deficit. Comparison of screening methods by calculating drought
tolerance indices, showed that none of the indices were associated with
degree of yield losses. However, photosynthetic characters correlated
with RWC both at early growth stages and close to pollination. Hence,
further studies that evaluate profoundly the relationship between these
screening methods and degree of yield losses are needed.

1. Introduction

One means of increasing productivity in the future, and especially
stabilizing yield performance over various environments is improved
stress tolerance. Finnish growing conditions are generally favourable
for oat production due to long days and relatively cool growing seasons.
Early summer drought is, however, a recurrent problem and results in
decreased yield potential (Peltonen-Sainio, 1991). Comparison of crop
yields produced under favourable versus drought stressed environments is
the most accurate, but is a laborious, method for evaluating drought
tolerance in a breeding population. Extrapolations of drought tolerance
solely from long-term field experiments is uncertain because
environmental factors other than precipitation markedly affect yield
performance. In this context, rapid and reliable methods for measuring
drought tolerance are needed. The aim of this study was to compare the
sensitivity and suitability of several physiological screening methods

465

P. C. Struik et al. (eds.), Plant Production on the Threshold of a New Century, 465–467.
© 1994 *Kluwer Academic Publishers.*

for measuring differences in response of oat lines to water deficit.

2. Materials and methods

Three experiments were carried out in a greenhouse of the University of Helsinki. Daylength was 18 hours at a photosynthetically active radiation of 200 μmol m^{-2} s^{-1}, and day and night temperatures were 20 °C and 17 °C, respectively. Plant material consisted of 19 oat lines and three wild species. The experiments included two treatments; control and water stress (either moderate or severe water deficit).

Assimilation rate (CER, mg CO_2 m^{-2} s^{-1}), stomatal resistance (s m^{-1}), and disparity between air and leaf temperatures (°C) were measured with a LI-COR LI-6200 portable photosynthetic system. Endogenous ABA (pmol ml^{-1} extractant) were measured using Phytodetek Immunoassay Test Kits. Accumulation of proline (μmol cm^{-2}) and relative water content (RWC) were measured as described by Troll and Lindsley (1955) and Ritchie *et al.* (1990), respectively. Phytomass and vegetative phytomass (g plant^{-1}), panicle weight (g), and harvest index (%) were measured. Drought tolerance indices were calculated for each line by dividing each character under water deficit by that under adequate watering. Hence, the greater the deviation of each index from 1, the greater the response of a specific cultivar to drought stress. Differences in yield components and physiological characteristics between lines were tested by analysis of variance. Relationships between drought tolerance indices were established using linear regression analysis.

3. Results and discussion

In cereals, even short drought periods occurring at critical developmental stages may result in considerable yield reduction. In the study, both pre- and post-anthesis drought resulted in significant yield losses ($P \leq 0.001$). It reduced vegetative phytomass, panicle weight, and harvest index. There were, however, significant differences ($P \leq 0.001$) in response to drought between lines.

Comparison of physiological characteristics with grain yield is very difficult, as physiological measurements only give momentary information about the effect of water deficit without, e.g., indicating the relative ability of lines to recover from drought. Lack of significant correlation between panicle weight index and the physiological indices confirmed this, even though the effect of drought on each physiological parameter over all lines was significant ($P \leq 0.05$). Drought stress significantly increased stomatal resistance and accumulation of abscisic acid and proline, whereas disparity between air and leaf temperatures and RWC decreased due to water deficit. Our results indicated that RWC index correlated with photosynthetic parameters; positively with CER ($0.47* \leq r \leq 0.57**$), and negatively with stomatal resistance ($-0.53* \leq r \leq -0.75***$). Moreover, RWC is an inexpensive and rapid screening technique, which emphasizes the practicability of this method, in particular. However, supplementary data from field experiments from several contrasting environments are required to enable ranking of the lines.

4. References

Peltonen-Sainio, P. (1991) 'Effect of moderate and severe drought stress on the pre-anthesis development and yield formation of oats'. J. Agric. Sci. Finl. 63, 379-389.

Ritchie, S.W., Nguyen, H.T. and Holaday, A.S. (1990) 'Leaf water content and gas-exchange parameters of two wheat genotypes differing in drought resistance'. Crop Sci. 30, 105-111.

Troll, W. and Lindsley, J. (1955) 'A photometric method for the determination of proline'. J. Biol. Chem. 215, 655-660.

THE EFFECT OF LINURON ON CHEMICAL COMPOSITION OF CARROT ROOTS

J. SOBIESZCZAŃSKI, J. SZYMAŃSKI*, R. STEMPNIEWICZ,
AND W. JURGIELEWICZ
*Departments of Biotechnology and Food Microbiology
and *Fruits and Vegetables Technology
Academy of Agriculture
50-375 Wroław
Poland*

ABSTRACT. Aim of the research was to examine, in the field, the effect of various doses of afalon 50W (linuron) on the chemical components in carrot roots. The results showed that the herbicide changed the plant metabolism, chemical composition and nutritive value of carrot roots.

1. Introduction

Herbicides reduce the costs of production but induce many fears, like plant contamination by the residue of these compounds (Orth, 1968). They can also influence the plant metabolism and change their nutritive value (Żechałko et al., 1971; Michalik et al., 1974). The goal of our research was to examine to which degree under our climatic and soil conditions afalon-linuron can influence the chemical composition of carrot roots.

2. Material and methods

In a field experiment (loamy-sand soil, pH 7.7) the carrot "Perfekcja" was cultivated. The plots: A - control without herbicide and plants, B - control without herbicide but with plants, C - treated 2 kg/ha, D - treated 4 kg/ha and E - 6 kg/ha of herbicide with plants. Afalon 50W (linuron) - N/3,4-dichlorophenyl/N-methoxy-N-methyl-urea (Hoechst-Farb Werke, FRG) was used. All objects in 5 fold repetition. The carrots were sown on March 24 and herbicide applied on April 2. The carrots were harvested October 27, 1991. The climatic conditions were rather dry (high temperature). The carrot roots were analysed just after the harvest (dry mass, refractometric dry mass, sugars, acidity, nitrates and carotenoides after Polish Norm PN-71/75101, PN-85/A-75101). The amino acids were analysed by using an automatic analyser AAA-339 "Microtechna", Czechoslovakia. For analyses 1 kg of roots was taken from various places of each plot. Next roots, after cleaning, were analysed, method of Moove and Stein (1954), each sample with four repetitions. For

P. C. Struik et al. (eds.), Plant Production on the Threshold of a New Century, 469–471.
© 1994 *Kluwer Academic Publishers.*

470

statistical calculations the Duncan's multiple range test was used.

3. Results

The herbicide clearly affected the chemical composition of the carrots;
especially the amino acids were strongly affected (Figures 1 and 2).

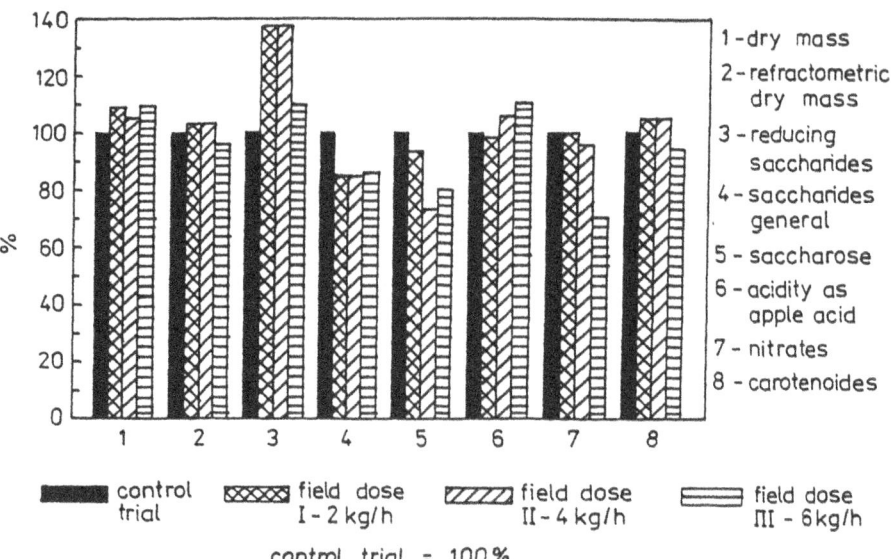

Figure 1. The effect of different doses of linuron on relative yield of
different chemical components of carrot roots.

4. Discussion

The results obtained showed that afalon-linuron changed the metabolism,
chemical composition and nutritive value of the carrot roots. Similar
effects on the amount of sugars in carrot roots by linuron were observed
by Żechałko et al. (1971), but Michalik et al. (1974) found that the
amount of sugars depended on the cultivar. "Perfekcja" contains a higher
amount of sugar than "Nantejska" from a field treated with herbicides
and it is supposed that cultivars can differently react to the
herbicides. Płoszyński et al. (1975) found that monuron and monolinuron
decreased the protein nitrogen and increased the amount of some free
amino acids. The intensity of these changes depends on the phosphor
fertilization. In our research, by higher doses, the afalon-linuron
increased the amounts of amino acids in the carrots by normal dose of
phosphor fertilization. Changes of the chemical composition in the

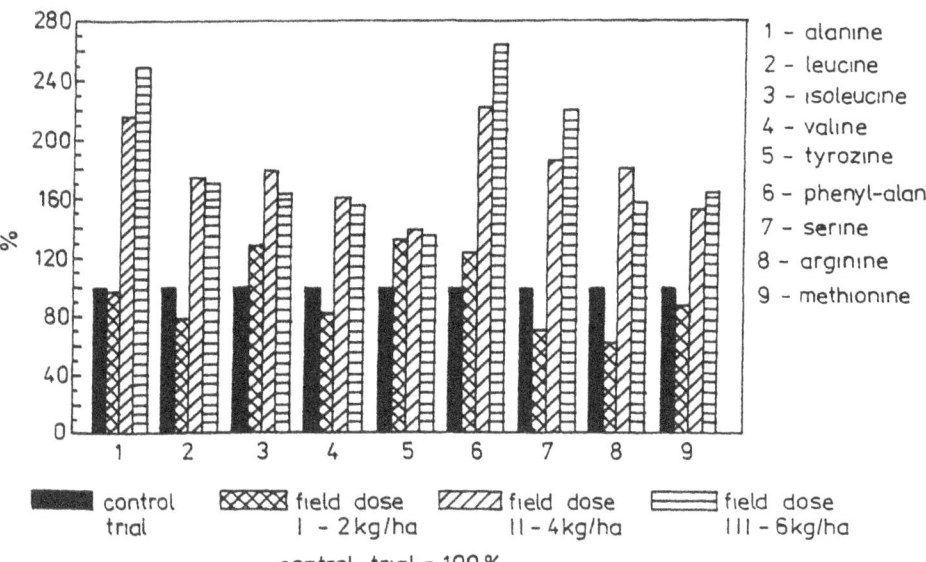

Figure 2. The effect of different doses of linuron on relative yields of various amino acids.

carrots observed in our research confirm a necessity of carrying out more versatile research in order to establish the most profitable conditions of herbicide application in the carrot cultivation.

5. References

Michalik, H. and Elkner, K. (19740 'Wpływ stosowania herbicidów na wartość biologiczną marchwi. Biuletyn Warzywniczy XVI, Instytut Warzywnictwa, Skierniewice, pp. 358-365.

Move, S. and Stein, L.H. (1954), Biol. Chem., pp. 211-223.

Orth, H. (1968) 'Herbicide im Gemüsebau'. Z. Pflkrank. Piltschutz, Sonderheft IV, pp. 97-108.

Płoszyński, M., Runowska-Hryńczuk (1975). 'Działanie monuronu i monolinuronu na dynamike wzrostu, plony i zmiany jakościowe żyta ozimego na tle różnych dawek nawożenia azotowego, fosforowego i potasowego'. Roczn. Nauk Roln. 10, pp. 295-313.

żechałko, A., Rola, J., Jaasińska, M., and Biernat, J. (1971) 'Badania nad wplywem herbicidów na plony marchwi i jej wartość odżwcza'. Pamietnik Puławski, Prace IUNG, pp. 228-242.

The work was supported by US Dept. of Agric. Grant FG-Po 384.

DEGRADATION OF PHENOLIC ACIDS BY BACTERIA FROM LIQUID HYDROPONIC CULTURES OF TOMATO

P. SUNDIN AND B. WAECHTER-KRISTENSEN
Department of Horticulture
P.O. Box 55
S-230 53 Alnarp
Sweden

ABSTRACT. Bacterial isolates from liquid hydroponic cultures of tomato were able to tolerate high concentrations of phenolic acids, both on agar plates and in liquid culture. None of the isolates tested, though, did cause any degradation of 2 mM of p-hydroxybenzoic, vanillic and caffeic acids, or 50 μM of p-hydroxybenzoic acid, in liquid culture. Nutrient solutions from liquid hydroponic cultures of tomato, however, contained agents capable of degrading 50 μM of p-hydroxybenzoic, vanillic and caffeic acids.

1. Introduction

Hydroponic systems, in which the nutrient solution is circulated and supplemented with nutrients, may facilitate the greenhouse cultivation of vegetables with a low environmental impact. In circulating nutrient solutions, however, accumulation of phytotoxic organic compounds, like phenolic acids, may occur.

In hydroponic tomato cultures with circulating nutrient solutions, different bacterial populations inhabit the root surface (rhizoplane) and the nutrient solution. This is possibly due to differences in the availability of attractive carbon sources, such as amino acids and low molecular weight carbohydrates exuded by roots. In the nutrient solution bacteria with the ability to use less valuable carbon and energy sources are probably competitive. Bacteria that are capable to degrade potentially phytotoxic phenolic acids would be desired in this environment. Such bacteria are known to be present in the soil environment (Vaughan *et al.*, 1983; Blum and Shafer, 1988).

2. Materials and methods

Cellulose test pads ($\varnothing$ 12.7 mm) were saturated with 3 mM aqueous solutions of benzoic, caffeic, ferulic, p-coumaric, salicylic, vanillic and p-hydroxybenzoic acids. Bacterial inocula, obtained from roots or

P. C. Struik et al. (eds.), Plant Production on the Threshold of a New Century, 473–475.
© 1994 *Kluwer Academic Publishers.*

nutrient solutions of hydroponic tomato cultures, were spread on TSA dishes, followed by placing of the test pads. Pads saturated with sterile water served as controls.

Bacterial inocula of isolates from roots or nutrient solutions of hydroponic tomato cultures were aseptically added to flasks containing sterile nutrient media amended with a) 2 g sucrose/ml and 2 mM of p-hydroxybenzoic, vanillic and caffeic acids, b) 2 g sucrose/ml, c) 50 μM of p-hydroxybenzoic acid and 2 g sucrose/ml. The flasks were incubated in triplicate, on a rotary shaker in the dark and samples were removed for determination of viable counts, and phenolic acids (using HPLC). In control flasks sterile water was added instead of bacterial inocula.

Nutrient solution samples from two other hydroponic tomato culture systems, and the corresponding volumes of deionized water, were supplemented with solutions of p-hydroxybenzoic, vanillic and caffeic acids to a final concentration of 50 μM. The flasks were incubated, and phenolic acids were determined as above.

3. Results

The majority of the bacterial isolates were able to grow on agar dishes in the presence of 3 mM of phenolic acids. The isolates incubated in liquid culture, with 2 mM of a phenolic acid mixture or 50 μM of p-hydroxybenzoic acid, remained viable during the incubation time (up to five days) but did not increase in numbers (data not shown). Instead, bacteria isolated from roots, in particular, tended to decrease in numbers.

None of the isolates tested were able to degrade any of the phenolic acids during the incubation time.

The incubation of phenolic acids with nutrient solutions from liquid hydroponic tomato cultures resulted in an initial reduction in the concentration of caffeic acid, but did not cause any other change during the first 20 h of incubation (Figure 1). After 28 h, however, the concentrations of all three acids had decreased in comparison with the chemical controls, and after 49 h only traces remained in the nutrient solution samples.

4. Discussion

The incapability of bacterial isolates from liquid hydroponic tomato cultures to degrade phenolic acids, may reflect the absence of such compounds in the nutrient solutions sampled. However, the hydroponic tomato cultures from which complete nutrient solutions were used for the degradation studies obviously contained agents - probably bacteria - capable to degrade phenolic acids. Similar results have been obtained earlier, in studies of nutrient solutions from hydroponically cultured lettuce (P. Sundin, unpublished). The time lag before the phenolic acid concentration started to decrease, may indicate that enzymes had to be induced. The initial reduction in the concentration of caffeic acid may indicate that a potential for degradation of this compound was already at hand in these culture systems.

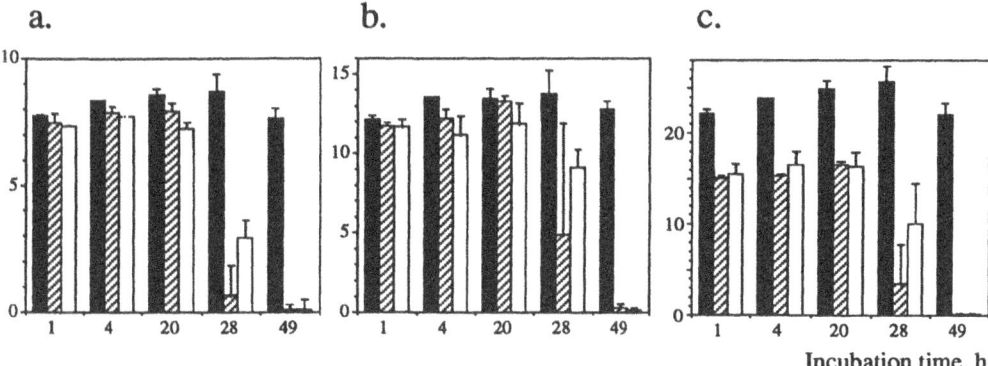

Figure 1. Degradation of 50 μM of a) p-hydroxybenzoic, b) vanillic and c) caffeic acids in mixture by incubation with nutrient solutions from two hydroponic tomato cultures (dashed and open columns) in comparison with deionized water (filled columns). The amounts of phenolic acids (x-axis) are given as integrator area units (x10^5) after HPLC analysis of 10 μl of solution.

5. References

Blum, U. and Shafer, S.R. (1988) 'Microbial populations and phenolic acids in soil'. Soil Biology and Biochemistry 20, 793-800.
Vaughan, D., Sparling, G.P. and Ord, B.G. (1983) 'Amelioration of the phytotoxicity of phenolic acids by some soil microbes'. Soil Biology and Biochemistry 15, 613-614.

DIVERSITY AND ADAPTATION IN SARDINIAN BARLEY (*HORDEUM VULGARE* L.) LANDRACES

R. PAPA[*], G. ATTENE[**] AND F. VERONESI[*]
[*] *Dipartimento di Biotecnologie Agrarie ed Ambientali*
   Via Brecce Bianche, 60131 Ancona, Italy
[**] *Istituto di Agronomia generale e Coltivazioni erbacee*
   Via E. De Nicola, 07100 Sassari, Italy

ABSTRACT. Twelve Sardinian six-row barley populations were evaluated in the field for several agronomic and morphological traits and eight genotypes of the Sinis 0 population were tested in four contrasting environments. High phenotypic variation among and within populations together with a significant GE-Interaction for the response of the Sinis 0 genotypes was found. These results highlight the importance of landraces not only as source of genetic variation but also as valuable materials to study the environmental adaptation of plant populations.

1. Introduction

In Sardinia barley is widely cultivated and many farmers prefer to grow local populations which are well adapted to the local environmental conditions. Characterization of morphological and agronomic traits as well as the salinity tolerance of one of these populations (Sinis 0) demonstrated the presence of a large genetic variation (Attene and Veronesi, 1991; Papa et al., 1991). The aim of the present research was to study the phenotypic variation of twelve Sardinian six-row barley populations collected in 1990 and to asses the response in contrasting environments, of eight genotypes of the population Sinis 0 diverging for their salt tolerance.

2. Materials and methods

First experiment - Twelve local populations of the landrace Su Orgiu Sardu (Papa, 1993) collected in Sardinia in 1990 were evaluated in the field in 1992 (Sassari, Italy). Per population 20 pure lines were tested. Data of the following characters were collected: number of culm nodes (NCN), plant height (PH), rachis length (RL), number of rachis segments (NSR), length of the 10 central segments of the rachis (LSR), collar length (CL), days to heading (DH), 1000 kernel weight (KW), grain yield (GY), biomass yield (BY) and harvest index (HI).

P. C. Struik et al. (eds.), Plant Production on the Threshold of a New Century, 477–479.

Second experiment - Eight genotypes extracted from the Sinis O population were evaluated in the field, one location in Italy (Sassari) and three locations in Syria (Tel Hadya, Breda and Bouider), with an average annual rainfall of 596 mm, 328 mm, 281 mm and 212 mm, respectively. At maturity grain yield was measured and GE-Interaction analysed.

3. Results

ANOVA results showed high significance for the component "among populations" ("AP") for all the characters evaluated except KW. High significance was also found for the component "within population" ("WP") for six traits except NCN, PH, GY and BY. The analysis of variance components showed that the component "WP" was higher than the component "AP" for four traits: CL, RL, LSR and NSR. For DH the two components were almost similar, 26.8 % and 22.1 % "WP". The among replications component was ranging from 42 % (LSR) to 61.7 % (HI). The analysis of the results of the Sinis O genotypes showed highly significant GE-Interaction for GY.

4. Discussion and conclusions

The study of the phenotypic variation in the Sardinian barley populations shows that the parameters measured vary considerably both among and within populations. A consistent variation within population was found for several traits, and in some cases, like for DH, KW and HI, with high adaptive and agronomic value. Moreover, the different behaviour of the Sinis O genotypes in response to contrasting environmental conditions highlight the role of population polymorphism as an adaptive tool.

As already suggested by Allard (1968) high variation is essential to the survival of populations. The maintenance of a stable non trivial polymorphism within a population depends on a complex set of interactions between genetic factors, mating system and ecological factors.

The results of the present research highlights the importance of landraces not only as a source of genetic variation but also as valuable materials to study the adaptation mechanisms of plant populations, viewed as complex organisms, to the environment.

5. References

Allard, R.W. (1968) 'The genetics of inbreeding populations'. Advances in Genetics 14, 55-131.

Attene, G. and Veronesi, F. (1991) 'Osservazioni su una popolazione locale sarda di orzo polistico'. Rivista di Agronomia, XXV, 1, 54-59.

Papa, R. (1993) 'Diversità e adattamento in germoplasma sardo di orzo (*Hordeum vulgare* L.)'. Tesi di Dottorato die Ricerca, Università degli Studi di Sassari, Italy.

Papa, R., Attene, G. and Veronesi, F. (1991) 'Evaluation of genetic variability for agronomical traits and salt tolerance of barley genotypes from a Sardinian landrace', in Barley Genetics VI. July 22-27, Helsingborg, Sweden, 536-538.

This research was partially carried out as part of the research project "Barley breeding for the Mediterranean environment", financially supported by the Italian Government, Ministry of University and Scientific Research, MURST.

THE EFFECTS OF BIOTIC ENVIRONMENTAL FACTORS ON THE YIELD OF DIFFERENT WINTER WHEAT (*T. AESTIVUM* L.) GENOTYPES

M. CSÓSZ, L. CSEUZ, Z. KERTÉSZ AND J. PAUK
Cereal Research Institute
P.O. Box 391
6701 Szeged
Hungary

ABSTRACT. Fifty five genotypes of winter wheat were investigated under different levels of infection with leaf rust, powdery mildew and stem rust. Eleven traits, related to yield components, were evaluated by principal component analysis for each level of infection. Level of infection significantly influenced the relations between traits.

1. Introduction

In Hungary the most important fungal leaf diseases of winter wheat are leaf rust, powdery mildew and stem rust. These three diseases can cause approximately 5-40 % of yield loss depending on the resistance of the cultivar, the level of the epidemic and the efficiency of plant protection.

The growing conditions of winter wheat are significantly different in the different regions of Hungary, so it is an important question how the different yield components affect the yield under changing environments.

2. Materials and methods

To study this question, different traits of 55 genotypes were studied under three different conditions:
1. *Protected by fungicides (control),*
2. *Infected by a mixture of stem rust races,*
3. *Grown under natural conditions.*

3. Results

The correlation system of the studied traits was evaluated by ***principal component analysis***.

481

P C Struik et al (eds), Plant Production on the Threshold of a New Century, 481–483
© 1994 *Kluwer Academic Publishers*

1. Fungicide protected

Eleven traits were evaluated by principal component analysis (total mass of sample, thousand kernel mass, harvest index, yield-, seed number-, spikelet number-, head mass of main head, the mass of the main head, whole yield, the total mass of the heads, number of heads, plant height). The first three components explained 75 % of the variation. We found a close correlation between the main head yield and the whole mass of the main head. Between these traits and the number of spikes a negative correlation was found.

2. Infected condition

Powdery mildew and stem rust scoring data at three different growth stages were evaluated together with the 11 traits by the principal component analysis. The first three components explained 62 % of the variation. Harvest index, total yield, and the total mass of the heads were closely related. Between these traits and the second scoring for powdery mildew infection a negative correlation was found. Close relations were found between stem rust scoring data of different life stages.

3. Naturally grown

The evaluated 11 traits were completed by the powdery mildew and leaf rust scoring values. The first three components explained 68 % of the variation. We found close, positive correlations between the data of total mass of sample, the mean head's yield, the number of seeds in the mean head, the total mass of mean head, the total yield and total mass of spikes. Plant height and leaf rust infection data showed a negative correlation.

According to these results, we can state that the different environmental criteria changed the relationship among the characteristics evaluated.

Multiple regression analysis

By the multiple regression analyses we could determine how the yield reaction was affected by the different variables. Here we evaluated only nine traits because we did not use the closely related variables and calculated values. Yield under infected and normal conditions was expressed as a percentage of the chemically protected yield. 42.4 % of the variance of yield reaction could be interpreted by the nine studied variables ($R=0.65$, $P=0.1$ %). Change of yield could be interpreted (within the 42.4 %) by the yield of mean head (45.2 %), the number of seeds in the mean head (31.6 %) and the number of heads (28.5 %). The effect of the remaining variables was not significant. Under stem rust infected conditions 51 % of yield reaction could be interpreted by the nine variables used ($R=0.71$, $P=0.1$ %). The change in yield was due to the decrease of the thousand kernel mass (15.9 %), infection level of stem rust (15.8 %) and number of spikes (13.8 %).

4. Conclusion

According to these results, we can establish that in the role of certain traits significant changes can occur due to the effects of environmental factors. Under normal conditions diseases had a much lower effect on the quantity of yield, in spite of the fact that the infection level was 56.1 % (powdery mildew) and 48.2 % (leaf rust). Under infected conditions yield was significantly affected by stem rust which caused damages at early stages.

SUBJECT INDEX

The manufacturer's authorised representative in the EU is Springer
Nature Customer Service Centre GmbH, Europaplatz 3, 69115 Heidelberg,
Germany. If you have any concerns regarding our products, please
contact ProductSafety@springernature.com

Printed and bound by CPI Group (UK) Ltd, Croydon, CR0 4YY

23/04/2026

02095630-0001